Multitrophic Level Interactions

The multitrophic level approach to ecology addresses the complexity of food-webs much more realistically than the traditional focus on simple systems and interactions. Only in the last twenty years have ecologists become interested in the nature of more complex systems, including tritrophic interactions between plants, herbivores, and natural enemies. Plants may directly influence the behavior of their herbivores' natural enemies, ecological interactions between two species are often indirectly mediated by a third species, landscape structure directly affects local tritrophic interactions, and below-ground food-webs are vital to above-ground organisms. The relative importance of top-down effects (control by predators) and bottom-up effects (control by resources) must also be determined. These interactions are explored in this exciting new volume by expert researchers from a variety of ecological fields. This book provides a much-needed synthesis of multitrophic level interactions and serves as a guide for future research for ecologists of all descriptions.

TEJA TSCHARNTKE is Professor of Agroecology at the University of Göttingen, Germany. His research focus is on plant–herbivore–enemy interactions including parasitism, predation and pollination, insect communities and food-webs on a landscape scale, and temperate–tropical comparisons. He is editor-in-chief of *Basic and Applied Ecology* and a member of the editorial board of *Oecologia*.

BRADFORD A. HAWKINS is an Associate Professor in the Department of Ecology and Evolutionary Biology at the University of California, Irvine. His research focus is on the biology and ecology of insect parasitoids, insect community ecology, food-webs, and energy-diversity theory. He is the author of *Pattern and Process in Host–Parasitoid Interactions* (1994, ISBN 0 521 46029 8), and editor of *Parasitoid Community Ecology* (1994) with William Sheenan and *Theoretical Approaches to Biological Control* (1999, ISBN 0 521 57283 5) with Howard V. Cornell.

Multitrophic Level Interactions

Edited by

TEJA TSCHARNTKE

Universität Göttingen

and

BRADFORD A. HAWKINS

University of California, Irvine

CAMBRIDGE UNIVERSITY PRESS
Cambridge, New York, Melbourne, Madrid, Cape Town, Singapore, São Paulo, Delhi

Cambridge University Press
The Edinburgh Building, Cambridge CB2 8RU, UK

Published in the United States of America by Cambridge University Press, New York

www.cambridge.org
Information on this title: www.cambridge.org/9780521791106

First published 2002
This digitally printed version 2008

A catalogue record for this publication is available from the British Library

Library of Congress Cataloguing in Publication data

Tscharntke, Teja, 1952–
Multitrophic level interactions / Teja Tscharntke, Bradford A. Hawkins.
p. cm.
Includes bibliographical references (p.).
ISBN 0 521 79110 3 (hb)
1. Multitrophic interactions (Ecology) I. Hawkins, Bradford A. II. Title.
QH541.T78 2002
577′.1 – dc21 2001037338

ISBN 978-0-521-79110-6 hardback
ISBN 978-0-521-08418-5 paperback

Contents

Contributors

PEDRO BARBOSA
Department of Entomology, Plant Science Building, University of Maryland, College Park, MD 20742, USA

JUDITH L. BRONSTEIN
Department of Entomology and Evolutionary Biology, University of Arizona, Tucson, AZ 85721, USA

VALERIE K. BROWN
CABI Bioscience Environment, Silwood Park, Ascot, Berks SL7 5PY, UK

THOMAS L. BULTMAN
Division of Science, Truman State University, Kirksville, MO 63501, USA

JÉRÔME CASAS
University of Tours, Institut de Recherche sur la Biologie de l'Insecte, IRBI-CNRS ESA 6035, F-37200, Tours, France

PHYLLIS D. COLEY
Biology Department, University of Utah, Salt Lake City, UT 84112, USA

THOMAS DEGEN
Institute of Zoology, University of Neuchâtel, CH-2007, Neuchâtel, Switzerland

IMEN DJEMAI
University of Tours, Institut de Recherche sur la Biologie de l'Insecte, IRBI CNRS ESA 6035, F-37200, Tours, France

LEE A. DYER
Biology Department, Department of Ecology and Evolutionary Biology, Tulane University, New Orleans, LA 70118, USA

Stanley H. Faeth
Department of Biology, Arizona State University, Tempe, AZ 85287, USA

Maria Elena Fritzsche-Hoballah
Institute of Zoology, University of Neuchâtel, CH-2007, Neuchâtel, Switzerland

Alan C. Gange
School of Biological Sciences, Royal Holloway College, University of London, Egham Hill, Surrey TW20 0EX, UK

Sandrine Gouinguené
Institute of Zoology, University of Neuchâtel, CH-2007, Neuchâtel, Switzerland

Ilkka Hanski
Department of Ecology and Systematics, Division of Population Biology, University of Helsinki, FIN-00014, Helsinki, Finland

J. Daniel Hare
Department of Entomology, University of California, Riverside, CA 92521, USA

Bradford A. Hawkins
Department of Ecology and Evolutionary Biology, University of California, Irvine, CA 92697, USA

Stefan Scheu
Technische Universität Darmstadt, Fachbereich 10, Biologie, D-64287, Darmstadt, Germany

Heikki Setälä
Department of Biological and Environmental Science, University of Jyväskylä, FIN-40351, Jyväskylä, Finland

Teja Tscharntke
Agroecology, University of Göttingen, D-37073, Göttingen, Germany

Ted C. J. Turlings
Institute of Zoology, University of Neuchâtel, CH-2007, Neuchâtel, Switzerland

Saskya van Nouhuys
Department of Ecology and Systematics, Division of Population Biology, University of Helsinki, FIN-00014, Helsinki, Finland

TEJA TSCHARNTKE AND BRADFORD A. HAWKINS

1

Multitrophic level interactions: an introduction

Terrestrial ecosystems are characterized by a huge diversity of species and a corresponding diversity of interactions between these species, but community ecology has historically been dominated by interactions between two trophic levels; in particular, plant–herbivore and predator–prey interactions. Only more recently have ecologists become interested in the nature of more complex interactions involving three or more trophic levels (e.g., Price *et al.*, 1980; Bernays and Graham, 1988; Barbosa *et al.*, 1990; Hawkins, 1994; Gange and Brown, 1997; Olff *et al.*, 1999; Pace *et al.*, 1999; Dicke, 2000; Schmitz *et al.*, 2000). It has quickly become clear that a multitrophic level approach addresses the complexity of food-webs much more realistically than does the simpler approach. Our reasons for generating this book are to provide an overview of progress that has been made in demonstrating how research on more realistic models of food webs has enriched our understanding of complex biological systems, and to highlight new and particularly exciting avenues of future research in this area.

In the past two decades there has been intense interest in tritrophic interactions between plants, herbivores, and natural enemies, driven by the need both to integrate host plant resistance and biological control in the management of arthropod pests and to understand the relative importance of direct and indirect interactions in ecological communities. Many examples document the direct effects of physical, chemical, and nutritional qualities of plants on the attack rate, survival and reproduction of natural enemies. In addition, it is well known that in some cases these same plant qualities have indirect effects on natural enemies by influencing the distribution, abundance, and vulnerability of herbivores. Even so, there is a need to specify conditions where multitrophic interactions are important and to determine the habitat characteristics that influence the relative importance of top-down effects (control by predators) and bottom-up effects (control by resources). This latter problem remains an important part of community ecology and is answerable only when we utilize multitrophic level thinking.

This book provides an overview and current perspectives on the field of multitrophic interactions. The book comprises ten chapters, the topics of which have been selected by the editors to include what we feel represent the most important aspects of multitrophic interactions. We have selected several standard topics that should be included in a book with this theme, but we have also selected newly emerging topics that should receive greater attention in the coming years. Consequently, the book will very much focus on the future rather than on the history of the field, and the authors provide critical reviews of the areas encompassed by their chapters, as well as an assessment of the most important areas for further research. Hence, this edited volume, without being overly long, provides an update of the field and serves as a guide for future research. It will become obvious that we restrict coverage to terrestrial systems. This represents a conscious choice to keep the book focused and relatively short.

The concept of multitrophic interactions implies that evolved plant traits enhance the success of natural enemies as mortality agents of herbivores. Hare (chapter 2) focuses on the question of whether or not enemy impact exerts sufficiently strong selection pressure to modify plant traits. He develops a useful criteria set to test when natural enemies represent significant agents of natural selection for plants. After a short overview on the diverse ways in which morphological and chemical plant characteristics may affect the parasitoids and predators of herbivores, Hare summarizes the literature on the compatibility of plant resistance and biological control. Development of transgenic crop plants as well as conventional plant breeding may be used to manipulate and exploit plant traits for improved biological control. Due to the wide range of variation in herbivore–enemy interactions, predictability is low and may be practical only on a case-by-case basis. Hare proposes four critical areas for future research: the need to measure tritrophic effects on plant fitness, the identification of the mechanisms involved, whether results from applied systems can be generalized to natural systems, and a call for the study of natural tritrophic systems in a true coevolutionary context.

Bronstein and Barbosa (chapter 3) emphasize that mutualisms may be multitrophic/multispecies in nature and that this complexity has infection with endophytic fungi, with special emphasis on grass endophytes. Evidence is accumulating that the mycotoxins of grass endophytes not only negatively affect herbivores, but consistently also affect insect parasitoids. Since contents of alkaloid mycotoxins tend to be lower in native than agronomic grasses, negative effects on herbivores and their parasitoids should also be lower in native systems. Whether the increased host grass resistance is counterbalanced by decreased natural enemy impact remains untested in these little-studied natural systems. Given that endophytic infections in woody plants are probably neutral or have, at best, weak and indirect impacts on herbivores, Faeth and Bultmann predict even weaker effects of endophytes on the third trophic level.

Van Nouhuys and Hanski (chapter 6) argue that an organism's spatial distribution may be considered to be a further attribute that potentially affects its trophic interactions. The addition of landscape structure and of spatial population dynamics to the analysis of multitrophic level interactions is becoming increasingly important in fragmented natural landscapes. The inclusion of space may show how local trophic interactions affect regional dynamics and how large-scale population dynamics, such as extinction–colonization dynamics in metapopulations, affect local interactions. Van Nouhuys and Hanski review the theory of metapopulation dynamics extended to several interacting species and trophic levels. They then discuss empirical findings from the literature. A basic conclusion from metapopulation models is that specialist food chains are constrained in fragmented landscapes, and the degrees of specialization and dispersal rates are essential for understanding the resulting patterns. Empirical findings from intensively studied plant–butterfly–parasitoid–hyperparasitoid interactions confirmed the theoretical predictions that poorly dispersing parasitoids and hyperparasitoids were absent from a small-patch network. The truncation of food chains as a consequence of habitat fragmentation is probably a common occurrence, although empirical evidence is currently limited.

Turlings, Gouinguené, Degen, and Fritzsche-Hoballah (chapter 7) review chemically mediated tritrophic level interactions. They emphasize three aspects relevant to the evolution of herbivore-induced plant signals, the factors inducing the emission of these signals, the specificity and reliability of the signals for natural enemies, and the benefits that plants may derive from attracting enemies. The importance of herbivore-induced plant volatiles in the location of hosts and prey by parasitoids and predators was not demonstrated until the 1980s, and these fascinating interactions turned out to vary greatly depending on the species involved.

received little attention in the past. They provide a variety of examples of mutualisms with a third species involved, either of the same or another trophic level. In many cases, mutualism depends on the impact of natural enemies or competitors. Ant–lycaenid caterpillar mutualisms may be only important in case of high rates of parasitism and predation. Mycorrhizae–plant mutualisms may depend on the impact of competing plants and promote species coexistence in terrestrial vegetation. A review of the evidence of third-species mediation of mutualism or "apparent" mutualism is the basis of a proposed set of working hypotheses. They hypothesize that the conditional nature of many mutualisms may depend on the behavior of a third species and discuss both the ecological and evolutionary implications.

Although the latitudinal gradient in species diversity is the oldest known pattern in ecology, many of the major hypotheses attempting to explain differences between tropical and temperate communities remain untested. Dyer and Coley (chapter 4) address the assumptions underlying, for example, the hypothesis that species are more specialized and the impact of predation is greater in the tropics, with special emphasis on the relative importance of bottom-up versus top-down forces across the latitudinal gradient. They review the evidence suggesting that, in the tropics, plants are better defended and herbivory as well as natural enemy impacts are higher. They further present a meta-analysis testing whether herbivores have adapted to selection pressures from bottom-up and top-down forces. Dyer and Coley found enhanced levels of plant defenses (mechanical, biotic, and chemical effects), but not a more negative effect on tropical herbivores, presumably due to better adaptations to plant defences. In contrast, top-down effects of predators on herbivores and herbivores on plants were significantly stronger in the tropics. Clearly, more empirical studies are needed to test latitudinal differences in top-down cascades on plant communities and bottom-up cascades on consumer communities, to reduce the still great number of speculative points in the discussion on tropical–temperate differences.

The sheer number of endophytic fungi and insect taxa associated with plants implies that plant–fungi–insect interactions are much more common and important than the relatively few studies conducted to date suggest. Plant pathogens may change host resistance to herbivores and herbivore–enemy interactions in diverse ways. Faeth and Bultman (chapter 5) focus on changes in host plant resistance in response to an

Signals are very specific in only some cases, and the mechanisms that allow for this specificity within the constraints of genotypic plant variation remain to be shown. The direct benefits of plants derived from the impact of natural enemies appear to be equally variable. From the applied point of view, manipulation of traits of crop plants to enhance pest control is a major challenge. Future research can be expected to show whether genetic modifications of volatile emissions in conventionally bred and transgenic plants may have the potential to largely attract beneficial insects and to reduce pest damage.

Plant architecture defines the physical environment in which most herbivores and their natural enemies move, but only rarely has the geometry of the environment been incorporated in predictions of the outcome of prey–predator and host–parasitoid interactions. Casas and Djemai (chapter 8) review the available information on the role of plant geometry for the distribution of herbivores and the intrinsic movement rules of predators and parasitoids. Decision rules of parasitoids, such as giving-up times, may be influenced by the structural complexity of the environment, including the surface area within which predators have to forage, as well as the structural heterogeneity and connectivity of plant parts. The authors describe the latest developments in the modeling of plant canopies in relation to random walking. Simple random walks in homogeneous environments, and the approximating diffusion equations, appear to be poor guides for understanding search strategies of predators and successful prey location in plant canopies. They are best replaced by a concept of random walks in randomly or deterministically determined, geometrically structured environments. In the future, an integration of modeling of canopy architecture with carefully designed field experiments encompassing detailed observations of prey and predator movements will lead to significant progress in this newly developing field.

Indirect interactions between spatially separated organisms are common, due to changes in plant growth or other mediating mechanisms. Even more complex interactions occur between below- and aboveground organisms, which belong to traditionally separated research areas, but which affect each other in manifold ways. Differential effects on plant life-history groups modify competition between species and drive ecological successions of plants and plant–insect interactions, as shown by Brown and Gange (chapter 9). The authors focus on plant–mycorrhizal fungal–herbivore interactions in grasslands and present a simple model of community structure. For example, the exclusion of foliar-feeding, but not the root-feeding insects leads to a high grass–forb ratio, and this effect is particularly strong when mycorrhizal fungi are absent. There are very many species involved in the below- and above-ground food-web, and analyses of the possible combinations of species in both laboratory experiments and field studies are a challenge for the future. Plant succession is not solely a domain of plant ecologists. There are still many gaps in our understanding of how succession in plant communities is influenced by pathogens, mycorrhizae, nematodes, decomposers, vertebrates, and their interactions.

Concepts in food-web ecology largely rely on aquatic systems and above-ground terrestrial systems, while the trophic interactions in the soil have received little attention. Reinforcing the potential importance of below-ground components of food-webs discussed by Brown and Gange, Scheu and Setälä (chapter 10) stress that soil ecosystems form the basis of virtually all terrestrial life, and an appreciation of their many unique features may significantly change the way we perceive nature. Large soil invertebrates, in particular earthworms, function as ecosystem engineers and affect other soil organisms via habitat modifications. Due to the dense species packing, the prevalence of generalist feeders and the ubiquity of omnivory, soil communities are exceptionally complex. Trophic cascades are presumably of limited importance, but top-down control appears to be widespread. For example, fungivores, nematodes, and detritivores are under certain conditions controlled by predators, and the interactions between the bacterivorous microfauna (mainly Protozoa and Nematoda) and bacteria regulate the nutrient acquisition by plants. The many unresolved questions include how mineralization of nutrients by decomposer activity influences above-ground plant growth and the associated food-web, and how plant foliar–herbivore interactions modify soil communities.

In conclusion, the focus in ecology is changing from the traditional study of simple systems and interactions to approaches that consider the spatio-temporal variability of direct and indirect interactions among multiple trophic levels. This turning-point in ecology includes the realization that plants directly influence the behavior of their herbivores' natural enemies (chapters 1, 7, and 8), that ecological interactions between two species are often indirectly mediated by a third species (chapters 3 and 5), that landscape structure directly affects local tritrophic interactions and community structure (chapter 6), and that below-ground food-webs are extremely complex and vital to above-ground organisms (chapters 9 and 10). No one book can cover all the complexity found in nature. However, it is our hope that this volume will facilitate further development of the study of a range of ecological phenomena and patterns encompassed within the concept of "multitrophic level interactions." Integrating carefully designed field studies and mathematical models will be a necessary precondition for further development in this field. We have only recently started on a most exciting path to find out the main mechanisms and driving forces of ecosystems typically determined by multitrophic level interactions.

REFERENCES

Barbosa, P., Kirschik, L. and Jones, E. (eds.) (1990) *Multitrophic Level Interactions among Microorganisms, Plants, and Insects*. New York: John Wiley.

Bernays, E. and Graham, M. (1988) On the evolution of host specificity in phytophagous arthropods. *Ecology* **69**: 886–892.

Dicke, M. (2000) Chemical ecology of host plant selection by herbivorous arthropods: a multitrophic perspective. *Biochemical Systematics and Ecology* **28**: 601–617.

Gange, A. C. and Brown, V. K. (eds.) (1997) *Multitrophic Interactions in Terrestrial Systems*, 36th Symposium of the British Ecological Society. Oxford: Blackwell Science.

Hawkins, B. A. (1994) *Pattern and Process in Host–Parasitoid Interactions*. Cambridge: Cambridge University Press.

Olff, H., Brown, V. K. and Drent, R. H. (eds.) (1999) *Herbivores: Between Plants and Predators*, 38th Symposium of the British Ecological Society. Oxford: Blackwell Science.

Pace, M. L., Cole, J. J., Carpenter, S. R. and Kitchell, J. F. (1999) Trophic cascades revealed in diverse ecosystems. *Trends in Ecology and Evolution* **14**: 483–488.

Price, P. W., Bouton, C.E., Gross, P., McPheron, B.A., Thompson, J.N., and Weis, A.E. (1980) Interactions among three trophic levels: influence of plants on interactions between herbivores and natural enemies. *Annual Review of Ecology and Systematics* **11**: 41–65.

Schmitz, O. J., Hambäck, P. A. and Beckerman, A. P. (2000) Trophic cascades in terrestrial ecosystems: a review of the effects of carnivore removal on plants. *American Naturalist* **155**: 141–153.

J. DANIEL HARE

2

Plant genetic variation in tritrophic interactions

Introduction

The host plants of herbivores are not neutral substrates upon which interesting herbivores–natural enemies occur. Both the dynamics as well as the outcome of particular herbivore–natural enemy interactions may vary with the herbivore's host plant species, or genotype within species, and understanding such variation is central to the study of tritrophic interactions. The theory of tritrophic interactions implies that plant characteristics that enhance the success of the natural enemies have evolved; plants with traits that encourage the success of natural enemies should have a selective advantage over plants that do not, thus the trait should spread through the plant population (Price *et al.*, 1980; Fritz, 1992, 1995; Hare, 1992). The primary genetic question underlying this chapter is whether the impact of natural enemies on herbivores is sufficiently strong and systematic to cause changes in gene frequencies in plant traits affecting the impact of natural enemies on those herbivores.

In this chapter, I will provide a brief review of several studies showing how genetic variation in plants may affect the outcomes of herbivore–natural enemy interactions. Most of these studies originated from applied studies in biological control. The strengths and weaknesses of those data in developing a general expectation of tritrophic interactions will be evaluated. I will conclude with a list of specific research objectives that may facilitate the development of a more predictive theory of tritrophic interactions.

The main goal in research on tritrophic interactions in applied systems is to determine whether biological control can be combined with host plant resistance in developing a more highly integrated pest management program. Studies on the interaction between genetically determined host plant resistance and biological control agents are useful for the development of theories of tritrophic interactions because they clearly demonstrate the potential for natural enemies to vary in effectiveness on different plant genotypes. What must not be forgotten, however, is that genetically based variation in any plant trait that affects biological control agents is most likely fortuitous; plant breeders have not, as yet, explicitly incorporated such traits into plant breeding programs. Applied systems therefore are primarily valuable in showing how pre-existing host plant variation may affect herbivore–natural enemy interactions. Such systems are less useful to determine if natural enemies might impose natural selection on plants for traits that enhance the success of those natural enemies.

Overview of effects

The growth of studies showing effects of plant traits on herbivore–natural enemy interactions has grown considerably since the reviews by Bergman and Tingey (1979) and Price *et al.* (1980). It is impossible to provide an exhaustive review of those studies here, but several recent reviews appear elsewhere (e.g., Vet and Dicke, 1992; Tumlinson *et al.*, 1992; Marquis and Whelan, 1996; Bottrell *et al.*, 1998; Turlings and Benrey, 1998; Cortesaro *et al.*, 2000).

Bottrell *et al.* (1998), for example, provide a table that lists 74 selected references on how various plant traits may directly or indirectly affect arthropod parasitoids and predators. Some of the morphological features include plant size, aspects of shape of whole plants and plant parts, aspects of color, phenological differences, and surface characteristics such as pubescence. Semiochemical features acting directly on natural enemies include the production of various attractants, repellents, mimics, sticky substances, and plant toxins. Plant population traits affecting natural enemies directly include variation in plant density, host patch size, and vegetation diversity. In addition to these direct effects, Bottrell *et al.* (1998) also list several indirect plant effects, including the release of semiochemicals following plant attack by herbivores, the sequestration of plant toxins by herbivores, the effect of nutritional and resistance factors affecting herbivore quality for utilization by natural enemies, and the effect of microbial symbionts on herbivore–natural enemy interactions.

Similarly, Cortesaro *et al.* (2000) provide another table listing 148 citations to many plant traits similar to the above but also including how plants may provide food and shelter for natural enemies. The fact that these two tables contain only 15 citations in common speaks to the volume of research on the various ways in which plant characteristics may affect the natural enemies of herbivores.

Some general effects of plants on herbivore–natural enemy interactions include the following. Plant morphological features may either enhance or reduce natural enemy activity. Among the more beneficial traits are "domatia," which are small structures that provide food, shelter, or a hospitable environment for natural enemies (Agrawal and Karban, 1997). Other domatia may house ants that may protect the plants from herbivores (Beattie, 1985). In general, plant pubescence (i.e., plant trichomes) interferes with the movement of natural enemies, and this often reduces their effectiveness (Kauffman and Kennedy, 1989). In some cases, however, the presence of plant trichomes reduces the walking speed of parasitoids, thereby causing them to search more thoroughly (van Lenteren and de Ponti, 1991). The presence of wax on leaves can reduce the efficiency of searching by natural enemies because the leaves are too slippery for the insects to grip (see Eigenbrode and Espelie, 1995 for a review). Increasing complexity of leaf shape also interferes with the foraging efficiency of some predators (Kareiva and Sahakian, 1990) and parasitoids (Andow and Prokrym, 1990).

Plant chemical compounds can have diverse effects on natural enemies. Volatile semiochemicals often serve as attractants, not only for herbivorous insects, but also for their natural enemies. Some of these compounds may be constitutive plant products that are produced whether plants are damaged or not. Others may be released after mechanical damage from diverse agents, and still others may be released only after feeding by particular herbivore species (reviewed by Tumlinson *et al.*, 1992; Vet and Dicke 1992; Turlings *et al.*, 1995; Turlings and Benrey, 1998). Plant allelochemicals that are taken up by herbivores often may be deleterious to natural enemies, either through active sequestration of toxins to which the herbivore is well adapted, or indirectly by causing reductions in feeding that lead to reductions in the size or quality of herbivores as hosts for natural enemies. Alternatively, plant toxins can also be advantageous to the natural enemies and indirectly to the plant if the toxins weaken the defenses or prolong the vulnerability of the herbivores (reviewed by Turlings and Benrey, 1998).

Host plant resistance and biological control

Historically, pest managers assumed that host plant resistance and biological control were compatible and largely independent pest management strategies (Adkisson and Dyck, 1980; Kogan, 1982). An early theoretical approach reflecting this view was derived from deterministic mathematical models of host–parasitoid population dynamics. Those models show that effective control by natural enemies would be enhanced if the rate of increase of the host population were reduced (van Emden, 1966; Beddington *et al.*, 1978; Hassell, 1978; Lawton and McNeil 1979; Hassell and Anderson 1984). Such a prediction assumes that the host plant affects only the growth rate of the prey population and not the attractiveness or quality of prey individuals for discovery and utilization by natural enemies. One can argue that the development of tritrophic interactions as a separate field of inquiry grew from tests of this basic assumption. By now, it is well known that prey on different host plants are often not of uniform quality (reviewed by Bergman and Tingey, 1979; Price *et al.*, 1980; Boethel and Eikenbary, 1986; Duffey and Bloem, 1986; Price, 1986; Vinson and Barbosa, 1987; Barbosa and Letourneau, 1988; Fritz, 1992; Hare, 1992; Vet and Dicke, 1992; Turlings and Benrey, 1998).

In order to better understand the consequences of host plant variation on population dynamics of herbivore and natural enemies, I previously developed five models of responses based upon the statistical form of the interaction between host plant resistance and biological control on equilibrium pest density (Hare, 1992). Since then, these models have been useful in resolving some of the confusion in the use of the term "compatibility" between host plant resistance and biological control. Four of these models are reviewed below.

A purely additive relationship between host plant resistance and biological control exists when the incremental numerical reduction in equilibrium herbivore density caused by natural enemies is independent of that caused by host plant resistance, and uniform at all levels of host plant resistance. Therefore, the expected equilibrium pest density due to host plant resistance and natural enemies can be predicted simply from the combined effects of both acting independently. An additive relationship is the "null" hypothesis and precludes any biological or statistical interactions.

In a simple synergistic model, the incremental reduction in equilibrium herbivore density caused by natural enemies is relatively greater at high host plant resistance levels than at low. This form of interaction is obviously compatible in a pest management program and the ideal model to be sought.

An antagonistic model specifies that the reduction in equilibrium herbivore density due to host plant resistance and natural enemies is less than would be calculated if the interaction were additive. Mildly antagonistic interactions also may be "compatible" interactions in pest management because the reduction in equilibrium pest densities due to both host plant resistance and biological control is still greater than that expected from either tactic alone.

When the antagonism is more severe, then host plant resistance replaces mortality once caused by natural enemies in reducing equilibrium herbivore density. This was termed a disruptive interaction (Hare, 1992). Such an interaction would be expected when natural enemies are more susceptible than herbivores to plant resistance mechanisms (e.g., Campbell and Duffey, 1979; Obrycki and Tauber, 1984; Farrar *et al.*, 1994). This form of interaction would be incompatible from low to intermediate host plant resistance levels, and there would be essentially no interaction at high host plant resistance levels due to the high mortality suffered by natural enemies.

An important point to recognize is that all models except the disruptive model show qualitative compatibility between host plant resistance and biological control. The only difference between the additive, synergistic, or mildly antagonistic models is whether the magnitude of pest population reduction differs from that predicted assuming host plant resistance and biological control imposed independent sources of mortality.

A summary of 61 studies exploring the potential interaction between resistant crop varieties and natural enemies in a number of crops is shown in Table 2.1. With regard to parasitoids, antagonistic interactions were found in eight of 30 unambiguous cases (27%), while a synergistic relationship was found in only three (10%). Additive interactions were found in 16 of the cases (53%), and three clearly disruptive cases also were found (10%). A similar summary was developed about ten years ago from a smaller number of studies (Hare, 1992). In that summary, antagonistic interactions were slightly more frequent than additive interactions, while synergistic interactions were rare and no disruptive interactions were identified. Much of the more recent literature tends to show a higher frequency of additive interactions since the earlier review.

Table 2.1. *Effect of host plant resistance in cultivated crops on life history characteristics of natural enemies of selected insect pests*

Pest	Natural enemy	Resistant plant	Mode of resistance	Effect on natural enemy	Comments	Reference
Parasitoids						
Pseudoplusia includens	*Copidosoma truncatellum*	Soybean (*Glycine max*)	Reduced survival and growth	Antagonistic	Increased parasitoid development time	Orr & Boethel, 1985
Pseudoplusia includens	*Microplitis demolitor*	Soybean	Lengthened development time, reduced weight	Antagonistic	Reduced parasitoid survival, reduced number of available prey	Yanes & Boethel, 1983
Pseudoplusia includens	*Voria ruralis*	Soybean	Reduced survival	Synergistic/ antagonistic, varied with resistance level	Reduced parasitoid puparia per host, reduced pupal weight, increased host death prior to parasitoid emergence	Grant & Shepard, 1985
Epilachna varivestis	*Pediobius foveolatus*	Soybean	Lengthened development time, reduced survival, reduced egg production	Antagonistic	Possibly a net compatibility of biological control with host plant resistance	Kauffman & Flanders, 1985
Epilachna varivestis	*Pediobius foveolatus*	Soybean	Reduced survival, growth and weight	Additive/ antagonistic	Male-biased sex ratio on one cultivar, no effects on others	Dover *et al.*, 1987
Heliothis virescens, *Helicoverpa zea*	*Microplitis croceipes*	Soybean	Antibiosis, reduced growth	Synergistic/ antagonistic, varied with resistance level	Increased development time, reduced parasitoid adult weight	Powell & Lambert, 1984

Table 2.1. *(cont.)*

Pest	Natural enemy	Resistant plant	Mode of resistance	Effect on natural enemy	Comments	Reference
Parasitoids *(cont.)*						
Nezara viridula	*Telenomus chloropus*	Soybean	Mild antibiosis and antixenosis	Antagonistic	Reduced survival, reduced total fecundity of egg parasitoid	Orr *et al.*, 1985
Four Lepidoptera species	*Cotesia marginiventris*, and others	Soybean	Reduced survival, reduced growth, lengthened development time	Additive	No differences in parasitism levels across cultivars	McCutcheon & Turnipseed, 1981
Spodoptera frugiperda	*Campoletis sonorensis*	Maize (*Zea mays*)	Reduced growth	Antagonistic	Increased parasitoid development time	Isenhour & Wiseman, 1989
Spodoptera frugiperda	*Campoletis* sp.	Maize	Antixenosis	Synergistic	Increased parasitization rate	Pair *et al.*, 1986
Schizaphis graminum	*Lysiphlebus testaceipes*	Barley (*Hordeum vulgare*), sorghum (*Sorghum bicolor*)	Reduced population growth, reduced size	Antagonistic	Fewer and smaller mummies, but not disruptive	Starks *et al.*, 1972
Diuraphis noxia	*Aphelinus albipodus*, *A. asychis*	Barley	Tolerance, antibiosis	Additive	Approximately equal parasitization rates in resistant and susceptible cultivars	Brewer *et al.*, 1998, 1999
Schizaphis graminum	*Lysiphlebus testaceipes*	Oat (*Avena sativa*)	Reduced population growth	Additive	No effect on parasitoid survival, sex ratio, or development time	Salto *et al.*, 1983

Diuraphis noxia	*Diaeretiella rapae*	Wheat (*Triticum aestivum*)	Tolerance	Additive	Similar levels of aphid suppression on resistant and susceptible varieties	Reed *et al.*, 1991
Diuraphis noxia	*Diaeretiella rapae*	Wheat	Antibiosis, antixenosis	Additive	Survival, sex ratio, adult longevity, and wasp size were not affected by host plant resistance	Farid *et al.*, 1998
Oulema melanopus	*Anaphes flavipes*	Wheat	Antibiosis, antixenosis	Additive	Pubescent wheat had no effect on larval parasitoids	Lampert *et al.*, 1983
Oulema melanopus	*Tetrastichus julis, Diaparsis termoralis, Lemophagus curtus*	Wheat	Antibiosis, antixenosis	Additive	Pubescent wheat had no effect on larval parasitoids	Casagrande & Haynes, 1976
Diuraphis noxia	*Diaeretiella rapae*	Triticale (*Triticosecale*)	Antibiosis	Antagonistic	Reduced parasitoid size and growth rate	Reed *et al.*, 1991
Aphis gossypii	*Lysiphlebus testaceipes*, and others	Cantaloupe (*Cucumis melo*)	Antibiosis and/or antixenosis	Additive	Parasitization level independent of aphid density across cultivars	Kennedy *et al.*, 1975
Acyrthosiphon pisum	Various parasitoids and predators	Alfalfa (*Medicago sativa*)	Reduced population growth	Additive	No variation in parasitism rates or predator numbers	Pimentel & Wheeler, 1973
Myzus persicae	*Aphidius matricariae*	Chrysanthemum (*Chrysanthemum* sp.)	Reduced population growth	Additive	No effect of density on searching behavior or apparent effect on parasitoid population growth	Wyatt, 1970

Table 2.1. *(cont.)*

Pest	Natural enemy	Resistant plant	Mode of resistance	Effect on natural enemy	Comments	Reference
Parasitoids (*cont.*)						
Bruchus pisorum	*Eupteromalus leguminis*	Pea (*Pisum sativum*)	Antixenosis	Synergistic	Increased parasitization	Annis & O'Keefe, 1987
Helicoverpa zea	*Archytas marmoratus*	Tomato (*Lycopersicon hirsutum* f. *glabratum*)	Antibiosis	Disruptive	Plant toxic to tachinid larva	Farrar & Kennedy, 1993
Helicoverpa zea	*Eucelatoria bryani*	Tomato	Antibiosis	Additive	No effect on tachinid parasitoid	Farrar & Kennedy, 1993
Helicoverpa zea, *Heliothis virescens*	*Trichogramma* sp.	Tomato	Antibiosis	Disruptive	Parasitoid inhibited more by glandular trichomes than host	Farrar *et al.*, 1994
Manduca spp.	*Telenomus sphingis*	Tomato	Antibiosis	Disruptive	Parasitoid inhibited more by glandular trichomes than host	Farrar *et al.*, 1994
Helicoverpa zea	*Campoletis sonorensis*	Tomato	Antibiosis	Antagonistic	Parasitoid inhibited somewhat more by glandular trichomes than host	Farrar *et al.*, 1994
Helicoverpa zea, *Heliothis virescens*	*Cotesia marginiventris*	Tomato	Antibiosis	Additive	Parasitoid not strongly affected by host plant resistance	Farrar *et al.*, 1994
Heliothis virescens	*Cardiochiles nigriceps*	Tomato	Antibiosis	Additive	Parasitoid not affected by host plant resistance	Farrar *et al.*, 1994

Aphis gossypii	*Lysiphlebus testaceipes*	Cotton (*Gossypium hirsutum*)	Reduced population growth on smooth-leaved cultivars	Additive	No effect of pubescence on parasitization rates	Weathersbee & Hardee, 1994
Helicoverpa zea	*Trichogramma pretiosum*	Cotton	Antixenosis	Synergistic	Glabrous plants are non-preferred by the host and host eggs suffer greater parasitization	Treacy *et al.*, 1985
Heliothis virescens	*Campoletis sonorensis*	Cotton	Antibiosis, antixenosis	Additive	Similar degree of parasitization on resistant and susceptible cultivars	Lindgren *et al.*, 1978
Planococcus citri	*Leptomastix dactylopii*	Coleus (*Coleus blumei*)	Antibiosis	Synergistic / antagonistic, depends on resistance level	Parasitoid cannot completely compensate for plant-mediated variation in host population growth	Yang & Sadof, 1997
Predators						
Nephotettix virescens	*Lycosa pseudoannulata*	Rice (*Oryza sativa*)	Reduced oviposition and survival	Additive	Predatory spider	Myint *et al.*, 1986
Nephotettix virescens	*Cyrtorhinus lividipennis*	Rice	Reduced oviposition and survival	Additive	Predatory bug	Myint *et al.*, 1986
Sogatella furcifera	*Lycosa pseudoannulata*	Rice	Reduced survival	Additive	All predators cause *c.* 30% additional mortality on all varieties	Salim & Heinrichs, 1986

Table 2.1. *(cont.)*

Pest	Natural enemy	Resistant plant	Mode of resistance	Effect on natural enemy	Comments	Reference
Predators *(cont.)*						
Sogatella furcifera	*Cyrtorhinus lividipennis*	Rice	Reduced survival	Additive	All predators cause *c.* 30% additional mortality on all varieties	Salim & Heinrichs, 1986
Sogatella furcifera	*Harmonia octomaculata*	Rice	Reduced survival	Additive	All predators cause *c.* 30% additional mortality on all varieties	Salim & Heinrichs, 1986
Sogatella furcifera	*Paederus fucipes*	Rice	Reduced survival	Additive	All predators cause *c.* 30% additional mortality on all varieties	Salim & Heinrichs, 1986
Nilaparvata lugens	*Lycosa pseudoannulata*, and others	Rice	Mild antixenosis, reduced survival	Synergistic	Increased predation rate due to increased prey movement	Kartohardjono & Heinrichs, 1984
Anticarsia gemmatalis	*Geocoris punctipes*	Soybean	Reduced growth rate	Antagonistic	Reduced growth rate and survival	Rogers & Sullivan, 1986
Pseudoplusia includens	*Geocoris punctipes*	Soybean	Reduced growth rate	Antagonistic	Reduced growth rate	Rogers & Sullivan, 1986
Pseudoplusia includens	*Podisius maculiventris*	Soybean	Lengthened development time, reduced growth	Antagonistic	Similar effects on predator as on host	Orr & Boethel, 1986
Tetranychus urticae	*Phytoseiulus persimilis*	Soybean	Antibiosis, antixenosis	Additive	No interaction between host plant resistance and predation rate	Wheatley & Boethel, 1992

Spodoptera frugiperda	*Orius insidiosus*	Maize	Antibiosis, antixenosis	Synergistic	Increased predation	Isenhour *et al.*, 1989
Helicoverpa zea	*Orius insidiosus*	Maize	Reduced growth	Synergistic	Increased predation	Isenhour *et al.*, 1989
Helicoverpa zea	*Chrysopa rufilabris*	Cotton	Antixenosis	Synergistic	Glabrous plants are non-preferred by the host, and host eggs suffer greater predation by young predator larvae	Treacy *et al.*, 1985
Tetranychus spp.	Three predators	Cotton	Antibiosis, antixenosis	Additive	Similar suppression of mites on resistant and susceptible cultivars	Trichilo & Leigh, 1986
Acyrthosiphon pisum	*Hippodamia convergens*	Alfalfa	Reduced aphid reproduction	Additive	Aphid consumption and beetle larval development did not differ between cultivars	Karner & Manglitz, 1985
Bemisia argentifolii	*Delphastus pusillus*	Tomato	Antibiosis	Additive	Plant trichomes inhibited whiteflies but did not affect predation rates by the beetle	Heinz & Zalom, 1996
Pathogens						
Spodoptera exigua	*Bacillus thuringiensis*	Celery (*Apium graveolens*)	Antibiosis, reduced growth rate	Additive	No interaction between host plant resistance and Bt on larval mortality	Meade & Hare, 1995

Table 2.1. *(cont.)*

Pest	Natural enemy	Resistant plant	Mode of resistance	Effect on natural enemy	Comments	Reference
Pathogens (*cont.*)						
Trichoplusia ni	*Bacillus thuringiensis*	Celery	Antibiosis, reduced growth rate	Additive	No interaction between host plant resistance and Bt on larval mortality	Meade & Hare, 1995
Helicoverpa zea	*Bacillus thuringiensis*	Soybean	Reduced growth rate	Synergistic	Increased dose-specific mortality	Kea *et al.*, 1978
Helicoverpa zea	*Bacillus thuringiensis*	Soybean	Reduced survival	Synergistic	Death at earlier age	Bell, 1978
Helicoverpa zea	*Nomuraea rileyi*	Soybean	Reduced survival	Synergistic	Death at earlier age	Bell, 1978
Pseudoplusia includens	Nuclear polyhedrosis virus (NPV)	Soybean	Antibiosis, antixenosis	Additive	Similar population reductions by the virus on resistant and susceptible cultivars	Beach & Todd, 1988
Anticarsia gemmatalis	NPV	Soybean	Antibiosis, antixenosis	Additive	Similar population reductions by the virus on resistant and susceptible cultivars	Beach & Todd, 1988
Four Lepidoptera species	*Nomuraea rileyi*	Soybean	Antibiosis	Additive	No effect on other pathogens	Gilreath *et al.*, 1986
Helicoverpa zea	Cytoplasmic polyhedrosis virus	Maize	Antibiosis	Additive	Host plant resistance and the virus acted independently on insect survival	Bong *et al.*, 1991

Heliothis virescens	*Bacillus thuringiensis*	Cotton	Antixenosis	Synergistic	Increased host mortality	Schuster *et al.*, 1983
Aphis gossypii	*Neozygites fresenii*	Cotton	Reduced population growth on smooth-leaved cultivars	Additive	No effect of pubescence on fungal infection rates	Weathersbee & Hardee, 1994

The three cases of synergism all involved greater parasitization rates of hosts on resistant plants. For *Spodoptera frugiperda* being attacked by *Campoletis* spp., the greater parasitization rate may have been the result of greater movement of host larvae on the antixenotic maize varieties, thereby increasing their conspicuousness and time of exposure to foraging natural enemies (Pair *et al.*, 1986). For the egg parasitoid *Trichogramma pretiosum* attacking eggs of *Helicoverpa zea*, it is likely that the absence of leaf pubescence increased the foraging efficiency of this small wasp on the cotton cultivars that also were antixenotic to *H. zea* larvae (Treacy *et al.*, 1985). No specific mechanism was presented in the third example for increased parasitization of bruchid beetle larvae on partially resistant genotypes of pea (Annis and O'Keefe, 1987).

Host plant resistance in soybean was responsible for four of the eight clearly antagonistic interactions. These antagonistic interactions may reflect the absence of any long-term coevolutionary history between soybean, its North American pests, and those pests' natural enemies, for this group of species has been associated for less than a century (see also Orr and Boethel, 1986; Boethel, 1999).

The disruptive interactions involved parasitoids attacking hosts on wild tomato plants (*Lycopersicon hirsutum* f. *glabratum*) with two resistance mechanisms. The first is a trichome-based mechanism that confers resistance to *Manduca sexta* and other insect species, and the second is a non-trichome-based mechanism conferring partial resistance to *Helicoverpa zea* and *Heliothis virescens*. In a number of studies, Kennedy and co-workers showed that the methyl ketones causing the trichome-based resistance were toxic or deleterious to several species of insect parasitoids (Farrar and Kennedy, 1993; Farrar *et al.*, 1994; and references therein). In the field, the deleterious impact on the natural enemies of *Helicoverpa zea* and *Heliothis virescens* was sufficiently great that the densities of these two insect pests were similar, if not higher, on the partially resistant plants (Farrar *et al.*, 1994).

The general trend is for host plant resistance to be compatible with biological control by generalist predators (59% of the cases involving predators). The only clearly antagonistic interactions involve two predators of pests of soybean. One of these, *Geocoris punctipes*, may acquire deleterious allelochemicals from resistant soybean cultivars through direct host plant feeding (Rogers and Sullivan, 1986), although this is apparently not the case for the other negatively affected predator, *Podisius maculiventris* (Orr and Boethel, 1986).

Three synergistic interactions involving predators are probably the result of an increased rate of discovery of prey by predators due to increased prey movement on antixenotic cultivars (Kartohardjono and Heinrichs, 1984; Isenhour *et al.*, 1989). In the fourth synergistic interaction noted, foraging efficiency of chrysopid larvae increased as plant trichome density declined (Treacy *et al.*, 1985).

Although a number of laboratory studies have shown both positive and negative effects of selected plant chemicals on the efficacy of insect pathogens (reviewed by Reichelderfer, 1991; Schultz and Keating, 1991), most studies using susceptible and resistant plant cultivars generally show an overall compatibility between host plant resistance and biological control by pathogens (64% additive interactions and 36% synergistic interactions). For the synergistic interactions, the susceptibility of host larvae to pathogens was inversely related to the growth and vigor of the larvae, which itself was directly related to the level of host plant susceptibility. In none of the 11 cases evaluated were there any antagonistic or disruptive interactions.

The implications of many of the results in Table 2.1 must be accepted with caution, for most of them simply examine the impact of host plant resistance on selected life history parameters of natural enemies. There could have been other effects opposite to those listed in other, unmeasured natural enemy parameters. Without explicit population studies, it may be difficult to translate results of short-term studies showing changes in such life history parameters of natural enemies into changes in herbivore–natural enemy population dynamics. Reductions in survivorship, growth, or fecundity of a particular natural enemy would not result in a reduction in biological control if the population growth rate of the prey population were reduced even more.

For example, while plant resistance led to a reduction in the size and number of parasitoids from greenbugs, overall plant damage was least and greenbug populations were smallest on resistant varieties in the presence of parasitoids (Starks *et al.*, 1972). Similarly, several life history parameters of *Pediobius foveolatus*, a parasitoid of the Mexican bean beetle, *Epilachna varivestis*, were reduced when reared on hosts that were themselves reared on resistant soybean cultivars. However, the population growth potential of the parasitoid was reduced less than was that of the host by host plant resistance, so that the intrinsic rate of increase of the parasitoid was greatest relative to that of its host on the resistant cultivar (Kauffman and Flanders, 1985).

More recently, Yang and Sadof (1997) compared population growth rates of the citrus mealybug, *Planococcus citri,* with those of its parasitoid, *Leptomastix dactylopii*, and concluded that most effective biological control would occur on a coleus cultivar (*Coleus blumei*) that expressed an intermediate level of resistance to citrus mealybugs. This was despite the fact that parasitoids reached greatest size and fecundity on mealybugs reared on the most susceptible *Coleus* cultivar. However, despite the increased parasitoid population growth rate on the most susceptible cultivar, the parasitoid could not compensate for the even greater population growth rate of the host on that cultivar. On the most resistant cultivar, the parasitoid's rate of population growth was reduced more than that of its host, so the host was expected to have a net advantage on the most resistant cultivar as well as the least resistant cultivar (Yang and Sadof, 1997). These three studies point out the continuing need for comparative research on population growth in order to better resolve ambiguities from laboratory studies on the actual impact of host plant resistance on herbivore and natural enemy population dynamics.

Biological control and transgenic plants

Transgenic crop plants expressing toxic proteins from *Bacillus thuringiensis* (hereafter Bt) are being grown commercially on ever-increasing acreage (Gould, 1998; Schuler *et al.,* 1998). Such releases have occurred without much concern on how those plants may affect natural levels of biological control, not only of the primary target pest of the crop, but also secondary pests within the agroecosystem (Schuler *et al.,* 1999a.)

Such effects could include substantial reductions in population densities of biocontrol agents resulting from: (1) reduction in densities of hosts, (2) feeding on hosts that have acquired the toxin, or (3) reductions in densities by feeding on hosts of smaller size or otherwise reduced quality. The latter factor may be more important for the dynamics of parasitoid populations than for predators, for predators simply may increase their feeding rate to compensate for reduced prey size (Schuler *et al.,* 1999a).

To date, the effects of transgenic Bt toxins on natural enemies seem to be relatively minor (Table 2.2). Part of the reason for this may be the selective toxicity of Bt proteins against Lepidoptera or Coleoptera but not Hymenoptera. Some indirect effects have been noted, however, when targeted pests ingest sublethal Bt doses. This results in a reduction or

Table 2.2. *Effect of transgenic* Bacillus thuringiensis *on life history characteristics of natural enemies*

Pest	Natural enemy	Crop	Effect on natural enemy	Comments	Reference
Pectinophora gossypiella, other leaf-feeding Lepidoptera	Various	Cotton	Additive	No effect on densities in small-plot field study	Flint *et al.*, 1995
Pectinophora gossypiella, other leaf-feeding Lepidoptera	Various	Cotton	Additive	No effect on densities in small-plot field study	Wilson *et al.*, 1992
Ostrinia nubilalis	*Orius insidiosus, Coleomegilla maculata* and other predators	Maize	Additive	No effect on parasitization rates, predation, or density of predators	Orr & Landis, 1997
None	*Coleomegilla maculata, Orius insidiosus, Chrysoperla carnea*	Maize	Additive	No effect of transgenic maize pollen on these predators or reductions in densities in the field	Pilcher *et al.*, 1997
Spodoptera frugiperda	*Chrysoperla carnea*	Maize	Antagonistic	Increased mortality	Hilbeck *et al.*, 1998
Ostrinia nubilalis	*Chrysoperla carnea*	Maize	Antagonistic	Increased mortality, increased development time	Hilbeck *et al.*, 1998
Myzus persicae	*Hippodamia convergens*	Potato (*Solanum tuberosum*)	No effect	Potatoes expressed Bt toxin used against the Colorado potato beetle. No effect on aphids or predatory beetle	Dogan *et al.*, 1996
Helicoverpa zea	*Nabis* spp.	Tobacco (*Nicotiana tabacum*)	Additive	No effect on densities in small-plot field study	Hoffmann *et al.*, 1992
Heliothis virescens	*Campletis sonorensis*	Tobacco	Synergistic	Greater than expected host mortality when attacked on Bt tobacco	Johnson & Gould, 1992
Plutella xylostella	*Cotesia plutellae*	Oilseed rape (*Brassica napus*)	Additive	No difference in survival of wasps in Bt-resistant hosts when fed transgenic or wild-type plants; no difference in attractiveness of Bt-resistant hosts on transgenic or wild-type plants	Schuler *et al.*, 1999b

termination of feeding and slower growth. Thus hosts may spend an increased time in life stages that are vulnerable to natural enemies (e.g., Johnson and Gould, 1992).

The reduced feeding, however, also may result in reduced emissions of feeding-related kairomones or synomones thus reducing the rate of discovery (Johnson *et al.*, 1997). Reduced consumption of Bt-transgenic oilseed rape foliage by Bt-susceptible diamondback moth, *Plutella xylostella*, was apparently responsible for the significantly lower attractiveness of damaged transgenic oilseed rape foliage compared to wild-type foliage to the parasitoid, *Coteslia plutellae*. The observation that the wasp was equally attracted to transgenic and wild-type foliage that were damaged equivalently by Bt-resistant *P. xylostella* (Schuler *et al.*, 1999b) ruled out a direct, differential response of the wasp to wild-type vs. transgenic foliage alone.

Both an increased time of vulnerability and reduced attractiveness of damaged transgenic foliage effects occurred in the system involving *H. virescens* and the parasitoid, *Campoletis sonorensis* and transgenic tobacco. The net result of these opposing effects was that *H. virescens* suffered reduced parasitism on transgenic tobacco plants (Johnson *et al.*, 1997). Thus, sublethal effects of consumption and ingestion of Bt-transgenic foliage on subsequent risk to natural enemies may affect the risk of herbivores to discovery and utilization via several potentially conflicting mechanisms, thereby making the overall outcome locally variable and difficult to predict.

Such conflicting mechanisms also introduce a variety of outcomes when predicting how the activity of natural enemies might affect the rate of adaptation of pests to resistant plants. Gould *et al.* (1991) developed mathematical models addressing this point that yielded straightforward conclusions: enemies that increase the difference in fitness between adapted and non-adapted plants increased the rate of adaptation to resistant plants, while enemies that decreased the difference in fitness between adapted and non-adapted plants retarded the rate of adaptation to resistant plants.

For example, in many natural enemies, host quality is a function of host size. Larger hosts provide greater resources for parasitoid utilization and are preferentially attacked. On resistant plants, better-adapted hosts are likely to reach a larger size than are less-adapted hosts. The natural enemy will preferentially attack these larger, better-adapted larvae, however, and the natural enemy will therefore preferentially impose additional mortality on the better-adapted hosts from the population. Thus, the natural enemy acts to decrease the difference in fitness between adapted and non-adapted hosts, thereby reducing the rate of adaptation of the host population.

In contrast, other natural enemies are limited to attacking small hosts within a limited size range. In this case, the better-adapted hosts may pass through the window of vulnerability more rapidly than may less-adapted hosts. If so, then the natural enemy will preferentially attack the less-adapted hosts, thereby imposing additional mortality on the less-adapted hosts. As a result, the natural enemy will increase the difference in fitness between adapted and less-adapted hosts, thereby increasing the rate of adaptation to resistant plants. Given the wide range of variation in herbivore–parasitoid interactions, predictions on the effect of natural enemies on pest adaptation to resistant plants may be practical only on a case-by-case basis (Gould *et al.*, 1991).

Plant breeding for attributes that improve biological control

With the growth of our understanding about how the effectiveness of natural enemies may vary as a function of specific host plant traits, it may be possible to manipulate and exploit those traits for improved biological control. There have been several recent reviews addressing this topic, (e.g., Poppy, 1997; Bottrell *et al.*, 1998; Verkerk *et al.*, 1998; Cortesero *et al.*, 2000). My purpose here is only to summarize some of the major prospects and problems concerning this novel approach toward improved pest management.

The deployment of plant trichomes in insect pest management illustrates an important problem in the breeding of natural-enemy-friendly plants. Glandular trichomes often directly defend plants against several insect pests (Dimock and Kennedy, 1983; Wagner, 1991; Duke, 1994; Juvik *et al.*, 1994), but they also may be as deleterious to particular natural enemies as to pest species (Farrar *et al.*, 1994). Thus, the plant breeder faces a dilemma of breeding for increased plant trichomes as a direct mechanism of plant defense, or breeding for reduced trichome numbers to encourage natural enemies. Although, in some cases, intermediate trichome densities may be an acceptable compromise (Obrycki, 1986), the final outcome may not be known without more detailed information on such factors as the herbivore mortalities imposed directly by trichomes, how much additional mortality might be provided by natural enemies on trichome-free plants, and which herbivore species are the most important to control.

In theory, breeding plants that produce greater quantities of herbivore-induced synomones that attract natural enemies should lead to enhanced biological control. Genetic variation in induced synomone production has been shown to exist among genotypes of cotton (Loughrin *et al.*, 1995) and maize (Gouinguené *et al.*, 2001; Turlings *et al.*, chapter 7, this volume). The effectiveness of this approach, however, may be uncertain and depend upon particular characteristics of the plant–herbivore–parasitoid system.

Turlings *et al.* (1995) developed three criteria to evaluate whether herbivore-induced volatiles can serve as effective signals of plant damage to natural enemies. First, the emitted signal must be clear enough so that it can be perceived and distinguished from background noise (e. g., normal plant volatiles). Second, the signal has to be specific enough to reliably indicate the presence of a suitable host or prey. Third, the signal must be emitted when natural enemies are foraging. For the volatile emissions of corn and cotton plants, Turlings *et al.* (1995) concluded that the criteria of clarity and appropriate timing were satisfied, but the criterion of specificity had limited support, especially when different herbivores damage the same plant and cause the release of similar, if not identical volatile compound blends. (See Dicke (1999) for a recent review of the evidence in favor or against the hypothesis of the specificity of herbivore-induced synomones.)

Parasitoid species also are known to vary in their response to plants with varying degrees and types of damage. Even closely related natural enemy species may differ in their response to damaged vs. undamaged plants and different sources of damage (Takabayashi *et al.*, 1998). Although some wasps respond only to specific volatile compounds that are released only in response to host feeding, sometimes by a particular life stage of the herbivore (e.g., Takabayashi *et al.*, 1998), other wasps appear to be as attracted to damage caused by non-host insects as host insects, (e.g., Geervliet *et al.*, 1996).

Additional factors in manipulating plant-produced synomones for pest management include complications caused by environmental variation in the ability of plants to emit such synomones as a function of variation among leaves within plants, among cultivars, or growing conditions within cultivars (Takabayashi *et al.*, 1994). The emission of synomone components that attract predatory mites differed qualitatively between young and old leaves, between leaves exposed to different light levels, and under different levels of water stress. In general, young leaves produced more attractive chemical blends than older leaves, leaves grown under high light levels produced more attractive blends than leaves under low light, and leaves from water-stressed plants produced more attractive volatile blends than leaves that were not water stressed (Takabayashi *et al.*, 1994). Such sources of variation in the strength and composition of herbivore-induced synomones may result in variable levels of natural-enemy enhancement under field conditions.

Bottrell *et al.* (1998) conclude their chapter on manipulation of natural enemies for pest management by listing several important considerations. The first is to carefully choose which natural enemies require manipulation. If a suite of natural enemies is necessary for successful pest control, then plant cultivars that only enhance the effectiveness of one may be of little value. Bottrell *et al.* (1998) also warn against unwanted side-effects, in which characteristics that enhance the effectiveness of natural enemies of one pest have deleterious consequences for other pests. The "frego bract" characteristic of cotton is an example of such a trait. Biological control of cotton boll weevils, *Anthonomus grandis*, by *Bracon mellitor* is enhanced on frego bract cotton, but frego bract cotton also is more susceptible to *Lygus* spp. and other homopterans that feed on floral buds (Lincoln *et al.*, 1971). Thus the deployment of a trait of cotton that enhances the success of a natural enemy of the boll weevil would occur at the expense of increasing the susceptibility of cotton to other pests.

Although there is substantial information on the ways in which plants may influence biological control agents, often we still lack an appropriate ecological context within which to interpret that information. The potential interactions among the pests of different crops, the dynamics of single pests in different crop systems, and interactions among natural enemies all need to be understood before the value of manipulating a particular natural enemy of a single pest species can be determined.

Finally, if natural-enemy enhancement is to succeed as a commercially viable pest management strategy, then growers also need to be convinced of its effectiveness. This requires that studies be carried out to demonstrate the values of natural-enemy manipulations on crop yield and crop value. Simple demonstrations of more immediate phenomena, such as greater densities of natural enemies following manipulation, are encouraging, but they are far from convincing proof that such manipulations have a role in contemporary agricultural production systems.

Tritrophic interactions in natural systems

One substantial gap in our understanding of how to exploit tritrophic interactions in agricultural systems is an almost complete lack of an understanding of the dynamics of tritrophic interactions in natural systems. This situation has changed little since a previous review of this topic (Hare, 1992). As with the applied systems reviewed earlier, most available studies on natural systems demonstrate that host plant variation can influence the interaction between predators and natural enemies. The studies, however, were not designed to address whether the host plant effects on natural enemies are evolved responses or simply fortuitous consequences of pre-existing plant genetic variation.

In goldenrod (*Solidago altissima*), the size of galls induced by *Eurosta solidaginis* is a heritable character of both insect and plant genotype (Weis and Abrahamson, 1986). In this system, gall-makers inhabiting large galls are less susceptible to parasitoids due to limitations on the length of the parasitoids' ovipositors. To the extent that gall diameter is under partial genetic control of the host plant, there exists the possibility that plants, their gall-makers, and the gall-makers' parasitoids may impose competing selection on plants for gall size. Variation among plant genotypes for gall size (and ultimate parasitoid success) would not, in general, lead to differential reproduction of plant genotypes (Weis and Abrahamson, 1985). The effect of parasitoids as selective agents favoring gall-makers that induce larger galls was more significant than was the effect of parasitoids reducing herbivore damage to plants. An additional complicating factor in this system is the fact that avian predators more readily attack large galls, and the advantage for gall-makers to induce large galls appears to be partially (but not completely) offset by increasing their risk of predation by birds (Weis and Abrahamson, 1986).

Price and Clancy (1986) and Clancy and Price (1987) similarly implicated genetic variation among clones of willow *Salix lasiolepis* in the size of galls induced by stem-galling sawflies. In the first study, mean gall diameter induced by *Euura lasiolepis* differed consistently among willow clones over a five-year study. Sawflies in smaller galls were more susceptible to parasitization by *Pteromalus* sp., and parasitism by *Pteromalus* was inversely related to gall diameter. In the second study, mean gall diameter induced by another tenthredinid, *Pontania* sp., also differed significantly among willow clones. In this case, the sawflies in the more rapidly growing and larger galls were more susceptible to parasitism. Individuals of another leaf-folding sawfly (*Phyllocolpa* sp.) attacking these willows also consistently differed in their susceptibility to parasites and other sources of mortality depending upon which plant clone they attacked (Fritz and Nobel, 1990). The question as to whether the variation in susceptibility of sawflies to their natural enemies due to variation among clones ultimately influences clone fitness has not yet been examined, however.

More recently, the effect of genetic variation among two willow species, *S. sericea* and *S. eriocephala*, and their hybrids on interactions between a leaf miner, *Phyllonorycter salicifoliella* and their eulophid parasitoids were studied over a two-year period (Fritz *et al.*, 1997). Survival of the leaf miner varied among taxa as well as among individual plants within taxa. Differences in rates of parasitization were not consistent between years, however, which may suggest that the genetic differences among and within taxa may be modified by environmental conditions.

In none of these studies was it possible to examine whether the differences in natural-enemy activity had any measurable impact on plant fitness. Nevertheless these studies clearly document that genetic variation among plants influences the interaction between herbivores and their natural enemies.

Criteria to demonstrate that herbivores' natural enemies are agents of natural selection on plants

In order to test for the significance of herbivores' natural enemies of agents of natural selection on plants for traits that enhance the success of those natural enemies, I propose the following set of four testable criteria that should be satisfied (Table 2.3). These criteria can be tested by a series of experiments that sequentially protect plants from both herbivores and natural enemies, then expose them to herbivores only, and finally to herbivores and natural enemies. The criteria are as follows:

1. Plant populations polymorphic for a natural-enemy-enhancing trait must be identified and the polymorphism must be controlled at least in part by additive genetic mechanisms.

Table 2.3 *Schematic design of a series of experiments to answer a sequence of questions in order to determine that natural enemies exclusively impose natural selection on plants for plant traits that enhance the success of those natural enemies*

Question		Conclusion
1. Is there additive genetic variation in a plant trait that enhances the effectiveness of natural enemies?	No →	There is no variation that selection can act upon to increase the effectiveness of natural enemies.
Yes ↓		
2. When both herbivores and natural enemies are excluded, is the fitness of plants expressing the trait EQUAL or LESS than plants not expressing the trait?	No →	Plants expressing the trait are intrinsically more fit than those that do not. Therefore, the trait directly benefits plant fitness independently of its role in herbivore suppression.
Yes ↓		
3. When natural enemies are excluded but herbivores are not, is the fitness of plants expressing the trait EQUAL or LESS than plants not expressing the trait?	No →	Plants expressing the trait are more fit than those that do not when both are exposed to herbivores alone. Therefore, the trait has a direct benefit on herbivore resistance irrespective of its effect on natural enemies.
Yes ↓		
4. In the presence of both herbivores and natural enemies, is the fitness of plants expressing the trait GREATER than plants expressing the trait?	No →	Plants expressing the trait have EQUAL or LESS fitness than plants not expressing the trait. If the plants have equal fitness, then the trait is neutral. If plants expressing the trait are less fit, then the trait is detrimental to plant fitness in the presence of natural enemies and natural selection will eliminate the trait from the population.
Yes ↓		
The plant trait is beneficial only in the presence of herbivores and natural enemies. Natural selection will act to spread the trait throughout the population in the presence of both herbivores and natural enemies.		

2. In the absence of both herbivores and natural enemies, plant fitness should be equal or lower for plants expressing the natural-enemy-enhancing trait than for those that do not.
3. In the presence of herbivores but not natural enemies, plant fitness should be equal or lower for plants expressing the natural-enemy-enhancing trait than for those that do not.
4. In the presence of both herbivores and natural enemies, plant fitness should be greater for plants expressing the natural-enemy-enhancing trait than for those that do not.

The significance of these criteria are as follows: criterion 1 must be satisfied simply to demonstrate the existence of genetic variation upon which natural selection can act. In the absence of this criterion, then there is little reason to pursue research on the particular plant–herbivore–natural enemy association. In addition, criterion 2 must be satisfied in order to rule out any benefit of the plant trait on plant fitness via direct increases in plant reproduction. In addition to these two, criterion 3 must be satisfied in order to rule out any effects of the trait acting upon herbivores directly as a plant resistance factor. Criterion 4 must be satisfied in order to demonstrate that the trait is beneficial to plants. Criterion 4 may be satisfied in the absence of criteria 2 and 3. If so, then additional research may be needed to determine the importance of the tritrophic aspects of the trait relative to the aspects that increase plant fitness directly through increased growth (criterion 2) or through increased levels of direct resistance to herbivores (criterion 3).

Critical areas for future research

Measure plant fitness

There are two general models for the development of tritrophic interactions. The most exciting model assumes that tritrophic interactions are coevolved mutualisms between plants and natural enemies attacking the plant's herbivores. A more mundane model assumes that observable tritrophic interactions only reflect evolutionary responses of natural enemies to plant traits that coincidentally improve the fitness of natural enemies; any benefit to plants is entirely fortuitous. If plant traits that enhance the success of natural enemies are indeed evolved traits, then they evolved because plants possessing such traits were more fit than plants that did not (see also van der Meijden and Klinkhamer, 2000). The only way to demonstrate that such traits are beneficial to plants is to measure plant fitness under experimental conditions where such traits can be manipulated experimentally (see above). Without such measurements, then the question of whether the beneficial plant effects on herbivore–natural enemy interactions are the result of plant evolution, or simply fortuitous, can never be resolved.

It is not sufficient to assume that plant fitness will be improved simply through the result of increased rates of parasitization of the plant's herbivores. Such observed differences must be manifest in terms of differential reproduction of plant genotypes that differentially express traits affecting the susceptibility of herbivores to natural enemies.

One particularly troublesome point in relying only on inferences about plant fitness is that the benefits of increased parasitization rates of herbivores may not be obvious. Although parasitized herbivores often consume less foliage than unparasitized caterpillars, in some cases, the parasitized hosts consume more foliage than unparasitized ones (Slansky, 1986). An immediate benefit on plant fitness of plants enhancing parasitization of their herbivores may not exist (e.g., Coleman *et al.*, 1999; see Turlings and Benrey, 1998 for additional examples). There is simply no alternative to careful measurements on plant fitness in order to demonstrate that plant traits that enhance successful parasitization are the results of natural selection.

Recently, Vinson (1999) developed an alternative model of tritrophic interactions. In his model, specialist tritrophic interactions are the result of evolved mutualisms. Vinson (1999) postulates that plants benefit from having a *few* specialized herbivores that facilitate chemical cycling and /or displace competitively more opportunistic herbivore species. Specialist natural enemies have coevolved with plants to keep these particular herbivores in check. This model also could be tested with specific measurements of plant fitness under different combinations of herbivore and natural enemy manipulations.

Consider all aspects of the mechanisms involved

Unfortunately, much of our appreciation of tritrophic interactions is based on an incomplete understanding of many of the underlying mechanisms. For example, the compound that causes corn to release volatile compounds that attract the parasitoid *Microplitis croceipes* is volicitin (Turlings *et al.*, 2000). This compound is synthesized by the herbivore *Spodoptera exigua* but is comprised of a fatty acid portion acquired from the plant and glutamine that is of insect origin (Paré *et al.*, 1998). The fact that the elicitor is synthesized by the insect and not the plant leads to two important and related points. The first is that if *M. croceipes* is so effective in reducing damage by volicitin-producing herbivores, then natural selection would favor herbivores that produce less volicitin. Alternatively, volicitin must have some other more important role in herbivore biology that overrides its deleterious effect in herbivore–parasitoid interactions if herbivores continue to produce it. Such additional functions of volicitin and related compound from other insect species (Pohnert *et al.*, 1999) need to be identified before the evolution and maintenance of the interaction between corn, *S. exigua*, and *M. croceipes* can be fully understood.

Recognize the limitations of applied systems

Tritrophic interactions in applied systems should be studied mainly to determine how host plant resistance and biological control can be best utilized for pest management. Inferences about how tritrophic systems evolved in general should be drawn with extreme caution from applied systems, for such systems are often far removed from their coevolutionary context. Applied studies may clearly illustrate how tritrophic interactions function in ecological time. Applied studies also can be used to develop better expectations about the effectiveness of a particular natural enemy attacking its host on different cultivars (e.g., Hare and Luck, 1991; Hare and Morgan, 2000). It may be problematic, however, to attempt to infer from applied systems how natural tritrophic interactions evolve because the interactions observed in the applied system may be far removed from their original ecological context.

Investigate more natural systems

Natural tritrophic systems require several attributes for profitable study. Ideally, in order to carry out controlled experiments, the systems must be tractable at all three levels. Plants must be easily grown, and both the herbivores and their natural enemies must be easily reared in the laboratory. In addition, plants must be polymorphic for traits that influence natural enemies and sufficiently short-lived that plant fitness can be measured (see above). Because these conditions must be met simultaneously, it is probably not surprising that so little work has been done in natural systems. Nevertheless, such work is necessary in order to determine the evolutionary trajectories of tritrophic interactions. Without such research, we will continue to be unable to determine to what extent the tritrophic effects that we commonly observe are evolved or merely fortuitous.

Acknowledgments

I thank B. A. Hawkins, T. Tscharntke, T. C. J. Turlings, and an anonymous reviewer for their comments on a previous draft of this manuscript. Partial support for research leading to this chapter was provided by US Department of Agriculture NRICGP award no. 94–37302 and National Science Foundation award DEB 96–15134.

REFERENCES

Adkisson, P. L. and Dyck, V. A. (1980) Resistant varieties in pest management systems. In *Breeding Plants Resistant to Insects*, ed. F. G. Maxwell and P. R. Jennings, pp. 233–251. New York: John Wiley.

Agrawal, A. A. and Karban, R. (1997) Domatia mediate plant–arthropod mutualism. *Nature* **387**: 562–563.

Andow, D. A. and Prokrym, D. R. (1990) Plant structural complexity and host-finding by a parasitoid. *Oecologia* **82**: 162–165.

Annis, B. and O'Keefe, L. E. (1987) Influence of pea genotype on parasitization of the pea weevil, *Bruchus pisorum* (Coleoptera: Bruchidae) by *Eupteromalus leguminis* (Hymenoptera: Pteromalidae). *Environmental Entomology* **16**: 653–655.

Barbosa, P. and Letourneau, D. K. (1988) *Novel Aspects of Insect–Plant Interactions*. New York: John Wiley.

Beach, R. M. and Todd, J. W. (1988) Discrete and interactive effects of plant resistance and nuclear polyhedrosis virus for suppression of soybean looper and velvetbean caterpillar (Lepidoptera: Noctuidae) on soybean. *Journal of Economic Entomology* **81**: 684–691.

Beattie, A. J. (1985) *The Evolutionary Ecology of Ant–Plant Mutualisms*. Cambridge: Cambridge University Press.

Beddington, J. R., Free, C. A. and Lawton, J. H. (1978) Characteristics of successful natural enemies in models of biological control of insect pests. *Nature* **273**: 513–519.

Bell, J. V. (1978) Development and mortality in bollworms fed resistant and susceptible soybean cultivars treated with *Nomuraea rileyi* or *Bacillus thuringiensis*. *Journal of the Georgia Entomological Society* **13**: 50–55.

Bergman, J. M. and Tingey, W. M. (1979) Aspects of interaction between plant genotypes and biological control. *Bulletin of the Entomological Society of America* **25**: 275–279.

Boethel, D. J. (1999) Assessment of soybean germplasm for multiple insect resistance. In *Global Plant Genetic Resources for Insect-Resistant Crops*, ed. S. L. Clement and S. S. Quisenberry, pp. 101–129. Boca Raton, FL: CRC Press.

Boethel, D. J. and Eikenbary, R. D. (eds.) (1986) *Interactions of Host Plant Resistance and Parasitoids and Predators of Insects*. New York: Halsted Press.

Bong, C. F., Sikorowski, P. P. and Davis, F. M. (1991) Effects of a resistant maize genotype and cytoplasmic polyhedrosis virus on growth and development of the corn earworm (Lepidoptera: Noctuidae). *Environmental Entomology* **20**: 1200–1206.

Bottrell, D. G., Barbosa, P. and Gould, F. (1998) Manipulating natural enemies by plant variety selection and modification: a realistic strategy? *Annual Review of Entomology* **43**: 347–367.

Brewer, M. J., Struttmann, J. M. and Mornhinweg, D. W. (1998) *Aphelinus albipodus* (Hymenoptera: Aphelinidae) and *Diaeretiella rapae* (Hymenoptera: Braconidae) parasitism on *Diuraphis noxia* (Homoptera: Aphididae) infesting barley plants differing in plant resistance to aphids. *Biological Control* **11**: 255–261.

Brewer, M. J., Mornhinweg, D. W. and Huzurbazar, S. (1999) Compatibility of insect management strategies: *Diuraphis noxia* abundance on susceptible and resistant barley in the presence of parasitoids. *BioControl* **43**: 479–491.

Campbell, B. C. and Duffey, S. S. (1979) Tomatine and parasitic wasps: potential incompatibility of plant antibiosis with biological control. *Science* **205**: 700–702.

Casagrande, R. A. and Haynes, D. L. (1976) The impact of pubescent wheat on the population dynamics of the cereal leaf beetle. *Environmental Entomology* **5**: 153–159.

Clancy, K. M. and Price, P. W. (1987) Rapid herbivore growth enhances enemy attack: sublethal plant defenses remain a paradox. *Ecology* **68**: 733–737.

Coleman, R. A., Barker, A. M. and Fenner, M. (1999) Parasitism of the herbivore *Pieris brassicae* L. (Lep., Pieridae) by *Cotesia glomerata* L. (Hym., Braconidae) does not benefit the host plant by reduction of herbivory. *Journal of Applied Entomology* **123**: 171–177.

Cortesaro, A. M., Stapel, J. O. and Lewis, W. J. (2000) Understanding and manipulating plant attributes to enhance biological control. *Biological Control* **17**: 35–49.

Dicke, M. (1999) Are herbivore-induced plant volatiles reliable indicators of herbivory identity to foraging carnivorous arthropods? *Entomologia Experimentalis et Applicata* **91**: 131–142.

Dimock, M. B. and Kennedy, G. G. (1983) The role of glandular trichomes in the resistance of *Lycopersicon hirsutum* f. *glabratum* to *Heliothis zea*. *Entomologia Experimentalis et Applicata* **33**: 263–268.

Dogan, E. B., Berry, R. E., Reed, G. L. and Rossignol, P. A. (1996) Biological parameters of convergent lady beetle (Coleoptera: Coccinellidae) feeding on aphids (Homoptera: Aphididae) on transgenic potato. *Journal of Economic Entomology* **89**: 1105–1108.

Dover, B. A., Noblet, R., Moore, R. F. and Shepard, B. M. (1987) Development and emergence of *Pediobius foveolatus* from Mexican bean beetle larvae fed foliage from *Phaseolus lunatus* and resistant and susceptible soybeans. *Journal of Agricultural Entomology* **4**: 271–279.

Duke, S. O. (1994) Glandular trichomes: a focal point of chemical and structural interactions. *International Journal of Plant Sciences* **155**: 617–620.

Duffey, S. S. and Bloem, K. A. (1986) Plant defense–herbivore–parasite interactions and biological control. In *Ecological Theory and Integrated Pest Management*, ed. M. Kogan, pp. 135–183. New York: John Wiley.

Eigenbrode, S. D. and Espelie, K. E. (1995) Effects of plant epicuticular lipids on insect herbivores. *Annual Review of Entomology* **40**: 171–194.

Farid, A., Johnson, J. B., Shafii, B. and Quisenberry, S. S. (1998) Tritrophic studies of Russian wheat aphid, a parasitoid, and resistant and susceptible wheat over three parasitoid generations. *Biological Control* **12**: 1–6.

Farrar, R. R. and Kennedy, G. (1993) Field cage performance of two tachinid parasitoids of the tomato fruitworm on insect resistant and susceptible tomato lines. *Entomologia Experimentalis et Applicata* **67**: 73–78.

Farrar, R. R., Barbour, J. D. and Kennedy, G. G. (1994) Field evaluation of insect resistance in a wild tomato and its effects on insect parasitoids. *Entomologia Experimentalis et Applicata* **71:** 211–226.

Flint, H. M., Henneberry, T. J., Wilson, F. D., Holguin, E., Parks, N. and Buchler, R. E. (1995) The effects of transgenic cotton, *Gossypium hirsutum* L., containing *Bacillus thuringiensis* toxin genes for the control of the pink bollworm *Pectinophora gossypiella* (Saunders) and other arthropods. *Southwestern Entomologist* **20:** 281–292.

Fritz, R. S. (1992) Community structure and species interactions of phytophagous insects on resistant and susceptible hosts. In *Plant Resistance to Herbivores and Pathogens: Ecology, Evolution And Genetics*, ed. R. S. Fritz and E. L. Simms, pp. 240–277. Chicago, IL: University of Chicago Press.

Fritz, R. S. (1995) Direct and indirect effects of plant genetic variation on enemy impact. *Ecological Entomology* **20:** 18–26.

Fritz, R. S. and Nobel, J. (1990) Host plant variation in mortality of the leaf-folding sawfly on the arroyo willow. *Ecological Entomology* **15:** 25–35.

Fritz, R. S., McDonough, S. E. and Rhoads, A. G. (1997) Effects of plant hybridization on herbivore–parasitoid interactions. *Oecologia* **110:** 360–367.

Geervliet, J. B. F., Vet, L. E. M. and Dicke, M. (1996) Innate responses of the parasitoids *Cotesia glomerata* and *C. rubecula* (Hymenoptera, Braconidae) to volatiles from different plant–herbivore complexes. *Journal of Insect Behavior* **9:** 525–538.

Gilreath, M. E., McCutcheon, G., Carner, G. R. and Turnipseed, S. G. (1986) Pathogen incidence in noctuid larvae from selected soybean genotypes. *Journal of Agricultural Entomology* **3:** 213–226.

Gouinguené, S., Degen, T. and Turlings, T. C. J. (2001) Variability in herbivore-induced odour emissions among maize cultivars and their wild ancestors (Teosinte). *Chemoecology* **11:** 9–16.

Gould, F. (1998) Sustainability of transgenic insecticidal cultivars: integrating pest genetics and ecology. *Annual Review of Entomology* **43:** 701–726.

Gould, F., Kennedy, G. G. and Johnson, M. T. (1991) Effects of natural enemies on the rate of herbivore adaptation to resistant host plants. *Entomologia Experimentalis et Applicata* **58:** 1–14.

Grant, J. F. and Shepard, M. (1985) Influence of three soybean genotypes on development of *Voria ruralis* (Diptera: Tachinidae) and on foliage consumption by its host, the soybean looper (Lepidoptera: Noctuidae). *Florida Entomologist* **68:** 672–677.

Hare, J. D. (1992) Effects of plant variation on herbivore–natural enemy interactions. In *Plant Resistance to Herbivores and Pathogens: Ecology, Evolution and Genetics*, ed. R. S. Fritz and E. L. Simms, pp. 278–298. Chicago, IL: University of Chicago Press.

Hare, J. D. and Luck, R. F. (1991) Indirect effects of citrus cultivars on life history parameters of a parasitic wasp. *Ecology* **72:** 1576–1585.

Hare, J. D. and Morgan, D. J. W. (2000) Chemical conspicuousness of an herbivore to its natural enemy: effect of feeding site selection. *Ecology* **81:** 509–519.

Hassell, M. P. (1978) *The Dynamics of Arthropod Predator–Prey Systems*. Princeton, NJ: Princeton University Press.

Hassell, M. P. and Anderson, R. M. (1984) Host suceptibility as a component in host–parasitoid systems. *Journal of Animal Ecology* **53:** 611–621.

Heinz, K. M. and Zalom, F. G. (1996) Performance of the predator *Delphastus pusillus* on *Bemisia* resistant and susceptible tomato lines. *Entomologia Experimentalis et Applicata* **81:** 345–352.

Hilbeck, A., Baumgartner, M., Fried, P. M. and Bigler, F. (1998) Effects of transgenic *Bacillus thuringiensis* corn-fed prey on mortality and development time of immature *Chrysoperla carnea* (Neuroptera: Chrysopidae). *Environmental Entomology* **27**: 480–487.

Hoffman, M. P., Zalom, F. G., Wilson, L. T., Smilanick, J. M., Malyj, L. D., Kiser, J., Hilder, V. A. and Barnes, W. M. (1992) Field evaluation of transgenic tobacco containing genes encoding *Bacillus thuringiensis* delta-endotoxin or cowpea trypsin inhibitor: efficacy against *Helicoverpa zea* (Lepidoptera: Noctuidae). *Journal of Economic Entomology* **85**: 2516–2522.

Isenhour, D. J. and Wiseman, B. R. (1989) Parasitism of the fall armyworm (Lepidoptera: Noctuidae) by *Campoletis sonorensis* (Hymenoptera: Ichneumonidae) as affected by host feeding on silks of *Zea mays* L. cv. Zapalote Chico. *Environmental Entomology* **18**: 394–397.

Isenhour, D. J., Wiseman, B. R. and Layton, R. C. (1989) Enhanced predation by *Orius insidiosus* (Hemiptera: Anthocoridae) on larvae of *Heliothis zea* and *Spodoptera frugiperda* (Lepidoptera: Noctuidae) caused by prey feeding on resistant corn genotypes. *Environmental Entomology* **18**: 418–422.

Johnson, M. T. and Gould, F. L. (1992) Interaction of genetically engineered host plant resistance and natural enemies of *Heliothis virescens* (Lepidoptera: Noctuidae) in tobacco. *Environmental Entomology* **21**: 586–597.

Johnson, M. T., Gould, F. and Kennedy, G. G. (1997) Effects of natural enemies on relative fitness of *Heliothis virescens* genotypes adapted and not adapted to resistant host plants. *Entomologia Experimentalis et Applicata* **82**: 219–230.

Juvik, J. A., Shapiro, J. A., Young, T. E. and Mutschler, M. A. (1994) Acylglucoses from wild tomatoes alter behavior and reduce growth and survival of *Helicoverpa zea* and *Spodoptera exigua* (Lepidoptera, Noctuidae). *Journal of Economic Entomology* **87**: 482–492.

Kareiva, P. and Sahakian, R. (1990) Tritrophic effects of a simple architectural mutation in pea plants. *Nature* **345**: 433–434.

Karner, M. A. and Manglitz, G. R. (1985) Effects of temperature and alfalfa cultivar on pea aphid (Homoptera: Aphididae) fecundity and feeding activity of convergent lady beetle (Coleoptera: Coccinellidae). *Journal of the Kansas Entomological Society* **58**: 131–136.

Kartohardjono, A. and Heinrichs, E. A. (1984) Populations of the brown planthopper, *Nilaparvata lugens* (Stahl) (Homoptera: Delphacidae) and its predators on rice varieties with different levels of resistance. *Environmental Entomology* **13**: 359–365.

Kauffman, W. C. and Flanders, R. V. (1985) Effects of variabily resistant soybean and lima bean cultivars on *Pediobius foveolatus* (Hymenoptera: Eulophidae), a parasitoid of the Mexican bean beetle, *Epilachna varivestis* (Coleoptera: Coccinellidae). *Environmental Entomology* **14**: 678–682.

Kauffman, W. C. and Kennedy, G. G. (1989) Toxicity of allelochemicals from wild insect-resistant tomato *Lycopersicon hirsutum* f. *glabratum* to *Campoletis sonorensis*, a parasitoid of *Heliothis zea*. *Journal of Chemical Ecology* **15**: 2051–2060.

Kea, W. C., Turnipseed, S. G. and Carner, G. R. (1978) Influence of resistant soybeans on the susceptibility of lepidopterous pests to insecticides. *Journal of Economic Entomology* **71**: 58–59.

Kennedy, G. G., Kishaba, A. N. and Bohn, G. W. (1975) Response of several pest species to *Cucumis melo* L. lines resistant to *Aphis gossypii* Glover. *Environmental Entomology* **4**: 653–657.

Kogan, M. (1982) Plant resistance in pest management. In *Introduction to Pest Management*, 2nd edn, ed. R. L. Metcalf and W. H. Luckman, pp. 93–134. New York: John Wiley.

Lampert, E. P., Haynes, D. L., Sawyer, A. J., Jokinen, D. P., Wellso, S. G., Gallun, R. L. and Roberts, J. J. (1983) Effects of regional releases of resistant wheats on the population dynamics of the cereal leaf beetle (Coleoptera: Chrysomelidae). *Annals of the Entomological Society of America* **76**: 972–980.

Lawton, J. H. and McNeil, S. (1979) Between the devil and the deep blue sea: on the problem of being a herbivore. *Symposium of the British Ecological Society* **20**: 223–244.

Lincoln, C., Dean, G., Waddle, B. A., Yearian, W. C., Phillips, J. R. and Roberts, L. (1971) Resistance of frego-type cotton to boll weevil and boll worm. *Journal of Economic Entomology* **64**: 1326–1327.

Lindgren, P. D., Lukefahr, M. J., Diaz, M., Jr. and Hartstack, A., Jr. (1978) Tobacco budworm control in caged cotton with a resistant variety, augmentative releases of *Campoletis sonorensis*, and natural control by other beneficial species. *Journal of Economic Entomology* **71**: 739–745.

Loughrin, J. H., Manukian, A., Heath, R. P. and Tumlinson, J. H. (1995) Volatiles emitted by different cotton varieties damaged by feeding beet armyworm larvae. *Journal of Chemical Ecology* **21**: 1217–1227.

Marquis, R. J. and Whelan, C. (1996) Plant morphology, and recruitment of the third trophic level: subtle and little-recognized defenses. *Oikos* **75**: 330–334.

McCutcheon, G. S. and Turnipseed, S. G. (1981) Parasites of lepidopterous larvae in insect resistant and susceptible soybeans in South Carolina. *Environmental Entomology* **10**: 69–74.

Meade, T. and Hare, J. D. (1995) Integration of host plant resistance and *Bacillus thuringiensis* insecticides in the management of lepidopterous pests of celery. *Journal of Economic Entomology* **88**: 1787–1794.

Myint, M. M., Raupsas, H. R. and Heinrichs, E. A. (1986) Integration of varietal resistance and predation for the management of *Nephotettix virescens* (Homoptera: Cicadellidae) population on rice. *Crop Protection* **5**: 259–265.

Obrycki, J. J. (1986) The influence of foliar pubescence on entomophagous insects. In *Interactions of Plant Resistance and Parasitoids and Predators*, ed. D. J. Boethel and R. K. Eikenbary, pp. 61–83. New York: Halsted Press.

Obrycki, J. J. and Tauber, M. J. (1984) Natural enemy activity on glandular pubescent potato plants in the greenhouse: an unreliable predictor of effects in the field. *Environmental Entomology* **13**: 679–683.

Orr, D. B. and Boethel, D. J. (1985) Comparative development of *Copidosoma truncatellum* (Hymenoptera: Encyrtidae) and its host, *Pseudoplusia includens* (Lepidoptera: Noctuidae) on resistant and susceptible soybean genotypes. *Environmental Entomology* **14**: 612–616.

Orr, D. B. and Boethel, D. J. (1986) Influence of plant antibiosis through four trophic levels. *Oecologia* **70**: 242–249.

Orr, D. B. and Landis, D. A. (1997) Oviposition of European corn borer (Lepidoptera: Pyralidae) and impact of natural enemy populations in transgenic versus isogenic corn. *Journal of Economic Entomology* **90**: 905–909.

Orr, D. B., Boethel, D. J. and Jones, W. A. (1985) Biology of *Telenomus chloropus* (Hymenoptera: Scelionidae) from eggs of *Nezara viridula* (Hemiptera: Pentatomidae) reared on resistant and susceptible soybean genotypes. *Canadian Entomologist* **117**: 1137–1142.

Pair, S. D., Wiseman, B. R. and Sparks, A. N. (1986) Influence of four corn cultivars on fall armyworm (Lepidoptera: Noctuidae) establishment and parasitisation. *Florida Entomologist* **69**: 566–570.

Paré, P. W., Alborn, H. T. and Tumlinson, J. H. (1998) Concerted biosynthesis of an insect elicitor of plant volatiles. *Proceedings of the National Academy of Sciences, USA* **95**: 13971–13975.

Pilcher, C. D., Obrycki, J. J., Rice, M. E. and Lewis, L. C. (1997) Preimaginal development, survival, field abundance of insect predators on transgenic *Bacillus thuringiensis* corn. *Environmental Entomology* **26**: 446–454.

Pimentel, D. and Wheeler, A. G. (1973) Influence of alfalfa resistance on a pea aphid population and its associated parasites, predators, and competitors. *Environmental Entomology* **2**: 1–11.

Pohnert, G., Jung, V., Haukioja, E., Lempa, K. and Boland, W. (1999) New fatty acid amides from regurgitant of lepidopteran (Noctuidae, Geometridae) caterpillars. *Tetrahedron* **55**: 11275–11280.

Poppy, G. M. (1997) Tritrophic interactions: improving ecological understanding and biological control? *Endeavour* **21**: 61–65.

Powell, J. E. and Lambert, L. (1984) Effects of three resistant soybean genotypes on development of *Microplitis croceipes* and leaf consumption by its *Heliothis* spp. hosts. *Journal of Agricultural Entomology* **1**: 169–176.

Price, P. W. (1986) Ecological aspects of host plant resistance and biological control: interactions among three trophic levels. In *Interactions of Plant Resistance and Parasitoid and Predator of Insects*, ed. D. J. Boethel and R. D. Eikenbary, pp. 11–30. New York: Halsted Press.

Price, P. W. and Clancy, K. M. (1986) Interactions among three trophic levels: gall size and parasitoid attack. *Ecology* **67**: 1593–1600.

Price, P. W., Bouton, C. E., Gross, P., McPheron, B. A., Thompson, J. N. and Weis, A. E. (1980) Interactions among three trophic levels: influence of plants on interactions betweeen insect herbivores and natural enemies. *Annual Review of Ecology and Systematics* **11**: 41–65.

Reed, D. K., Webster, J. A., Jones, B. G. and Burd, J. D. (1991) Tritrophic relationships of Russian wheat aphid (Homoptera: Aphididae), a hymenopterous parasitoid (*Diaeretiella rapae* McIntosh), and resistant and susceptible small grains. *Biological Control* **1**: 35–41.

Reichelderfer, C. F. (1991) Interactions among allelochemicals, some Lepidoptera, and *Bacillus thuringiensis* Berliner. In *Microbial Mediation of Plant–Herbivore Interactions*, ed. P. Barbosa, V. A. Krischik and C. G. Jones, pp. 507–524. New York: John Wiley.

Rogers, D. J. and Sullivan, M. J. (1986) Nymphal performance of *Geocoris punctipes* (Hemiptera: Lygaeidae) on pest-resistant soybeans. *Environmental Entomology* **15**: 1032–1036.

Salim, M. and Heinrichs, E. A. (1986) Impact of varietal resistance in rice and predation on the mortality of *Sogatella furcifera* (Horvath) (Homoptera: Delphacidae). *Crop Protection* **5**: 395–399.

Salto, C. E., Eikenbary, R. D. and Starks, K. J. (1983) Compatibility of *Lysiphlebus testaceipes* (Hymenoptera: Braconidae) with greenbug (Homoptera: Aphididae) biotype "C" and "E" reared on susceptible and resistant oat varieties. *Environmental Entomology* **12**: 603–604.

Schuler, T. H., Poppy, G. M., Kerry, B. R. and Denholm, I. (1998) Insect-resistant transgenic plants. *Trends in Biotechnology* **16:** 168–175.

Schuler, T. H., Poppy, G. M., Kerry, B. R. and Denholm, I. (1999a) Potential side effects of insect-resistant transgenic plants on arthropod natural enemies. *Trends in Biotechnology* **17:** 210–216.

Schuler, T. H., Potting, R. P. J., Denholm, I. and Poppy, G. M. (1999b) Parasitoid behaviour and *Bt* plants. *Nature* **400:** 825–826.

Schultz, J. C. and Keating, S. T. (1991) Host-plant-mediated interactions between the gypsy moth and a baculovirus. In *Microbial Mediation of Plant–Herbivore Interactions*, ed. P. Barbosa, V. A. Krischik and C. G. Jones, pp. 489–506. New York: John Wiley.

Schuster, M. F., Calvin, P. D. and Lanston, W. C. (1983) Interaction of high tannin with bollworm control by Pydrin and Dipel. In *Proceedings of the 1983 Beltwide Cotton Production Research Conference*, pp. 72–73. Memphis, TN: Sinauer Associates.

Slansky, F. (1986) Nutritional ecology of endoparasitic insects and their hosts: an overview. *Journal of Insect Physiology* **32:** 255–261.

Starks, K. J., Muniappan, R. and Eikenbary, R. D. (1972) Interaction between plant resistance and parasitism against greenbug on barley and sorghum. *Annals of the Entomological Society of America* **65:** 650–655.

Takabayashi, J., Dicke, M. and Posthumus, M. A. (1994) Volatile herbivore-induced terpenoids in plant mite interactions: variation caused by biotic and abiotic factors. *Journal of Chemical Ecology* **20:** 1329–1354.

Takabayashi, J., Sato, Y., Horikoshi, M., Yamaoka, R., Yano, S., Ohsaki, N. and Dicke, M. (1998) Plant effects of parasitoid foraging: differences between two tritrophic systems. *Biological Control* **11:** 97–103.

Treacy, M. F., Zummo, G. R. and Benedict, J. H. (1985) Interactions of host-plant resistance in cotton with predators and parasites. *Agricultural Ecosystems and Environment* **13:** 151–157.

Trichilo, P. J. and Leigh, T. F. (1986) The impact of cotton plant resistance on spider mite and their natural enemies. *Hilgardia* **20**(5): 1–20.

Tumlinson, J. H., Turlings, T. C. J. and Lewis, W. J. (1992) The semiochemical complexes that mediate insect parasitoid foraging. *Agricultural Zoology Reviews* **5:** 221–252.

Turlings, T. C. J. and Benrey, B. (1998) Effects of plant metabolites on the behavior and development of parasitic wasps. *Ecoscience* **5:** 321–333.

Turlings, T. C. J., Loughrin, J. H., McCall, P. J., Rose, U. S. R., Lewis, W. J. and Tumlinson, J. H. (1995) How caterpillar-damaged plants protect themselves by attracting parasitic wasps. *Proceedings of the National Academy of Sciences, USA* **92:** 4169–4174.

Turlings, T. C. J., Alborn, H. T., Loughrin, J. H. and Tumlinson, J. H. (2000) Volicitin, an elicitor of maize volatiles in oral secretion of *Spodoptera exigua*: isolation and bioactivity. *Journal of Chemical Ecology* **26:** 189–202.

van der Meijden, E. and Klinkhamer, G. L. (2000) Conflicting interests of plants and the natural enemies of herbivores. *Oikos* **89:** 202–208.

van Emden, H. F. (1966) Plant–insect relationships and pest control. *World Review of Pest Control* **5:** 115–123.

van Lenteren, J. C. and de Ponti, O. M. B. (1991) Plant-leaf morphology, host-plant resistance and biological control. *Symposia Biologica Hungarica* **39:** 365–386.

Verkerk, R. H. J., Leather, S. R. and Wright, D. J. (1998) The potential for manipulating crop–pest–natural enemy interactions for improved insect pest management. *Bulletin of Entomological Research* **88:** 493–501.

Vet, L. E. M. and Dicke, M. (1992) Ecology of infochemical use by natural enemies in a tritrophic context. *Annual Review of Entomology* **37**: 141–172.

Vinson, S. B. (1999) Parasitoid manipulation as a plant defense strategy. *Annals of the Entomological Society of America* **92**: 812–828.

Vinson, S. B. and Barbosa, P. (1987) Interrelationships of nutritional ecology of parasitoids. In *Nutritional Ecology of Insects, Mites, Spiders*, ed. F. Slansky and J. G. Rodriguez, pp. 673–695. New York: John Wiley.

Wagner, G. J. (1991) Secreting glandular trichomes more than just hairs. *Plant Physiology* **96**: 675–679.

Weathersbee, A. A. and Hardee, D. D. (1994) Abundance of cotton aphids (Homoptera: Aphididae) and associated biological control agents on six cotton cultivars. *Journal of Economic Entomology* **87**: 258–265.

Weis, A. E. and Abrahamson, W. G. (1985) Potential selective pressures by parasitoids on a plant–herbivore interaction. *Ecology* **66**: 1261–1269.

Weis, A. E. and Abrahamson, W. (1986) Evolution of host-plant manipulation by gall makers: ecological and genetic factors in the *Solidago–Eurosta* system. *American Naturalist* **127**: 681–695.

Wheatley, J. A. C. and Boethel, D. J. (1992) Populations of *Phytoseiulus persimilis* (Acari: Phytoseiidae) and its host, *Tetranychus urticae* (Acari: Tetranychidae), on resistant and susceptible soybean cultivars. *Journal of Economic Entomology* **85**: 731–738.

Wilson, F. D., Flint, H. M., Deaton, W. R., Fischhoff, D. A., Perlak, F. J., Armstrong, T. A., Fuchs, R. L., Berberich, S. A., Parks, N. J. and Stapp, B. R. (1992) Resistance of cotton lines containing a *Bacillus thuringiensis* toxin to pink bollworm (Lepidoptera: Gelechiidae) and other insects. *Journal of Economic Entomology* **85**: 1516–1521.

Wyatt, I. J. (1970) The distribution of *Myzus persicae* (Sulz.) on year-round chrysanthemums II. Winter season: the effect of parasitism by *Aphidius matricariae* Hal. *Annals of Applied Biology* **65**: 31–41.

Yanes, J. and Boethel, D. J. (1983) Effect of a resistant soybean genotype on the development of the soybean looper (Lepidoptera: Noctuidae) and an introduced parasitoid, *Microplitis demolitor* Wilkinson (Hymenoptera: Braconidae). *Environmental Entomology* **12**: 1270–1274.

Yang, J. and Sadof, C. S. (1997) Variation in the life history of the citrus mealybug parasitoid *Leptomastix dactylopii* (Hymenoptera: Encyrtidae) on three varieties of *Coleus blumei*. *Environmental Entomology* **26**: 978–982.

JUDITH L. BRONSTEIN AND PEDRO BARBOSA

3

Multitrophic/multispecies mutualistic interactions: the role of non-mutualists in shaping and mediating mutualisms

Introduction

Off the coast of Massachusetts (USA), the hermit crab *Pagurus longicarpus* is often found carrying a colonial hydroid, *Hydractinia*, on its shell. In some situations, this interaction is clearly mutualistic: hermit crabs transport hydroids to rich feeding sites, and hydroids in turn deter larger, damaging organisms from colonizing hermit crab shells. The outcome of this interaction shifts away from mutualism, however, under other ecological conditions. The hydroid tends to be positively associated with a burrowing marine worm that weakens hermit crab shells to the point where they are easily crushed by predatory blue crabs (*Callinectes sapidus*). The nature of the hermit crab–hydroid association thus varies depending on which other species are present, shifting from mutualism (when blue crabs and/or worms are scarce), to commensalism, to antagonism (Buckley and Ebersole, 1994).

In most introductory biology textbooks, mutualism is defined as an association between organisms of two species in which both species benefit (e.g., Starr and Taggart, 1998; Tobin and Dusheck, 1998; Krogh, 2000). However, the hermit crab–hydroid interaction clearly demonstrates that at least some mutualisms can only be understood within a broader community context. The influence of other species and other trophic levels on mutualism has received curiously little attention, particularly in contrast to other types of interactions (see Barbosa and Letourneau, 1988; Cardé and Bell, 1995; Barbosa and Benrey, 1998; Barbosa and Wratten, 1998; Olff *et al.*, 1999). Mutualisms have been discussed to a certain extent within the literature on indirect and tritrophic interactions (e.g., Vandermeer *et al.*, 1985; Bertness and Callaway, 1994; Wootton, 1994; Menge, 1995; Callaway and Walker, 1997;

Sabelis *et al.*, 1999). However, the full range of multi-trophic effects on mutualism remains incompletely defined (but see Thompson, 1988; Cushman, 1991; Cushman and Addicott, 1991; Bronstein, 1994), and consideration of their possible evolutionary consequences has been minimal (but see, e.g., Wilson and Knollenberg, 1987; West and Herre, 1994; Strauss, 1997).

In this chapter, we provide a review of multitrophic/multispecies mutualisms, that is, interactions in which other species and trophic levels influence the nature and outcome of a potentially beneficial pairwise interaction. We outline the conditions under which third species' influence will be an essential feature of a mutualism, in that a pairwise relationship will only be beneficial in the context of these species' influence. (The hermit crab–hydroid interaction is an example of one such mutualism.) We also document interactions that can in fact be considered mutualistic in a strictly pairwise context, but whose outcomes and effects will be substantially determined by the abundance or actions of other species. In particular, we show that variation in the abundance of third species leads the outcomes of many mutualisms to vary in space and in time, in some cases to the point where they are no longer beneficial to one or even both partners. Finally, based on the lessons gleaned from these examples, we develop a set of working hypotheses regarding the broader evolutionary implications of third species mediation of mutualism.

Table 3.1 summarizes the range of interactions that we identify and discuss. The examples we present were chosen to illustrate these phenomena from diverse habitats, taxa, and kinds of mutualism. Some of the examples are documented in studies explicitly designed to test hypotheses about third-species mediation of mutualism; others emerge from more inferential data collected for other purposes. Throughout this chapter, we will emphasize the need for further direct tests of all of the phenomena we identify.

Mutualisms that exist only in the presence of a third species

Mutualisms are commonly grouped according to the nature of the benefits that partners exchange (Boucher *et al.*, 1982). Three kinds of benefits are usually recognized: *transportation*, in which one species moves its partner or its partner's gametes to places they could not otherwise reach; *nutrition*, in which partners are provided with essential limiting nutrients; and *protection*, in which one species protects its partners from negative influences of their biotic or abiotic environments. This section

Table 3.1. *Types of multitrophic/multispecies mutualism discussed in this chapter*

- I. Mutualisms that exist only in the presence of other species
  - I.A. Mutualism in the presence of a natural enemy of one of the partners
    - I.A.1. Direct attack on partner's predators and parasites
    - I.A.2. Modification of partner traits so as to reduce its susceptibility to attack
    - I.A.3. Shared defense against a common enemy
  - I.B. Mutualism in the presence of a competitor of one of the partners
    - I.B.1. Direct attack on partner's competitors
    - I.B.2. Modification of partner traits so as to increase its competitive performance
- II. Mutualisms whose outcomes are influenced by other species
  - II.A. Mutualisms altered by antagonists of one of the partners
    - II.A.1. Reduction in benefit via depressed density of one partner
    - II.A.2. Reduction in benefit via effects mediated by one partner's natural enemies
    - II.A.3. Reduction in benefit via effects mediated by one partner's competitors
    - II.A.4. Augmentation in benefit via effects mediated by one partner's natural enemies
  - II.B. Mutualisms altered by other mutualists of one of the partners
    - II.B.1. Augmentation in benefit via shared attraction of mutualists
    - II.B.2. Augmention in benefit via facilitated colonization by mutualists
  - II.C. Mutualisms altered by exploiters of mutualism
    - II.C.1. Reduction in benefit via competition for mutualistic rewards and services
    - II.C.2. Reduction in benefit via deterrence of mutualistic partners
  - II.D. Mutualisms altered by incidental disruption by humans and other species
    - II.D.1. Reduction in benefit via introduction of natural enemies or competitors
    - II.D.2. Reduction in benefit via habitat modification

focuses on a subgroup of protective mutualisms in which one mutualist shields its partner from the effects of natural enemies or competitors. These mutualisms are of particular interest from a multitrophic/multispecies perspective because, in these interactions, a pair of interacting species can be defined as mutualists *only in the context of their association with a third species*. In the absence of that species, most of these pairwise interactions appear not to involve reciprocal benefits. The range of phenomena we discuss in this section is summarized under the first heading in Table 3.1.

Mutualisms dependent on third species: natural enemies

The best-known examples of protective mutualism, and some of the best-known mutualisms overall, are those in which one species deters its partner's predators, parasites, parasitoids, or herbivores. Deterrence can take at least three forms: one partner can directly attack the other's enemies, it can modify the other's traits in ways that reduce that partner's susceptibility to enemies, or it can share in defense against a common enemy. Below, we discuss each of these forms of deterrence. They are summarized in section I.A of Table 3.1.

Ants aggressively defend diverse plants, homopterans, and lycaenid caterpillars from their natural enemies in exchange for various food rewards (Pierce, 1987; Huxley and Cutler, 1991; Koptur, 1992). When enemies are abundant, these behaviors can translate into increased growth, survival, or reproduction for ant-tended individuals. For example, Pierce *et al.* (1987) showed that rates of predation and parasitism of larvae of the lycaenid *Jalmenus evagoras* are so high that individuals deprived of ant defenders cannot survive. Similarly, ant-defended *Maieta guianensis* plants (Melastomataceae) experience little herbivory and are able to produce 45 times more fruits than individuals denied ants for a year (Vasconcelos, 1991). Ant defense can be effective even when less than ferocious. Certain relatively timid, competitively inferior species benefit plants by removing very small herbivores and their eggs (Letourneau, 1983; Gaume *et al.*, 1997).

Another kind of active deterrence involves removal and consumption of the partner's ectoparasites. Interactions of this form are known from both marine and terrestrial habitats, and include cleaner fish and their hosts (Poulin and Grutter, 1996) and oxpeckers and oxen (Weeks, 2000). However, it has proven very difficult to show that cleaners significantly reduce parasite loads (but see Grutter, 1999), and no case is yet known in which cleaning unequivocally increases host fitness. It is possible that parasite loads are only rarely high enough to reduce host success, and thus for parasite removal to confer a measurable benefit (Grutter, 1997). In fact, cleaners commonly inflict significant damage to host tissues while feeding, and this cost may outweigh any benefit conferred. Thus, while cleaning is a multitrophic and multispecies interaction, it is debatable whether its outcome is ever truly mutualistic.

Some mutualists deter their partners' enemies by means other than direct attack. In general, these interactions involve some modification of the partner's traits in ways that reduce the partner's susceptibility to enemies. For example, certain endophytic (leaf-inhabiting) fungi produce secondary compounds that render their host plants distasteful or even fatal to herbivores (Clay, 1991; Saikkonen *et al.*, 1998; see also chapter 5). Decorator crabs gain protection from predator attack by adorning themselves with distasteful algae (Stachowicz and Hay, 1999a). Other invertebrates become covered with plants that effectively alter their hosts' colors or shapes, making them difficult for visually oriented predators to recognize (Gressit *et al.*, 1965; Espoz *et al.*, 1995). It should be pointed out, however, that the consequences to the plants in these associations remain generally uninvestigated. It is entirely possible that, like cleaning associations, they are rarely if ever mutualistic.

A final form of mutualistic protection is shared defense against a common enemy. For example, in Müllerian mimicry systems, different chemically defended species resemble one another and thus predators can learn more quickly and remember more effectively to avoid distasteful prey (Gilbert, 1983). Similarly, mixed-species foraging associations confer joint protection when the benefit of group predator vigilance outweighs competition for food (Metcalfe, 1989; Székely *et al.*, 1989).

Mutualists dependent on third species: competitors

All of the examples above involve situations in which one mutualist alters an interaction between its partner and its partner's enemies. In a second, overlapping category of mutualism, mutualists provide protection against competitors (Table 3.1, section I.B). As is true for mutualisms that modify predator–prey interactions, many of these relationships confer no other benefits to the protected partner, and often inflict some costs as well. Hence, the pairwise relationship tends to be mutualistic only when competitors are abundant.

Many examples are known in which one species directly attacks its partner's competitors, either by consuming or simply removing them. The best-documented cases involve removal of epibionts, species that use the external surface of another species as a habitat and thus compete with it for a key resource such as light and/or nutrients (Witman, 1987; Fiala *et al.*, 1989; Dudley, 1992; McQuaid and Froneman, 1993; Ellison *et al.*, 1996; Amsler *et al.*, 1999). For example, along the coast of North Carolina (USA), the coral *Oculina arbuscula* harbors the omnivorous crab *Mithrax forceps*, which feeds on seaweeds and invertebrates growing on or near the coral. Stachowicz and Hay (1999b) have shown that in certain habitats, corals from which crabs have been experimentally removed develop a dense cover of epibionts, resulting in reduced coral growth and increased mortality relative to crab-associated corals, which remain epibiont-free.

Other mutualists modify competition not by active removal of their partner's competitors, but by more indirect reductions of their competitive effects. For example, *Andropogon* grasses function as competitive dominants in tallgrass prairie communities only when colonized by mutualistic mycorrhizae, which provide them with a resource-access advantage (Hartnett *et al.*, 1993). Endophytic fungi can confer sufficient resistance to natural enemies that infected plants outperform competitors that would otherwise displace them (Marks *et al.*, 1991; Clay *et al.*, 1993).

By augmenting the performance of otherwise competitively inferior species, mutualists like these can promote species coexistence, particularly in communities where competition is intense (Wilson and Hartnett, 1997; Stachowicz and Hay, 1999b). Thus, it is possible that multitrophic/multispecies mutualisms that modify competitive environments play critical roles in the maintenance of biological diversity at the community scale. The role of mutualisms in generating and maintaining community and ecosystem-level diversity is attracting increasing attention (e.g., see Bever, 1999; Traveset 1999; Wall and Moore, 1999), and this currently untested hypothesis deserves close scrutiny in this context.

Do these mutualisms confer multiple benefits?

The examples we have cited here for third-species mediation of mutualism are based on studies ranging from simple observations to complex manipulative experiments. We strongly emphasize the importance of further studies, particularly those that use experimental approaches, to explore these phenomena further. In some cases, mutualistic third-species mediation, although present, plays a fairly insignificant role in the association between two species. This appears to be the case for many plant–endophyte interactions, as Faeth and Bultman describe in chapter 5. In other protective mutualisms, additional, more direct benefits have been identified. For example, Morales (2000) explored an ant–treehopper mutualism via experimental manipulations of both ant and predator densities. He demonstrated that ants do benefit treehoppers by protecting them from predators, but also showed that they must provide other benefits as well, since treehoppers perform better in the presence of ants even when predators are absent; some evidence suggests that ant-tending may increase treehopper feeding rates. Multiple benefits like these may prove to be common within mutualisms, but as yet have barely been explored.

Mutualisms whose outcomes are influenced by third species

From the perspective of each participant, the net effect of any interaction can be thought of as its gross benefits to that partner minus its costs. It is now clear that the magnitudes of both gross benefits and costs, and thus of net effects as well, are highly variable for virtually all mutualisms (Thompson, 1988; Bronstein, 1994, in press). At times and places where the costs of mutualism exceed its benefits, the net effect can even be antagonistic. The idea that the effects of a single interaction can range from mutualistic to antagonistic at ecological scales of time and space has only been considered relatively recently; hence, we know very little about the conditions (e.g., mutualist densities, presence of third species, and abiotic factors) that produce observed patterns of variation. In this section, we discuss variation in the intensity and outcome of mutualisms that is mediated by species external to the mutualism. These effects (which are summarized in the second section of Table 3.1) can be found in all forms of mutualism, not only in the set of protective mutualisms discussed in the previous section of this chapter.

Mutualisms altered by third species: antagonists

Natural enemies not only reduce the fitness of their prey, but can also indirectly reduce the fitness of their prey's mutualists (Table 3.1, section II.A). In the simplest case, densities of mutualistic species might be expected to rise and fall together depending on the abundance of natural enemies, but it is surprisingly difficult to find good examples of this in nature (but see Dyer and Letourneau (1999) for a suggestive case). Tightly linked density fluctuations may only be common in relatively obligate and species-specific mutualisms. In more generalized mutualisms, the reduction of any one partner species can have a barely discernible effect on partner density. For example, declines in one species can spur increases in the abundance of species that were previously excluded or suppressed competitively; these resurgent species can function as equally effective, or even more effective, mutualists (e.g., Young *et al.*, 1997).

More complex antagonist-mediated effects, in which the benefits of mutualism are reduced in the presence of antagonists even though mutualist density *per se* is not, have been documented in a wide variety of systems. For example, the spider *Dipoena banksii*, an ant predator, interferes with protective mutualisms between the ant *Pheidole bicornis* and *Piper* plants (Piperaceae) by building webs at the base of new leaves. Since the ant can detect and will avoid these webs, plants with spiders suffer more herbivory than those without spiders (Gastreich, 1999). Herbivores interfere with pollination and seed dispersal mutualisms by making plants less attractive or accessible to their partners (Christensen and Whitham, 1993; Strauss, 1997). For example, effects of herbivory on pollination occur via reductions in resource allocation to flowers, nectar, and pollen grains (Aizen and Raffaele, 1996, 1998; Lohman *et al.*, 1996; Lehtila and Strauss, 1997; Mothershead and Marquis, 2000), as well as via modifications in plant architecture and flowering phenology (Juenger and Bergelson, 1997). Parasites can also interfere with pollination. Schmid-Hempel and Stauffer (1998) describe a case in which the quality of bumblebees as pollinators deteriorates when they are infected with larvae of an endoparasitic conopid fly: parasitized individuals show reduced fidelity to any one plant species. Similarly, pathogenic fungi and bacteria that colonize nectar, flowers, and fruit can make plants less attractive to pollinators and dispersers, leading to reduced visitation rates to infected plants (Alexander, 1987; Borowicz, 1988; Buchholz and Levey, 1990; Ehlers and Olesen, 1997).

Competitors of mutualists can similarly reduce the benefits of mutualism. Plant species that flower simultaneously can compete for pollinators, reducing reproductive success of one or more of them relative to their performance when flowering alone (Rathcke, 1988). Different pollinator species, in turn, can compete for access to plants, with inferior competitors left with inferior resources (Johnson and Hubbell, 1975). Pollinators also compete with non-pollinators for flowers. For example, wasp species that feed on sterile tissue within developing figs compete for resources with the offspring of pollinator wasps that also develop there. In some cases, this reduces pollinator maturation success, and hence the potential success of the fig tree as a pollen donor (West and Herre, 1994; Kerdelhué and Rasplus, 1996).

In some cases antagonists can actually enhance mutualistic benefits, at least for one of the two partners. One type of unidirectional enhancement takes place in interactions in which mutualistic rewards are only produced, or are produced in greater quantities, when attack by a natural enemy increases the rewarding species' need for defenders. Leimar and Axén (1993), for example, demonstrated that the rate of reward secretion by one lycaenid caterpillar increases after simulated predator attacks. Similarly, in some plants, extrafloral nectar is an inducible defense primarily produced in response to herbivore damage (Agrawal and Rutter,

1998). In these interactions, the rewarded species is the one that benefits from the presence of its mutualist's antagonists; the species under attack clearly does worse. Attack by herbivores occasionally leads to increases in flower production, however, potentially benefiting pollinators and plants alike (e.g., Lennartson *et al.*, 1998).

Mutualisms altered by third species: mutualists

In so-called "friend's friend" interactions (Boucher *et al.*, 1982), the success of one mutualism enhances the success of another (Table 3.1, section II.B). Two species that share a mutualist species are sometimes able to attract so many more partner individuals when they co-occur that the positive effects of sharing partners outweigh the negative effects of having to compete for their attention. For example, certain plant species that flower at the same place and time attract disproportionately high numbers of pollinators, facilitating each other's reproductive success (Rathcke, 1983; Laverty, 1992). Similarly, plants may attract more seed dispersers when neighboring species that share those dispersers bear fruit at the same time (Sargent, 1990). Different plant species may also share root symbionts via underground connections, resulting in bidirectional nutrient transfer beneficial to both plants (Simard *et al.*, 1997). Another form of mutualistic enhancement of mutualism occurs when an individual is able to obtain more mutualists once it has established another kind of mutualism with a third species. Mycorrhizal inoculation can lead to an increase in nodulation of the same plant by *Rhizobium*, as a consequence of the positive effect of mycorrhizae on plant size (Cluett and Boucher, 1983). Such mutually facilitated colonization has been observed in other pairs of root symbionts as well (Sempavalan *et al.*, 1995).

Mutualisms altered by third species: exploiters

Exploiter species and individuals (often called "cheaters") interfere with mutualisms by reaping the benefits that mutualists offer their partners while providing no benefits in return (Table 3.1, section II.C; Soberon and Martinez del Rio, 1985; Bronstein, 2001). For example, species requiring transport may advertise rewards but deliver none. Many orchids, for instance, display nectarless flowers that mimic those of rewarding species, and they receive sufficient visits to set seed (Dafni, 1984). Exploitation can occur on the transport side of a transportation mutualism as well, by visitors that collect rewards but do not transport or even destroy the associate or its gametes. For example, nectar-robbers consume nectar but neither pick up nor deposit pollen (Maloof and Inouye, 2000). Exploiters in protective mutualisms include species that consume plant and insect rewards but never protect the reward-provider from its enemies (DeVries and Baker, 1989; Gaume and McKey, 1999), as well as Batesian mimics associated with Müllerian mimicry complexes, which gain protection from enemies without producing the expensive chemical defenses shared by the Müllerian species (Gilbert, 1983). In nutritional mutualisms, exploiters include strains of mycorrhizae (Smith and Smith, 1996) and *Rhizobium* (Batzli *et al.*, 1992) that obtain fixed carbon from their host plants but transport no nutrients back to them. Lichens are attacked by diverse parasitic fungi that enslave the algae and confer no benefit in return (Richardson, 1999). As these examples might suggest, exploiters can be found within essentially every kind of mutualism (Bronstein, 2001).

Exploiters do not always reduce the benefits that mutualists accrue from their interaction (e.g., Maloof and Inouye, 2000). However, the negative impact of exploitation can sometimes be dramatic. Letourneau (1990) has documented such an effect by a clerid beetle associated with an obligate ant–plant mutualism in Costa Rica. The plant, a species of *Piper*, produces food rewards once it detects the presence of its specific ant-defender species. However, the beetle invades the plant, consumes the ants, then somehow induces the plant to continue producing food bodies for it. Thus, the plant loses its biotic defense system and its ant mutualists lose their lives. Ant defensive mutualisms also can be disrupted by non-predatory exploiters. Non-mutualistic ants have recently been shown to prune buds and flowers from the ant-plants with which they are associated, increasing vegetative plant growth but reducing sexual reproduction to near zero (Yu and Pierce, 1998; Stanton *et al.*, 1999). In some cases, monopolization of a reward by a non-mutualistic ant species deters visits by more mutualistic ants (Gaume and McKey, 1999).

Mutualisms altered by third species: incidental and anthropogenic disruptions

Finally, some species disrupt mutualisms simply as by-products of other behaviors. Prominent among these disruptions are the activities of humans (Table 3.1, section II.D). Anthropogenic effects on mutualisms can be dramatic and are increasing at an alarming rate (e.g., see Smith and Buddemeier, 1992; Buchmann and Nabhan, 1996; Kearns *et al.*, 1998; Richardson *et al.*, 2000). Notable among these effects is the intentional or accidental introduction of mutualists' predators, parasites, and competitors. For example, invasion by the Argentine ant (*Linepithema humile*) has been altering mutualisms worldwide. In South Africa, it outcompetes native ants, resulting in reduced seed dispersal of native myrmecochorous plants (Bond and Slingsby, 1984), while in Hawaii it destroys nests of endemic bees that are essential pollinators of the endemic flora (Cole *et al.*, 1992). Humans also alter the environment in a number of ways that lead mutualisms to break down, including habitat destruction and fragmentation (Aizen and Feinsinger, 1994), generation of inferior edge habitats (Jules and Rathcke, 1999), introduction of pollutants (Kevan *et al.*, 1985), and carbon dioxide enrichment (Smith and Buddemeier, 1992; but see Staddon and Fitter (1998) for a case in which increased CO_2 may *enhance* a set of mutualisms).

Variation in third-species effects

In each of the examples discussed in this section, a third species has been shown to have either a positive or negative effect on one or both mutualists, and thus on the mutualism as a whole. However, it is not uncommon for third-species effects on a given mutualism to vary in direction. For example, consider the effects that plants experience from certain antagonists of their biotic pollinators and seed dispersers. The benefits of both pollination and seed dispersal depend not only upon mutualists arriving to the plants, but upon whether they depart with the plant's gametes. Visits that are too long are therefore less beneficial, since they can result in selfing in the case of pollination, and deposition of seeds directly under the parent plant in the case of seed dispersal. For this reason, predators and other antagonists whose presence leads mutualists to shorten their visits therefore might actually enhance plant reproductive success (Pratt and Stiles, 1983; Maloof and Inouye, 2000). However, if these antagonists are either too common or too successful, mutualist visits might well become limiting to plant success. In other words, depending on their abundance and actions, certain antagonists might either increase or decrease the benefits of the mutualism they disrupt.

The most thoroughly studied cases in which species external to a mutualism have varying effects involve interactions that are mutualistic only in the presence of a third species (Table 3.1, section I). In these cases, it is the presence and identity of a *fourth* species that affects the direction of the outcome. One group of mutualisms in which these effects are seen are those in which ants defend other insect species from their natural enemies. The fourth species in these situations are the plants on which the insects feed while they are defended. At least two kinds of plant traits affect the magnitude of benefits that tended insects receive from their defenders. First, the nutritional status and water content of the host plant influences the quality and quantity of rewards produced by ant-tended herbivores for their ants, and hence the degree of protection that ants provide to them in return (Fiedler, 1990; Pierce *et al.*, 1991; Burghardt and Fiedler, 1996). Baylis and Pierce (1991), for example, showed that larvae of the lycaenid *Jalmenus evagoras* that feed on fertilized plants attract more ant defenders, and consequently experience higher survivorship. Not surprisingly, adult lycaenids preferentially oviposit on fertilized plants (Pierce and Elgar, 1985). Second, whether or not the plant directly offers rewards to ants can potentially affect the success of ant–insect defensive mutualisms taking place on its surface. Plant rewards could either shift ant attention away from the insects they tend, disrupting ant–insect mutualisms (Becerra and Venable, 1989), or else contribute to the support of a larger population of defenders, enhancing ant–insect mutualisms (Del-Claro and Oliveira, 1993). Converse effects are also possible: if ants prefer insect rewards over plant rewards, ant–plant mutualisms can be disrupted (Buckley, 1983; DeVries and Baker, 1989).

Evolutionary implications of multitrophic/multispecies mutualisms

Cushman and Addicott (1991) have suggested three general sources of variation in the outcomes of mutualism: (1) variation in the kinds of ecological "problems" species experience; (2) variation in the solutions that partners can provide to these problems; and (3) variation in the availability of mutualists. The direct and indirect influences of other species are key determinants of all of these sources of variation. Although this point has been developed previously (Thompson, 1988; Bronstein, 1994) and elaborated upon for one kind of mutualism (Cushman, 1991; Cushman and Addicott, 1991), its broader effects, particularly in an evolutionary context, have been underappreciated. In this final section, we speculate upon and offer some hypotheses regarding the evolutionary dimension of multitrophic/multispecies mutualisms.

It is clear that species from various trophic levels not only mediate the interactions between mutualists in ecological time, but may act as selective forces in the evolution of mutualism. That is, *interactions between mutualists are mediated by species of the same and other trophic levels not only at ecological but evolutionary time-scales.* The evolutionary influence of a third species on mutualists is likely to be difficult to demonstrate, particularly in circumstances in which the influence of third species is indirect. In the (usual) absence of opportunities to demonstrate ongoing evolution, attempts to prove evolutionary pathways have generally entailed demonstrations that the mechanisms and conditions that would facilitate the proposed evolutionary scenario do at least exist.

Following this approach, we suggest that (1) *traits mediating mutualisms that exist only in the presence of external species are likely to indicate the evolutionary influence of those external species.* For example, it has proven difficult to explain the evolution of certain complex plant and insect structures and exudates (e.g., Pierce, 1987; Letourneau, 1990; Folgarait and Davidson, 1994) outside the context of soliciting ant protection against natural enemies. Put another way, in the absence of selective pressure imposed by natural enemies, those ant rewards would not have arisen. (Of course, the natural enemy species important within these interactions today are not necessarily the same ones that exerted selection at the time those traits evolved. Furthermore, additional past or present functions of these traits might well be found some day.) At the same time, however, it should be remembered that many of the critical traits that mediate protective mutualisms probably did not evolve within those interactions. For example, ants exhibit diverse behaviors that facilitate their functions as mutualistic protectors of plants and insects, but it is difficult to find evidence that any of these behaviors arose or changed after the establishment of those associations. Rather, other species evolved traits that co-opted and redirected pre-existing ant behaviors, such as interspecific aggression and nest-cleaning.

We further suggest that (2) *traits within mutualisms that serve multiple functions often indicate an evolutionary role of other species.* The existence of traits that serve multiple functions and that influence several species can be observed in all pairwise interactions, not only mutualisms. In predator–prey interactions, many anti-predator behaviors are exaptations, i.e., traits that originally evolved in response to a third species or other selective force but now serve additional functions (Sih, 1992; Yosef and Whitman, 1992; Kudo, 1996; Kudo and Ishibashi, 1996; Matsuda *et al.*, 1996; Rayor, 1996). For example, many lepidopteran larvae drop (balloon) on a strand of silk as a method of dispersal (McManus and Mason, 1983). This same behavior can also serve as an anti-predator defense. It is not effective against all types of predators and parasitoids (see Yeargan and Braman, 1989a, b), that is, the outcome of the expression of this behavior is conditional. One might speculate that the evolution of ballooning as an anti-predator defense may have been constrained by its importance in dispersal, or vice versa. Another example of single traits that function in multiple interactions involves guppies (*Poecilia reticulata*), in which mate selection by females is based on male color patterns. The evolution of interactions between males and females may have been constrained by the responses of other species, such as predators, to the color patterns. Thus, as predation intensity increases, color patterns become simpler, body size is reduced, and schooling intensifies (Endler, 1995).

In the case of mutualism, indirect influences like these might arise if a behavioral, physical, or physiological trait that originally evolved in response to a third species concurrently or subsequently acquired a role in the context of a mutualistic interaction. It might then come under selection in the context of that mutualism. In the clearest example to date, Armbruster (1997; Armbruster *et al.*, 1997) has provided extensive phylogenetic evidence documenting that floral compounds in *Dalechampia* vines (Euphorbiaceae) are exaptations; they apparently originated as chemical defenses against herbivores, and were subsequently co-opted and further modified in the context of attracting and rewarding pollinators.

Considerably more complex evolutionary scenarios can also be envisioned. We offer one hypothetical example to illustrate this point. Many flowering plants lose nectar to animals that provide no pollination service (Maloof and Inouye, 2000). Floral morphologies of some of these species, such as highly elongated corolla tubes, have been hypothesized to have evolved as mechanisms to deter such nectar-robbers. However, some of these same floral traits restrict the pool of pollinators to species that possess matching morphologies (e.g., hummingbirds with highly elongated bills), and may in fact have selected for those morphologies (McDade, 1992). At the same time, many robbers have evolved traits or adopted behaviors that allow them to feed on nectar regardless of corolla length. For instance, certain hummingbird species have evolved serrated bills that allow them to rip corollas open (Ornelas, 1994). It is possible that floral traits like these evolved in the context of robbing and subsequently influenced the evolution of the mutualism, particularly the evolution of specialization to pollinators. Alternatively, these floral traits may have evolved in the context of mutualism, and subsequently influenced the evolution of robbing. It is also conceivable that they have been selected simultaneously in the context of fostering mutualists and deterring robbers. To our knowledge, there is as yet no evidence available that would lend support any one of these scenarios over the other two. Whichever happened, it is clear that multiple evolutionary (and coevolutionary) processes can take place at one time within a given mutualism, and can interact when they do.

Traits that affect multiple species can impart remarkable ecological and evolutionary complexity to mutualistic relationships. For example, many plants possess extrafloral nectaries (EFNs), and the most common mutualists that feed from them are ants that defend those plants (Koptur, 1992). However, most plant defense mutualisms based on EFNs are highly generalized, involving interactions with many species other than or in addition to ants. Some of these EFN visitors benefit the plant, including predatory and fungivorous mites and parasitic wasps (Pemberton, 1993; Pemberton and Lee, 1996; van Rijn and Tanigoshi, 1999). Others are commensal or even antagonistic towards the plant (e.g., DeVries and Baker, 1989). The evolution of pairwise interactions between ants and plants is likely to have constrained and been constrained by the evolution of plants' relationships with many other species. It will be a challenge to tease apart and identify the evolutionary pressures exerted by single species in diffuse multitrophic/multispecies mutualisms like these.

This will be a particularly difficult (although interesting!) challenge when a single species functions either as a mutualist or antagonist of its partner, depending on the ecological context. For example, traits might evolve in response to selection on the mutualistic component of an interaction in one subset of ecological habitats, but in response to selection on its antagonistic component elsewhere. Such a process might give rise the kind of "geographic mosaic of coevolution" studied by John Thompson and his colleagues for strictly pairwise mutualisms (i.e., ones considerably simpler than those discussed in this chapter), in which a single species pair coevolves as mutualists in some patches but as antagonists in others (e.g., Nuismer *et al.*, 1999, 2000; Gomulkiewicz *et al.*, 2000; Hochberg *et al.*, 2000). An important step for future research is to delineate the kinds of mutualisms in which these and other evolutionary processes might be expected.

This review has barely touched on a number of other fascinating questions about the ecological and evolutionary distribution of multitrophic/multispecies effects. For instance, are they more common in symbiotic versus non-symbiotic, specialized versus generalized, or obligate versus non-obligate mutualisms? Are they more abundant in certain habitats? In symbiotic mutualisms, are third-party influences greater on the symbiont or the external partner? Is there a relationship between the evolutionary age of a mutualism and the likelihood that it is mediated by other species? Can multitrophic/multispecies mutualisms undergo coevolution?

It is quite clear that third species influence both the ecological nature and intensity of mutualistic relationships, and quite likely that they have played important roles in the evolution of mutualistic traits. It is equally apparent that major new insights stand to be gained by studying mutualisms in the context of the role that other species play within them both currently and in the past. Only a handful of unifying principles have yet been proposed that help us understand the nature of mutualisms, and the few that do exist have as yet generated even fewer testable hypotheses. The review and analysis of the literature on multitrophic/multispecies mutualisms that we have presented here generates several testable hypotheses that we hope ultimately will contribute to this process.

Acknowledgments

Some of the ideas expressed here were developed from research conducted by PB with the support of a USAID/Ministry of Agriculture grant and the University of Maryland, Agricultural Experiment Station Project MD-H-201. Writing of the manuscript was facilitated by funding from the National Science Foundation (DEB 99-73521) to JLB. Comments of Anurag Agrawal, Goggy Davidowitz, Stan Faeth, David Inouye, Mike Leigh, Deborah Letourneau, Manuel Morales, Jennifer Rudgers, and an anonymous reviewer greatly improved the manuscript.

REFERENCES

Agrawal, A. A. and Rutter, M. T. (1998) Dynamic anti-herbivore defense in ant-plants: the role of induced responses. *Oikos* **83**: 227–236.

Aizen, M. A. and Feinsinger, P. (1994) Forest fragmentation, pollination, and plant reproduction in a chaco dry forest, Argentina. *Ecology* **75**: 330–351.

Aizen, M. A. and Raffaele, E. (1996) Nectar production and pollination in *Alstroemeria aurea*: responses to level and pattern of flowering shoot defoliation. *Oikos* **76**: 312–322.

Aizen, M. A. and Raffaele, E. (1998) Flowering-shoot defoliation affects pollen grain size and postpollination pollen performance in *Alstroemeria aurea*. *Ecology* **79**: 2133–2142.

Alexander, H. M. (1987) Pollination limitation in a population of *Silene alba* infected by the anther-smut fungus, *Ustilago violacea*. *Journal of Ecology* **75**: 771–780.

Amsler, C. D., McClintock, J. B. and Baker, B. J. (1999) An antarctic feeding triangle: defensive interactions between macroalgae, sea urchins, and sea anemones. *Marine Ecology Progress Series* **183**: 105–114.

Armbruster, W. S. (1997) Exaptations link evolution of plant–herbivore and plant–pollinator interactions: a phylogenetic study. *Ecology* **78**: 1661–1672.

Armbruster, W. S., Howard, J. J., Clausen, T. P., Debevec, E. M., Loquvam, J. C., Matsuki, N., Cerendola, B. and Andel, F. (1997) Do biochemical exaptations link evolution of plant defense and pollination systems? Historical hypotheses and experimental tests with *Dalechampia* vines. *American Naturalist* **149**: 461–484.

Barbosa, P. and Benrey, B. (1998) The influence of plants on insect parasitoids: implications to conservation biological control. In *Conservation Biological Control*, ed. P. Barbosa, pp. 55–82. San Diego, CA: Academic Press.

Barbosa, P. and Letourneau, D. K. (eds.) (1988) *Novel Aspects of Insect–Plant Interactions*. New York: John Wiley.

Barbosa, P. and Wratten, S. D. (1998) Influence of plants on invertebrate predators: implications to conservation biological control. In *Conservation Biological Control*, ed. P. Barbosa, pp. 83–100. San Diego, CA: Academic Press.

Batzli, J. M., Graves, W. R. and Berkum, P. V. (1992) Diversity among rhizobia effective with *Robinia pseudoacacia* L. *Applied and Environmental Microbiology* **58**: 2137–2143.

Baylis, M. and Pierce, N. E. (1991) The effect of host-plant quality on the survival of larvae and oviposition by adults of an ant-tended butterfly, *Jalmenus evagoras*. *Ecological Entomology* **16**: 1–9.

Becerra, J. X. I. and Venable, D. L. (1989) Extrafloral nectaries: a defense against ant–homoptera mutualisms? *Oikos* **55**: 276–280.

Bertness, M. D. and Callaway, R. (1994) Positive interactions in communities. *Trends in Ecology and Evolution* **9**: 191–193.

Bever, J. D. (1999) Dynamics within mutualism and the maintenance of diversity: inference from a model of interguild frequency dependence. *Ecology Letters* **2**: 52–62.

Bond, W. and Slingsby, P. (1984) Collapse of an ant–plant mutualism: the Argentine ant (*Iridomyrmex humilis*) and myrmecochorous Proteaceae. *Ecology* **65**: 1031–1037.

Borowicz, V. A. (1988) Do vertebrates reject decaying fruit? An experimental test with *Cornus amomum* fruits. *Oikos* **53**: 74–78.

Boucher, D. H., James, S. and Keeler, K. H. (1982) The ecology of mutualism. *Annual Review of Ecology and Systematics* **13**: 315–347.

Bronstein, J. L. (1994) Conditional outcomes in mutualistic interactions. *Trends in Ecology and Evolution* **9**: 214–217.

Bronstein, J. L. (2001) The exploitation of mutualisms. *Ecology Letters* **4**: 277–287.

Bronstein, J. L. (in press) Mutualisms. In *Evolutionary Ecology: Perspectives and Synthesis*, ed. C. Fox, D. Fairbairn and D. Roff. Oxford: Oxford University Press.

Buchholz, R. and Levey, D. J. (1990) The evolutionary triad of microbes, fruits, and seed dispersers: an experiment in fruit choice by cedar waxwings, *Bombycilla cedrorum*. *Oikos* **59**: 200–204.

Buchmann, S. L. and Nabhan, G. P. (1996) *The Forgotten Pollinators*. Washington, DC: Island Press.

Buckley, R. (1983) Interaction between ants and membracid bugs decreases growth and seed set of host plant bearing extrafloral nectaries. *Oecologia* **58**: 132–136.

Buckley, W. J. and Ebersole, J. P. (1994) Symbiotic organisms increase the vulnerability of a hermit crab to predation. *Journal of Experimental Marine Biology and Ecology* **182**: 49–64.

Burghardt, F. and Fiedler, K. (1996) The influence of diet on growth and secretion behaviour of myrmecophilous *Polyommatus icarus* caterpillars (Lepidoptera: Lycaenidae). *Ecological Entomology* **21**: 1–8.

Callaway, R. M. and Walker, L. R. (1997) Competition and facilitation: a synthetic approach to interactions in plant communities. *Ecology* **78**: 1958–1965.

Cardé, R. T. and Bell, W. J. (eds.) (1995) *Chemical Ecology of Insects*, 2nd edn. New York: Chapman and Hall.

Christensen, K. M. and Whitham, T. G. (1993) Impact of insect herbivores on competition between birds and mammals for pinyon pine seeds. *Ecology* **74**: 2270–2278.

Clay, K. (1991) Fungal endophytes, grasses, and herbivores. In *Microbial Mediation of Plant–Herbivore Interactions*, ed. P. Barbosa, V. A. Krischik and C. G. Jones, pp. 199–226. New York: John Wiley.

Clay, K., Marks, S. and Cheplick, G. P. (1993) Effects of insect herbivory and fungal endophyte infection on competitive interactions among grasses. *Ecology* **74**: 1767–1777.

Cluett, H. C. and Boucher, D. H. (1983) Indirect mutualism in the legume–*Rhizobium*–mycorrhizal fungus interaction. *Oecologia* **59**: 405–408.

Cole, F. R., Madeiros, A. C., Loope, L. L. and Zuehlke, W. W. (1992) Effects of the Argentine ant on arthropod fauna of Hawaiian high-elevation shrubland. *Ecology* **73**: 1313–1322.

Cushman, J. H. (1991) Host-plant mediation of insect mutualisms: variable outcomes in herbivore–ant interactions. *Oikos* **61**: 138–144.

Cushman, J. H. and Addicott, J. F. (1991) Conditional interactions in ant–plant–herbivore mutualisms. In *Ant–Plant Interactions*, ed. C. R. Huxley and D. F. Cutler, pp. 92–103. Oxford: Oxford University Press.

Dafni, A. (1984) Mimicry and deception in pollination. *Annual Review of Ecology and Systematics* **15**: 259–278.

Del-Claro, K. and Oliveira, P. S. (1993) Ant–homoptera interaction: do alternative sugar sources distract tending ants? *Oikos* **68**: 202–206.

DeVries, P. J. and Baker, I. (1989) Butterfly exploitation of a plant–ant mutualism: adding insult to herbivory. *Journal of the New York Entomological Society* **97**: 332–340.

Dudley, T. L. (1992) Beneficial effects of herbivore on stream macroalgae via epiphyte removal. *Oikos* **65**: 121–127.

Dyer, L. A. and Letourneau, D. K. (1999) Trophic cascades in a complex terrestrial community. *Proceedings of the National Academy of Sciences, USA* **96**: 5072–5076.

Ehlers, B. K. and Olesen, J. M. (1997) The fruit-wasp route to toxic nectar in *Epipactis* orchids? *Flora* **192**: 223–229.

Ellison, A. M., Farnsworth, E. J. and Twilley, R. R. (1996) Facultative mutualism between red mangrove and root-fouling sponges in Belizean mangal. *Ecology* **77**: 2431–2444.

Endler, J. A. (1995) Multiple-trait coevolution and environmental gradients in guppies. *Trends in Ecology and Evolution* **10**: 22–29.

Espoz, C., Guzmán, C. and Castilla, J. C. (1995) The lichen *Thelidium litorale* on shells of intertidal limpets: a case of lichen-mediated cryptic mimicry. *Marine Ecology Progress Series* **119**: 191–197.

Fiala, B., Maschwitz, U., Pong, T. Y. and Helbig, A. J. (1989) Studies of a South East Asian ant–plant association: protection of *Macaranga* trees by *Crematogaster borneensis*. *Oecologia* **79**: 463–470.

Fiedler, K. (1990) Effects of larval diet on myrmecophilous qualities of *Polyommatus icarus* caterpillars (Lepidoptera: Lycaenidae). *Oecologia* **83**: 284–287.

Folgarait, P. J. and Davidson, D. W. (1994) Antiherbivore defenses of myrmecophytic *Cecropia* under different light regimes. *Oikos* **71**: 305–320.

Gastreich, K. R. (1999) Trait-mediated indirect effects of a theridiid spider on an ant–plant mutualism. *Ecology* **80**: 1066–1070.

Gaume, L. and McKey, D. (1999) An ant–plant mutualism and its host-specific parasite: activity rhythms, young leaf patrolling, and effects on herbivores of two specialist plant-ants inhabiting the same myrmecophyte. *Oikos* **84**: 130–144.

Gaume, L., McKey, D. and Anstett, M.-C. (1997) Benefits conferred by "timid" ants: active anti-herbivore protection of the rainforest tree *Leonardoxa africana* by the minute ant *Petalomyrmex phylax*. *Oecologia* **112**: 209–216.

Gilbert, L. E. (1983) Coevolution and mimicry. In *Coevolution*, ed. D. Futuyma and M. Slatkin, pp. 263–281. Sunderland, MA: Sinauer Associates.

Gomulkiewicz, R., Thompson, J. N., Holt, R. D., Nuismer, S. L. and Hochberg, M. E. (2000) Hot spots, cold spots, and the geographic mosaic theory of coevolution. *American Naturalist* **156**: 156–174.

Gressitt, J. L., Sediacek, J. and Szent-Ivany, J. J. H. (1965) Flora and fauna on backs of large papuan moss-forest weevils. *Science* **150**: 1833–1835.

Grutter, A. S. (1997) Effect of the removal of cleaner fish on the abundance and species composition of reef fish. *Oecologia* **111**: 137–143.

Grutter, A. S. (1999) Cleaner fish really do clean. *Nature* **398**: 672–673.

Hartnett, D. C., Hetrick, B. A. D., Wilson, G. W. T. and Gibson, D. J. (1993) Mycorrhizal influence on intra- and interspecific neighbour interactions among co-occurring prairie grasses. *Journal of Ecology* **81**: 787–795.

Hochberg, M. E., Gomulkiewicz, R., Holt, R. D. and Thompson, J. N. (2000) Weak sinks could cradle mutualistic symbioses: strong sources should harbour parasitic symbioses. *Journal of Evolutionary Biology* **13**: 213–222.

Huxley, C. R. and Cutler, D. F. (eds.) (1991) *Ant–Plant Interactions*. Oxford: Oxford University Press.

Johnson, L. K. and Hubbell, S. P. (1975) Contrasting foraging strategies and coexistence of two bee species on a single resource. *Ecology* **56**: 1398–1406.

Juenger, T. and Bergelson, J. (1997) Pollen and resource limitation of compensation to herbivory in scarlet gilia, *Ipomopsis aggregata*. *Ecology* **78**: 1684–1695.

Jules, E. S. and Rathcke, B. J. (1999) Mechanisms of reduced *Trillium* recruitment along edges of old-growth forest remnants. *Conservation Biology* **13**: 784–793.

Kearns, C. A., Inouye, D. W. and Waser, N. M. (1998) Endangered mutualisms: the conservation of plant–pollinator interactions. *Annual Review of Ecology and Systematics* **29**: 83–112.

Kerdelhué, C. and Rasplus, J.-Y. (1996) Non-pollinating Afrotropical fig wasps affect the fig–pollinator mutualism in *Ficus* within the subgenus *Sycomorus*. *Oikos* **75**: 3–14.

Kevan, P. G., Thomson, J. D. and Plowright, R. C. (1985) Matacil insecticide spraying, pollinator mortality, and plant fecundity in New Brunswick forests. *Canadian Journal of Botany* **63**: 2056–2061.

Koptur, S. (1992) Extrafloral nectary-mediated interactions between insects and plants.

In *Insect–Plant Interactions*, vol. 4, ed. E. Bernays, pp. 81–129. Boca Raton, FL: CRC Press.

Krogh, D. (2000) *Biology: A Guide to the Natural World*. Upper Saddle River, NJ: Prentice Hall.

Kudo, S.-I. (1996) Ineffective maternal care of a subsocial bug against a nymphal parasitoid: a possible consequence of specialization to predators. *Ethology* **102**: 227–235.

Kudo, S.-I. and Ishibashi, E. (1996) Maternal defense of a leaf beetle is not effective against parasitoids but is effective against predators. *Ethology* **102**: 560–567.

Laverty, T. M. (1992) Plant interactions for pollinator visits: a test of the magnet species effect. *Oecologia* **89**: 502–508.

Leimar, O. and Axén, A. H. (1993) Strategic behavior in an interspecific mutualism: interactions between lycaenid larvae and ants. *Animal Behavior* **46**: 1177–1182.

Lehtila, K. and Strauss, S. Y. (1997) Leaf damage by herbivores affects attractiveness to pollinators in wild radish, *Raphanus raphanistrum*. *Oecologia* **111**: 396–406.

Lennartson, T., Nilsson, P. and Tuomi, J. (1998) Induction of overcompensation in the field gentian, *Gentianella campestris*. *Ecology* **79**: 1061–1072.

Letourneau, D. K. (1983) Passive aggression: an alternative hypothesis for the *Piper–Pheidole* association. *Oecologia* **60**: 122–126.

Letourneau, D. K. (1990) Code of ant–plant mutualism broken by parasite. *Science* **248**: 215–217.

Lohman, D. J., Zangerl, A. R. and Berenbaum, M. R. (1996) Impact of floral herbivory by parsnip webworm (Oecophoridae: *Depressaria pastinacella* Duponchel) on pollination and fitness of wild parsnip (Apiaceae: *Pastinaca sativa* L.). *American Midland Naturalist* **136**: 407–412.

Maloof, J. E. and Inouye, D. W. (2000) Are nectar robbers cheaters or mutualists? *Ecology* **81**: 2651–2661.

Marks, S., Clay, K. and Cheplick, G. P. (1991) Effects of fungal endophytes on interspecific and intraspecific competition in the grasses *Festuca arundinacea* and *Lolium perenne*. *Journal of Applied Ecology* **28**: 194–204.

Matsuda, H., Hori, M. and Abrams, P. A. (1996) Effects of predator-specific defense on biodiversity and community complexity in two-trophic-level communities. *Evolutionary Ecology* **10**: 13–28.

McDade, L. A. (1992) Pollinator relationships, biogeography, and phylogenetics. *BioScience* **42**: 21–26.

McManus, M. L. and Mason, C. J. (1983) Determination of the settling velocity and its significance to larval dispersal of the gypsy moth (Lepidoptera: Lymantriidae). *Environmental Entomology* **12**: 270–272.

McQuaid, C. D. and Froneman, P. W. (1993) Mutualism between the territorial intertidal limpet *Patella longicosta* and the crustose alga *Ralfsia verrucosa*. *Oecologia* **96**: 128–133.

Menge, B. A. (1995) Indirect effects in marine intertidal interaction webs: patterns and importance. *Ecological Monographs* **65**: 21–74.

Metcalfe, N. B. (1989) Flocking preferences in relation to vigilance benefits in mixed-species shorebird flocks. *Oikos* **56**: 91–98.

Morales, M. A. (2000) Mechanisms and density dependence of benefit in an ant–membracid mutualism. *Ecology* **81**: 482–489.

Mothershead, K. and Marquis, R. J. (2000) Fitness impacts of herbivory through indirect effects on plant–pollinator interactions in *Oenothera macrocarpa*. *Ecology* **81**: 30–40.

Nuismer, S. L., Thompson, J. N. and Gomulkiewicz, R. (1999) Gene flow and geographically structured coevolution. *Proceedings of the Royal Society of London Series B* **266**: 605–609.

Nuismer, S. L., Thompson, J. N. and Gomulkiewicz, R. (2000) Coevolutionary clines across selection mosaics. *Evolution* **54**: 1102–1115.

Olff, H., Brown, V. K. and Drent, R. H. (eds.) (1999) *Herbivores: Between Plants and Predators*. Oxford: Blackwell Science.

Ornelas, J. F. (1994) Serrate tomia: an adaptation for nectar robbing in hummingbirds? *Auk* **111**: 703–710.

Pemberton, R. W. (1993) Observations of extrafloral nectar feeding by predaceous and fungivorous mites. *Proceedings of the Entomological Society of Washington* **95**: 642–643.

Pemberton, R. W. and Lee, J.-H. (1996) The influence of extrafloral nectaries on parasitism of an insect herbivore. *American Journal of Botany* **83**: 1187–1194.

Pierce, N. E. (1987) The evolution and biogeography of associations between lycaenid butterflies and ants. *Oxford Surveys in Evolutionary Biology* **4**: 89–116.

Pierce, N. E. and Elgar, M. A. (1985) The influence of ants on host plant selection by *Jalmenus evagoras*, a myrmecophilous lycaenid butterfly. *Behavioral Ecology and Sociobiology* **16**: 209–222.

Pierce, N. E., Kitching, R. L., Buckley, R. C., Taylor, M. F. J. and Benbow, K. F. (1987) The costs and benefits of cooperation between the Australian lycaenid butterfly, *Jalmenus evagoras*, and its attendant ants. *Behavioral Ecology and Sociobiology* **21**: 237–248.

Pierce, N. E., Nash, D. R., Baylis, M. and Carper, E. R. (1991) Variation in the attractiveness of lycaenid butterfly larvae to ants. In *Ant–Plant Interactions*, ed. D. Cutler and C. Huxley, pp. 131–142. Oxford: Oxford University Press.

Poulin, R. and Grutter, A. S. (1996) Cleaning symbioses: proximate and adaptive explanations. *BioScience* **46**: 512–517.

Pratt, T. K. and Stiles, E. W. (1983) How long fruit-eating birds stay in the plants where they feed: implications for seed dispersal. *American Naturalist* **122**: 797–805.

Rathcke, B. (1983) Competition and facilitation among plants for pollination. In *Pollination Biology*, ed. L. Real, pp. 305–329. New York: Academic Press.

Rathcke, B. (1988) Interactions for pollination among coflowering shrubs. *Ecology* **69**: 446–457.

Rayor, L. S. (1996) Attack strategies of predatory wasps (Hymenoptera: Pompilidae; Sphecidae) on colonial orb web-building spiders (Araneidae: *Metepeira incrassata*). *Journal of the Kansas Entomological Society* **69**: 67–75.

Richardson, D. H. S. (1999) War in the world of lichens: parasitism and symbiosis as exemplified by lichens and lichenicolous fungi. *Mycological Research* **103**: 641–650.

Richardson, D. M., Allsopp, N., D'Antonio, C., Milton, S. J. and Rejmánek, M. (2000) Plant invasions: the role of mutualisms. *Biological Reviews* **75**: 65–93.

Sabelis, M. W., Van Baalen, M., Bruin, J., Egas, M., Jansen, V. A. A., Janssen, A. and Pels, B. (1999) The evolution of overexploitation and mutualism in plant–herbivore–predator interactions and its impact on population dynamics. In *Theoretical Approaches to Biological Control*, ed. B. A. Hawkins and H. Cornell, pp. 259–282. New York: Cambridge University Press.

Saikkonen, K., Faeth, S. H., Helander, M. and Sullivan, T. J. (1998) Fungal endophytes: a continuum of interactions with host plants. *Annual Review of Ecology and Systematics* **29**: 319–343.

Sargent, S. (1990) Neighborhood effects on fruit removal by birds: a field experiment with *Viburnum dentatum* (Caprifoliaceae). *Ecology* **71**: 1289–1298.

Schmid-Hempel, P. and Stauffer, H. P. (1998) Parasites and flower choice of bumblebees. *Animal Behaviour* **55**: 819–825.

Sempavalan, J., Wheeler, C. T. and Hooker, J. E. (1995) Lack of competition between *Frankia* and *Glomus* for infection and colonization of roots of *Casuarina equisetifolia* (L.). *New Phytologist* **130**: 429–436.

Sih, A. (1992) Integrative approaches to the study of predation: general thoughts and a case study on sunfish and salamander larvae. *Annales Zoologici Fennici* **29**: 183–198.

Simard, S. W., Perry, D. A., Jones, M. D., Myrold, D. D., Durall, D. M. and Molina, R. (1997) Net transfer of carbon between ectomycorrhizal tree species in the field. *Nature* **388**: 579–582.

Smith, F. A. and Smith, S. E. (1996) Mutualism and parasitism: diversity in function and structure in the "arbuscular" (VA) mycorrhizal symbiosis. *Advances in Botanical Research* **22**: 1–43.

Smith, S. V. and Buddemeier, R. W. (1992) Global change and coral reef ecosystems. *Annual Review of Ecology and Systematics* **23**: 89–118.

Soberon, M. J. and Martinez del Rio, C. (1985) Cheating and taking advantage in mutualistic associations. In *The Biology of Mutualism*, ed. D. H. Boucher, pp. 192–216. Oxford: Oxford University Press.

Stachowicz, J. J. and Hay, M. E. (1999a) Reducing predation through chemically mediated camouflage: indirect effects of plant defenses on herbivores. *Ecology* **80**: 495–509.

Stachowicz, J. J. and Hay, M. E. (1999b) Mutualism and coral persistence: the role of herbivore resistance to algal chemical defense. *Ecology* **80**: 2085–2101.

Staddon, P. L. and Fitter, A. H. (1998) Does elevated atmospheric carbon dioxide affect arbuscular mycorrhizas? *Trends in Ecology and Evolution* **13**: 455–458.

Stanton, M. L., Palmer, T. M., Young, T. P., Evans, A. and Turner, M. L. (1999) Sterilization and canopy modification of a swollen thorn acacia tree by a plant-ant. *Nature* **401**: 578–581.

Starr, C. and Taggart, R. (1998) *Biology: The Unity and Diversity of Life*. Belmont, CA: Wadsworth.

Strauss, S. Y. (1997) Floral characters link herbivores, pollinators, and plant fitness. *Ecology* **78**: 1640–1655.

Székely, T., Szép, T. and Juhász, T. (1989) Mixed species flocking of tits (*Parus* spp.): a field experiment. *Oecologia* **78**: 490–495.

Thompson, J. N. (1988) Variation in interspecific interactions. *Annual Review of Ecology and Systematics* **19**: 65–87.

Tobin, A. J. and Dusheck, J. (1998) *Asking About Life*. Orlando, FL: W. B. Saunders.

Traveset, A. (1999) The importance of mutualisms for biodiversity conservation in insular ecosystems. *Revista Chilena de Historia Natural* **72**: 527–538.

Vandermeer, J., Hazlett, B. and Rathcke, B. (1985) Indirect facilitation and mutualism. In *The Biology of Mutualism*, ed. D. H. Boucher, pp. 326–343. Oxford: Oxford University Press.

van Rijn, P. C. J. and Tanigoshi, L. K. (1999) The contribution of extrafloral nectar to survival and reproduction of the predatory mite *Iphiseius degenerans on Ricinus communis*. *Experimental and Applied Acarology* **23**: 281–296.

Vasconcelos, H. L. (1991) Mutualism between *Maieta guianensis* Aubl., a myrmecophytic melastome, and one of its ant inhabitants: ant protection against insect herbivores. *Oecologia* **87**: 295–298.

Wall, D. H. and Moore, J. C. (1999) Interactions underground: soil biodiversity, mutualism, and ecosystem processes. *BioScience* **49**: 109–117.

Weeks, P. (2000) Red-billed oxpeckers: vampires or tickbirds? *Behavioral Ecology* **11**: 154–160.

West, S. A. and Herre, E. A. (1994) The ecology of the New World fig-parasitising wasps *Idarnes* and implications for the evolution of the fig–pollinator mutualism. *Proceedings of the Royal Society of London Series B* **258**: 67–72.

Wilson, D. S. and Knollenberg, W. G. (1987) Adaptive indirect effects: the fitness of burying beetles with and without their phoretic mites. *Evolutionary Ecology* **1**: 139–159.

Wilson, G. W. T. and Hartnett, D. C. (1997) Effects of mycorrhizae on plant growth and dynamics in experimental tallgrass prairie microcosms. *American Journal of Botany* **84**: 478–482.

Witman, J. D. (1987) Subtidal coexistence: storms, grazing, mutualism, and the zonation of kelps and mussels. *Ecological Monographs* **57**: 167–187.

Wootton, J. T. (1994) The nature and consequences of indirect effects in ecological communities. *Annual Review of Ecology and Systematics* **25**: 443–466.

Yeargan, K. V. and Braman, S. K. (1989a) Life history of the parasite *Diolcogaster facetosa* (Weed) (Hymenoptera: Braconidae) and its behavioral adaptation to the defensive response of a lepidopteran host. *Annals of the Entomological Society of America* **79**: 1029–1033.

Yeargan, K. V. and Braman, S. K. (1989b) Comparative behavioral studies of indigenous hemipteran predators and hymenopteran parasites of the green cloverworm (Lepidoptera: Noctuidae). *Journal of the Kansas Entomological Society* **62**: 156–163.

Yosef, R. and Whitman, D. W. (1992) Predator exaptations and defensive adaptations in evolutionary balance: no defense is perfect. *Evolutionary Ecology* **6**: 527–536.

Young, T. P., Stubblefield, C. H. and Isbell, L. A. (1997) Ants on swollen-thorn acacias: species coexistence in a simple system. *Oecologia* **109**: 98–107.

Yu, D. W. and Pierce, N. E. (1998) A castration parasite of an ant–plant mutualism. *Proceedings of the Royal Society of London Series B* **265**: 375–382.

LEE A. DYER AND PHYLLIS D. COLEY

4

Tritrophic interactions in tropical versus temperate communities

Introduction

The latitudinal gradient in diversity is one of the oldest (e.g., Wallace, 1878) and most obvious trends in ecology, and a wealth of literature is devoted to understanding both the causes and consequences of this gradient (Dobzhansky, 1950; also reviewed by Rohde, 1992). Given the enormous latitudinal differences in both diversity and productivity between temperate and tropical habitats, it is likely that relationships among trophic levels may also be fundamentally different. Although trophic interactions can be complex, a current research goal in community ecology is to determine which populations at different trophic levels are limited due to resource availability and which are limited due to consumption by higher trophic levels. In this chapter, we review the literature to determine if latitudinal trends exist for trophic controls. Identifying these patterns should help clarify whether ecological paradigms developed in temperate systems are useful for understanding tropical systems. Tropical ecologists, conservation biologists, and agricultural scientists have suggested that many ecological paradigms do not apply to tropical systems and should not be used to make management decisions or theoretical assumptions. Another advantage of identifying latitudinal gradients in tritrophic level interactions is that many of the hypotheses attempting to explain the latitudinal gradient in diversity are based on untested assumptions about the differences between tropical and temperate communities. For example, it is assumed that higher levels of specialization (for all consumers) in the tropics have allowed for greater numbers of species (Dobzhansky, 1950; Pianka, 1966; MacArthur and Wilson, 1967), but it is not at all clear that a latitudinal gradient in specialization exists (Price, 1991a; Marquis and Braker, 1994; Fiedler, 1998). Similarly, levels of predation are assumed to be higher in the tropics (Paine, 1966; Janzen, 1970), and these high levels are hypothesized as a factor that maintains higher levels of diversity (Pianka, 1966). Tests of these assumptions are an important part of understanding the latitudinal gradient in diversity.

In order to describe latitudinal gradients in terrestrial tritrophic interactions we focus on direct and indirect effects of predators and parasitoids on lower trophic levels, and effects of plant resources on upper trophic levels. Hairston *et al.*'s (1960) initial top-down hypothesis for herbivore regulation resulted in many theoretical and empirical studies on the effects of top-down and bottom-up forces on community structure (most recently reviewed by Pace *et al.*, 1999; Persson, 1999; Polis, 1999). However, there is still disagreement regarding which factors limit populations of different trophic levels. Currently, there are three prominent models that incorporate direct and indirect effects in tritrophic interactions (Fig. 4.1):

1. Top-down trophic cascades. In these models, predators and plants are resource-limited while herbivores are limited by their consumers. Thus, predators regulate their prey and indirectly benefit plants.
2. Bottom-up trophic cascades. These models suggest that both herbivores and enemies are regulated by plant biomass. Bottom-up hypotheses incorporate basic thermodynamics: energy is lost as it is transferred up the trophic chain, so the biomass of herbivores, then primary and secondary carnivores attenuates and is dependent on total primary productivity (Lindeman, 1942; Slobodkin, 1960).
3. The green desert. This also addresses bottom-up hypotheses but focuses on resource limitation as the factor determining community structure (Menge, 1992; Moen *et al.*, 1993). In this hypothesis it is assumed that herbivores cannot utilize most plant parts, either because they cannot digest the most common plant macromolecules (e.g., cellulose; Abe and Higashi, 1991) or because of toxic secondary metabolites (e.g., Murdoch, 1966; White, 1978).

Although the above models are not necessarily mutually exclusive, each one probably has better predictive power in specific ecosystems. Some authors have criticized these models and presented convincing arguments to dispose of trophic cascade theories (Polis and Strong, 1996) because of the ubiquity of factors such as omnivory and diet shifts and a general lack of demonstrable trophic structure in real communities. For example, many terrestrial predators eat both herbivores and plants,

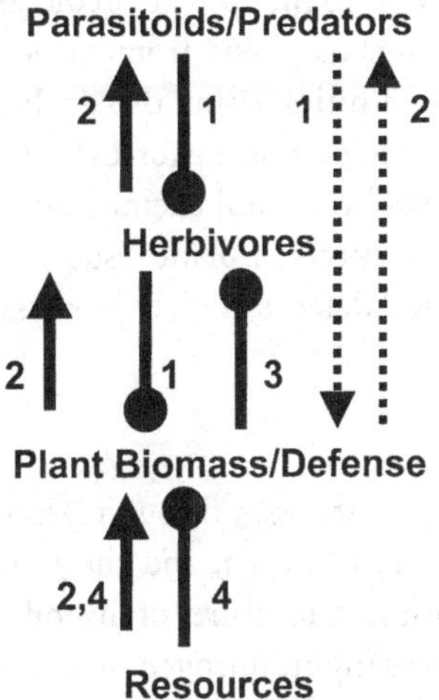

Fig. 4.1. Direct and indirect effects among three trophic levels and plant resources. Direct effects are indicated by a solid line between two trophic levels, and indirect effects (cascades) are indicated by a dashed line. A negative effect of one trophic level on the other is drawn with a bullet-head, and a positive effect is drawn with an arrowhead. The effect is on the trophic level nearest to the arrow- or bullet-head. The numbers closest to the lines refer to current models in ecology that examine trophic relationships: (1) top-down trophic cascades; (2) bottom-up trophic cascades; (3) the green desert model; and (4) resource availability models. The meta-analysis measured the strength of these interactions in tropical versus temperate systems.

potentially having no indirect positive effect on plants. Persson (1999) adds to these criticisms by pointing out that terrestrial studies of trophic cascades have not included appropriately scaled experiments with large vertebrate herbivores and predators and that there are many other indirect interactions that are equally important in structuring communities. Thus, the validity of these trophic models and their applicability to different habitats have been the target of much discussion. In this chapter, we compile information from the literature to assess the relative strength of top-down and bottom-up forces across a latitudinal gradient.

Specific predictions have been made about how aspects of tritrophic interactions differ between tropical and temperate systems. Below we review the evidence that suggests that in the tropics plants are better defended, herbivory is higher, and pressure from natural enemies is more intense. These patterns imply that tropical herbivore populations have adapted to pressures from intense bottom-up *and* top-down forces. In this chapter, we examine the literature relevant to the specific predictions of latitudinal differences and present a meta-analysis from 14 years of research in tropical and temperate communities. Using this analysis we evaluate the relative effects of top-down and bottom-up forces by directly comparing the suppression of herbivores by natural enemies versus by chemical compounds. We also assess the effects of plant resource availability on upper trophic levels via chemical defense or plant biomass.

Meta-analysis methods

The meta-analysis included data from January, 1985 through December, 1998. All papers in the journals *Oecologia*, *Biotropica*, and the *Journal of Tropical Ecology* were examined for quantitative measures of the following direct and indirect interactions: resources (light, nitrogen, phosphorus) on plant biomass or survivorship and on plant defenses; plant defenses (chemical defenses and leaf toughness) on percentage herbivory, herbivore biomass, or herbivore survivorship; herbivores (natural and artificial damage) on plant biomass or survivorship; natural enemies on prey biomass or survivorship; and natural enemies on plant biomass (see Fig. 4.1). The starting date was chosen because the first issue of the *Journal of Tropical Ecology* was published in that year. For the journal *Oecologia*, we used the same starting date but only included nine years of studies (1985–1993) because the work reported in that journal is mostly temperate, and we were attempting to collect a balanced sample of tropical and temperate work. A bibliography of the papers that were examined can be found on the internet along with the effect sizes from each study (http://www.caterpillars.org). Papers that were actually included in the meta-analysis were those that contained means, measures of dispersion, and sample sizes. We conducted a mixed model meta-analysis for temperate versus tropical systems to uncover potential latitudinal differences. We defined tropical studies as all those conducted in natural ecosystems below 2000 m within the tropics of Cancer and Capricorn or on organisms that live exclusively in those latitudes.

Equations in Gurevitch and Hedges (1993) were used to calculate combined effect sizes across all studies and 95% confidence intervals for the meta-analysis. Means and standard deviations were taken directly from tables or text, were calculated from other statistics, or were gleaned from figures (using a ruler). We calculated only one effect size per interaction per paper. If more than one effect size was available for an interaction, we randomly selected a value or used the last value in a series of measurements. In this chapter, we report all effect sizes along with the range of the 95% confidence intervals (after Gurevitch and Hedges, 1993); all other measures of dispersion reported here are ±1 standard error. Any effect sizes greater than 1.0 were considered to be large effects (Gurevitch and Hedges, 1993). We compared the strength of specific trophic interactions (Fig. 4.1) in tropical versus temperate systems by using the between class heterogeneity statistic, Q_B, which has approximately a χ^2 distribution (Gurevitch and Hedges, 1993).

Utilizing a meta-analysis for a review such as this one has notable advantages because the effect size calculated is independent of sample size, avoiding the problems arising from the positive correlation between sample size and likelihood of attaining a significant result. However, meta-analyses are subject to the same problems as any literature review based on vote-counting or more subjective narrative reviews of existing studies, including subjectivity of data collection from the literature, biases in collections of studies, and loss of system-specific details for the sake of generality (Gurevitch and Hedges, 1993). We attempted to minimize subjectivity of data collection by only including those studies that had distinct statistics reported in tables, figures, or text. The only obvious bias in the studies we examined was a tendency to examine specialist invertebrate herbivores when studying the effects of herbivory on plants. We discuss consequences of this bias below.

Latitudinal trends in plant defenses

Plant defenses are an important component of tritrophic interactions over both ecological and evolutionary time-scales. Latitudinal differences in defenses among plant communities should influence population dynamics of plants, herbivores, and natural enemies, and these interactions shape the evolution of defenses. Several reviews and empirical studies indicate that there is a strong latitudinal gradient in chemical defenses, with tropical plants being better defended than temperate plants (Crankshaw and Langenheim, 1981; Langenheim *et al.*, 1986; Miller and Hanson, 1989; Coley and Aide, 1991; Basset, 1994; Gauld and Gaston, 1994; Coley and Kursar, 1996, in press a). Alkaloids are more common and toxic in the tropics (Levin, 1976; Levin and York, 1978). About 16% of the temperate species surveyed in these studies contained alkaloids, compared to more than 35% of the tropical species. Simple phenolics do not seem to vary between latitudes, but condensed tannins in mature leaves are almost three times higher in tropical forests (Becker, 1981; Coley and Aide, 1991; Turner, 1995). The diversity of secondary compounds is also much higher in tropical than temperate forests (Miller and Hanson, 1989; Gauld and Gaston, 1994). This may occur because plant diversity is far greater in the tropics, but it is also true that many sympatric closely related plants have different chemical defenses (Waterman, 1983; Gauld and Gaston, 1994). For many herbivores, leaf toughness is the most effective feeding deterrent (Coley, 1983; Lowman and Box, 1983; Langenheim *et al.*, 1986; Aide and Londoño, 1989). This defense increases threefold in the tropics across four different forest types, being lowest in temperate plants. Indirect plant defenses, such as domatia and extrafloral nectaries are also more common in the tropics (Koptur, 1991).

Another striking difference between tropical and temperate plant defenses is that young, expanding tropical leaves have the highest levels of investment in secondary compounds, while temperate plants invest in higher levels of chemical defense in mature leaves. In tropical trees, young leaves contain much higher concentrations of simple phenolics, condensed tannins, terpenes, and alkaloids compared to the concentrations found in mature leaves (Coley and Kursar, in press a). In temperate trees, young leaves contain half the concentration of condensed tannins as mature leaves (Coley and Kursar, in press a).

While the above data strongly indicate that both young and mature leaves of tropical species are substantially better defended than leaves from temperate species, our meta-analysis suggests that the negative impact of defenses on herbivores is similar in temperate and tropical regions (Fig. 4.2). There were large negative effects of plant defenses on herbivores for tropical (−1.06) and temperate (−1.32) systems, and there were no significant differences between the latitudes ($Q_B = 1.18$, $df = 1$, $P > 0.5$). These results are not inconsistent with the documented latitudinal gradient in plant defenses. In this case, herbivore response is not an adequate measure of severity of plant defense, since many of these studies examined specialist herbivores that are adapted to the defenses of their hosts. Temperate and tropical studies alike have demonstrated that specialists have evolved adaptations to detoxify or sequester the defensive compounds that are unique to their restricted array of host plants (Krieger *et al.*, 1971; Whittaker and Feeny, 1971; Feeny, 1976; Dyer, 1995; Camara, 1997). So, the similar magnitude of the

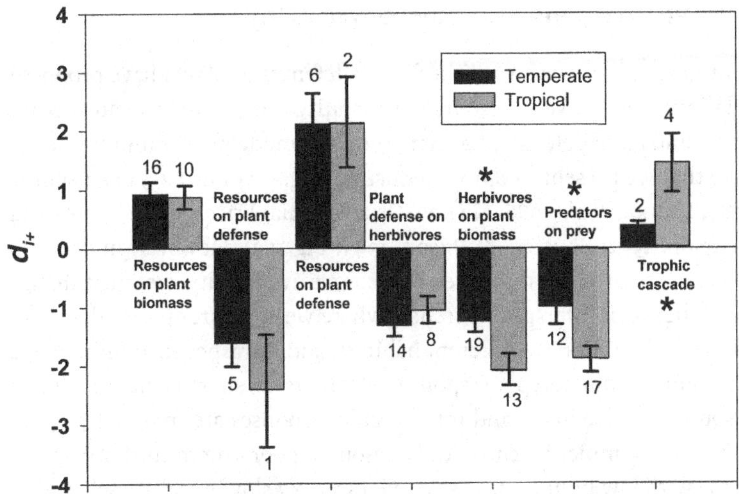

Fig. 4.2. Accumulated effect sizes across studies (d_{i+}) in tropical and temperate systems and 95% confidence intervals. The dependent variables included measures of biomass, defense, survivorship, and percentage damage. Any effect sizes greater than 1.0 were considered to be large effects. An asterisk indicates a significant difference ($P < 0.05$, based on the between class heterogeneity statistic, Q_B) for that interaction across lattitudes. The numbers above or below each bar indicate the number of studies included for the meta-analysis.

negative effect of defenses on herbivores across latitudes may result from coevolutionary interactions, where elevated defenses in the tropics are countered by elevated modes of tolerance or detoxification by specialist herbivores.

A more appropriate test of latitudinal differences in the effectiveness of plant defenses was recorded by Miller and Hanson (1989), who conducted experiments and literature reviews to compare development of a naïve generalist herbivore (*Lymantria dispar*) on 658 species of tropical and temperate food plants. Their results were consistent with the hypothesis that tropical plants are better defended: plant chemistry was a good predictor of suitability of host plants, and when tropical plants were added to their assay, the proportion of host plant rejections increased. A more extensive meta-analysis than the one reported here might allow for distinguishing the effects of plant defenses on adapted specialists versus generalists or naïve herbivores; in that case, we predict a greater negative effect of tropical versus temperate plants on the generalist or naïve herbivores.

Plant responses to resource availability

A number of studies on anti-herbivore defenses of plants have proposed relationships between resource availability, plant growth rates, plant vigor, and plant defense relevant to the three models of community structure that we present in our introduction (e.g., Bryant *et al.*, 1983; White, 1984; Larsson *et al.*, 1986; Nichols-Orians, 1991a; Price, 1991b; Herms and Mattson, 1992; Shure and Wilson, 1993; Fig. 4.1). However, in order to make sense of plant responses to resource availability, we must distinguish between interspecific trends, where we compare species that have evolved adaptations to different habitats, and intraspecific trends, where we compare phenotypic responses of plants to short-term changes in resources. These inter- and intraspecific responses are frequently opposite. For example, in chronically resource-poor communities, such as those with low light or poor soils, plants grow slowly and are selected to invest heavily in defenses (Janzen, 1974; Grime, 1979; Coley *et al.*, 1985). This in turn would limit herbivore populations, as predicted by the green desert hypothesis. However, in a given system, changing the availability of resources could either enhance or confound traditional hypotheses of bottom-up control. This is because plastic responses of plants reflect source–sink imbalances (not optimal solutions), and some resources increase growth, while others increase defenses. An example of enhancement of thermodynamic bottom-up control would be under lowered nitrogen conditions: levels of carbon-based defenses will increase, and herbivores will decline because of increased plant defense as well as lower plant biomass. The opposite situation (i.e., contradicting bottom-up predictions) could also result from variation in nitrogen or light availability. Kyto *et al.* (1996) found that despite predictions by bottom-up models, folivore populations did not increase in response to nitrogen additions, perhaps because of increases in nitrogen-based defenses. Similarly, under low light availability, herbivore populations might be expected to decline because of reduced plant productivity, but they are just as likely to increase because of lower levels of carbon-based defenses (Bryant *et al.*, 1983). Variation in light availability might also affect nitrogen-based defenses (Bryant *et al.*, 1983), which would alter effects of enhanced plant biomass on upper trophic levels. The few studies that have examined associations between resource availability, plant biomass, plant chemistry, and herbivory have yielded inconsistent results (Waterman *et al.*, 1984; Larsson *et al.*, 1986; Bryant *et al.*, 1987; Briggs, 1990; Dudt and Shure, 1994); thus the relationships between these variables need to be examined more closely. This type of work will enhance bottom-up models by improving our understanding of how communities adapt to different resource levels and how they respond to short-term fluctuations.

Plasticity in plant defenses

In an earlier section, we discussed evidence for a latitudinal trend in defenses that results from selection. The data suggest that the optimal level of defense is greater in the tropics. Here we examine plastic responses of plants to variation in light and mineral resources (Bryant *et al.*, 1983; White, 1984; Larsson *et al.*, 1986; Nichols-Orians, 1991a; Price, 1991b; Herms and Mattson, 1992; Shure and Wilson, 1993). Not surprisingly, data from our meta-analysis showed that plants respond to an increase in resources by increasing growth (Fig. 4.2). In addition, there were defense responses consistent with the theory of carbon–nutrient balance (Bryant *et al.*, 1983). This hypothesis suggests that resources in excess of baseline requirements for growth and defense are invested in defenses. Thus, under conditions of high light, carbon-based defenses (e.g., tannins and terpenes) should increase, whereas under nitrogen fertilization, nitrogen-containing compounds (e.g., alkaloids) should increase. In our analysis, increases in nitrogen, phosphorus, and light availability had strong effects on plant defenses. Depending on the resources and the defenses, both positive and negative effects were seen in approximately equal numbers of studies (Fig. 4.2). For example, Nichols-Orians (1991b) found that increased light availability was correlated with increased concentrations of condensed tannins (positive effect of resources), while Mihaliak and Lincoln (1985) found that increased levels of nitrate (from fertilizing) led to decreased concentrations of volatile terpenes (negative effect of resources). Although resource levels clearly influenced plant growth and levels of defense, there were no differences between tropical and temperate systems in the magnitude of effect (resources on plant biomass, $Q_B = 0.19$, df = 1, $P > 0.5$; resources negatively affecting plant defense, $Q_B = 2.16$, df = 1, $P > 0.1$; resources positively affecting plant defense, $Q_B = 0.0016$, df = 1, $P > 0.9$).

Herbivory

Levels of herbivory are variable at many different scales of time and space at all latitudes. For example, herbivores generally prefer young leaves over mature ones, but the difference is most dramatic in the tropics (Coley and Aide, 1991). In addition, within the tropics, leaf damage is significantly less in wet than in dry tropical forests (Barone, in press a), pioneer species have higher levels of herbivory than understory species (Coley, 1988; Nuñez-Farfan and Dirzo, 1989; Marquis and Braker, 1994), and understory plants suffer more herbivory than canopy plants (Lowman, 1985; Barone, in press b).

Despite this variation within latitudes, there is a detectable latitudinal pattern of herbivory. A review of herbivory in tropical versus temperate systems reported that mean folivory was 7% ($n = 13$ studies) in the temperate zone versus 16.6% ($n = 29$ studies) in the tropics (Coley and Barone, 1996). The effect sizes calculated in the meta-analysis support the hypothesis that herbivory is more intense in the tropics and has a greater negative effect on plant biomass and survivorship than herbivory on temperate plants (Fig. 4.2; $Q_B = 31.0$, df $= 1$, $P < 0.0001$). Despite this difference, the effects of herbivory on plants were large for both temperate (-1.25) and tropical (-2.1) studies.

Differences in herbivory on young versus mature leaves create a latitudinal pattern that mirrors the pattern of chemical defenses (Coley and Kursar, 1996). In the temperate zone, most of the damage occurs on mature leaves, while in the shade-tolerant species of the tropical wet forests, approximately 75% of the lifetime damage occurs during the short period of leaf expansion. The concentration of herbivores on ephemeral young leaves allows rapid herbivore development and might also select for efficient host-finding abilities in parasitoids.

Because physical and chemical defenses are higher in the tropics, the higher levels of herbivory suggest that herbivore pressure or specialized adaptations to specific plant defenses must also be greater. Our meta-analysis indicates that tropical herbivores probably are better adapted to defenses because the increased levels of tropical defenses do not have a greater negative effect on tropical herbivores when compared to the effect of weaker temperate plant defenses on their herbivores (Fig. 4.2). Some diversity hypotheses suggest that increased levels of specialized herbivory in the tropics help maintain the high diversity of trees (Janzen, 1970; Leigh, 1999). These authors suggest that if the herbivores are specialized, the intense levels of tropical herbivory will keep their host plant rare, allowing other species to coexist. Again, our meta-analysis supports this hypothesis since the tropical herbivores are more likely to suppress overall biomass of superior plant competitors. For example, one of the papers in our meta-analysis (Letourneau and Dyer, 1998b) uncovers a dramatic increase in the density of one understory plant (*Piper cenocladum*) when specialist herbivores are suppressed. Since *P. cenocladum* can occur at very high densities (Letourneau and Dyer, 1998b), forests where the plant is suppressed should be able to support higher species richness of understory plants.

Natural enemies

In addition to facing a diverse array of plant toxins, herbivores in the tropics may also be subjected to more intense pressure from natural enemies. It has long been thought that predation is more intense in tropical compared to temperate ecosystems (Paine, 1966; Elton, 1973; Rathcke and Price, 1976; Gauld and Gaston, 1994). There are some data that support this hypothesis (Jeanne, 1979) along with some indirect evidence, but very few appropriate comparisons have been made. The most cited indirect evidence that predation is more intense is that important predatory taxa are more diverse in the tropics. Ants provide a clear example of an important group of predators that are more species-rich and abundant in tropical versus temperate systems (Kusnezov, 1957; Fischer, 1960; Wilson, 1971). Jeanne (1979) tested the hypothesis of a latitudinal gradient in ant predation by offering wasp larvae to ants at five locations along a latitudinal gradient and found that rates of predation were significantly greater in the tropics. Our meta-analysis also confirms that natural enemies have strong negative effects on herbivores at all latitudes, but the magnitude of the effect is significantly higher in tropical (−1.89) versus temperate (−1.0) systems (Fig. 4.2; $Q_B = 21.3$, df $= 1$, $P < 0.0001$).

Overall levels of parasitism are either the same in tropical and temperate systems (Hawkins, 1994) or are slightly higher in tropical systems, despite the fact that for some parasitoid groups diversity is lower and assemblage sizes are smaller in the tropics compared to temperate systems. Hawkins (1994) examined levels of parasitism for over 1200 hosts all over the world and found no latitudinal gradient in mortality, and while he did document a positive relationship between parasitoid species richness and mean parasitism rates, the lower levels of diversity in the tropics were not associated with lower levels of parasitoid-induced mortality. Other rearing studies indicate that levels of parasitism are slightly higher in tropical versus temperate forests. G. Gentry and L. A. Dyer (unpublished data, but also see http://www.caterpillars.org and Dyer and Gentry, 1999) have compiled a five-year database of over 200 species of tropical Lepidoptera and have found that mean yearly levels of parasitism for 55 well-sampled species (17 families) were 32.5% ± 3%. In contrast, mean levels of parasitism across 98 species (13 families) of temperate caterpillars (from a long-term database published in Schaffner and Griswold, 1934 and used by Sheehan, 1991 then by Dyer and Gentry, 1999) were 17% ± 2%. Even if pressure from parasitoids is higher in the tropics than in the temperate zone, it is likely that predation is a more important source of mortality than parasitism in tropical systems while parasitism is more important source of mortality in temperate systems. Hawkins *et al.* (1997) quantified enemy-induced mortality for 78 species of herbivores and found that predators represent the dominant natural enemy in the tropics, whereas parasitoids are dominant in temperate systems.

An examination of latitudinal trends in plant defenses provides additional indirect evidence for higher pressure from natural enemies in tropical systems. Mature leaves of rainforest species have extremely high concentrations of condensed tannins as compared to temperate ones (Coley and Aide, 1991). Tannins as defenses present a paradox, because they cause herbivores to grow more slowly but also to consume more leaf tissue (Price *et al.*, 1980; Coley and Kursar, in press a). The paradox is solved if prolonged larval development makes herbivores susceptible to predation for longer, as the removal of larvae, particularly in the early instars, will reduce damage to the plant (Benrey and Denno, 1997). Therefore, we would only expect tannins to evolve as a defense if, by slowing herbivore growth, they made larvae more vulnerable to predators. The high tannin levels in mature tropical leaves, and the low abundance of mature leaf feeders, suggests that natural enemies may be quite effective in reducing herbivory in tropical forests (Coley and Kursar, in press a).

Herbivore defenses

The large negative effects of plant toxins on herbivores are attenuated by the fact that many specialized herbivores utilize these toxins for their own defense. Studies comparing different defensive mechanisms of herbivores have found chemical defenses to be the most effective against a diverse suite of natural enemies (Dyer, 1995, 1997). Chemical defenses of tropical versus temperate herbivores potentially mirror the defenses found in their host plants: tropical herbivores are generally more toxic than their temperate counterparts. Both direct and indirect evidence has been accumulated to support this generalization. Sime and Brower (1998) presented direct evidence that tropical Lepidoptera are more toxic than those in temperate latitudes. They demonstrated that the latitudinal gradient in species richness of unpalatable butterflies is greater than the gradient for the Papilionidae, which they use as an average (in terms of palatability) butterfly family. These results should be viewed with caution, since many supposedly toxic groups have never been investigated for toxicity (DeVries, 1987, 1997), and many groups that were thought to be toxic were not toxic to several different invertebrate predators (Dyer, 1995, 1997). In addition, the assumption that the immatures of entire families or subfamilies of butterflies are unpalatable (Sime and Brower, 1998) is unrealistic and has not been supported by empirical data (Dyer, 1995).

The "nasty host hypothesis" (Gauld *et al.*, 1992; Gauld and Gaston, 1994) provides further indirect evidence for the elevated toxicity of tropical herbivores. Many taxa of parasitoid Hymenoptera are not more diverse in the tropics, and one explanation for this could be that tropical hosts are more toxic than extra-tropical hosts. The parasitoid groups that are negatively affected by "nasty" compounds are less diverse in the tropics. Furthermore, diversity of tropical parasitoids is not lower for egg or pupal parasitoids because these stages are usually not chemically defended; likewise diversity is high for tropical parasitoids of herbivores that eat non-toxic plant tissue (Gauld *et al.*, 1992; Gauld and Gaston, 1994). Gauld *et al.* (1992) also pointed out that the proportion of aposematic insects is higher for many taxa in the tropics and that the tissues of most of these insects are likely to be toxic.

Chemically defended herbivores are often dietary specialists (Duffey, 1980; Bowers, 1990; Dyer, 1995), therefore it is possible that the gradient in herbivore unpalatability (if it does exist) is correlated with a latitudinal gradient in specialization. Limited evidence has been provided in support of such a gradient (Scriber, 1973, 1984; Basset, 1994; Scriber *et al.*, 1995; Sime and Brower, 1998), although there are notable exceptions where chemical and phylogenetic constraints minimize any latitudinal gradients in host plant specialization (Fiedler, 1998). For those groups for which diet breadths are narrower in the tropics, the increased specialization may be a result of plant chemistry (Ehrlich and Raven, 1964) or pressure from natural enemies (Bernays and Graham, 1988), or a combination of these top-down and bottom-up forces (Dyer and Floyd, 1993).

Tritrophic interactions and trophic cascades

Tropical ecosystems are generally considered to be more complex, containing longer trophic chains and trophic webs that exhibit more omnivory, intraguild predation, and unpredictable indirect effects. Convincing arguments have been made suggesting that top-down and bottom-up trophic cascades are unlikely to occur in such complex ecosystems. However, studies that have focused on top-down forces have discovered recipient control in terrestrial systems with high diversity that include omnivory and opportunistic diets (Spiller and Schoener, 1994; Dial and Roughgarden, 1995; Floyd, 1996; Moran *et al.*, 1996; Letourneau and Dyer, 1998b; Dyer and Letourneau, 1999a, b; Pace *et al.*, 1999). The concept of distinct trophic levels that exert statistically detectable forces on other levels (whether they be donors or recipients) is useful for community ecology; rather than discarding this concept, more empirical tests are needed to examine the role of omnivory with respect to mediating or mitigating top-down and bottom-up forces. Alternatively, the concept of "effective" trophic levels, in which trophic levels are fractional rather than discrete integers (e.g., 3 = a predator with a 100% diet of herbivores, 2.5 = an omnivore with a 50% herbivores and 50% plant diet), could be utilized to enhance the predictive power of the major trophic cascades models (Christian and Luczkovich, 1999).

Using either the traditional concept of trophic levels or the new concept of functional trophic levels, very few terrestrial studies have documented clear top-down cascades (as actual indirect effects) anywhere (Letourneau and Dyer, 1998a). This is because it is difficult to control for direct effects of predators and parasitoids on plants (or top predators on herbivores). For example, many of the ant–plant systems in the tropics, which have been used to demonstrate the positive effects of predators on plants, have not measured clear indirect effects because the ants may have considerable positive direct effects on the plant (nutrient procurement), considerable negative direct effects (costs of producing food), or other indirect effects (see Bronstein and Barbosa, chapter 3, this volume). With this caveat in mind, the limited numbers of studies that do exist suggest that top-down cascades occur in terrestrial systems (reviewed by Pace *et al.*, 1999). In fact, the strong negative effects of enemies on herbivores and negative effects of herbivores on plants uncovered by our meta-analysis (Fig. 4.2) support the idea that enemies can have indirect positive effects on plants even if they do shift diets, eat plants, or compete with other consumers. The very few studies in our meta-analysis that directly documented a top-down cascade also support this idea (Fig. 4.2). Effects of enemies on plants were positive for both tropical (1.44) and temperate (0.38) systems, but the effects were significantly greater for the tropics ($Q_B = 6.03$, df$=1$, $P<0.025$).

The strong top-down (direct and indirect effects) control demonstrated by tropical studies in our meta-analysis included large vertebrate predators and herbivores (e.g., Jedrzejewski *et al.*, 1992; Meserve *et al.*, 1993), which partially addresses Persson's (1999) criticism that trophic cascades studies have not been appropriately scaled. The results of these studies are also relevant to tropical conservation issues. Terborgh (1992) suggested that top-down cascades are important in Neotropical forests, and he hypothesized that the decline of large mammalian predators due to forest fragmentation and hunting could lead to an increase of mammalian seed predators and a decline in tree species with large seeds. Terborgh's specific predictions may be incorrect because a correlation between herbivore body size and seed size may not exist (Brewer *et al.*, 1997). However, it is clear that top-down control is important in tropical systems, and various cascading effects may cause tropical conservation problems similar to the negative cascading effects of disappearing coyotes (caused by habitat fragmentation) on bird diversity in temperate communities (Crooks and Soulé, 1999).

Conclusions

The main latitudinal trends noticed across the three trophic levels of plant, herbivore, and natural enemy indicate that with respect to temperate ecosystems, the tropics exhibit: (1) increased diversity for most taxa at all three trophic levels, with the exception of some parasitoids, (2) higher levels of plant defenses (mechanical, biotic, and chemical), (3) increased levels of herbivory, (4) more toxic herbivores, and (5) more intense pressure from natural enemies.

Examination of the effect sizes in the meta-analysis revealed that strong top-down and bottom-up forces were detectable in both temperate and tropical systems (Fig. 4.2). Despite the complex trophic structure of tropical communities, distinct trophic levels exert statistically detectable forces on other levels. There was no latitudinal difference in the effect of plant defenses on herbivores, however, top-down effects of predators on herbivores and herbivores on plants were significantly stronger in the tropics. Thus, if one looks at the relative importance of these forces on community structure, we see quite surprising and distinct patterns in the different systems. In temperate systems, plant chemistry appears to have a stronger ecological impact on herbivores than do natural enemies, even though levels of defense are relatively low. On the other hand, in tropical systems natural enemies seem to be more important than plant defenses. Thus, controls on community organization may follow different rules along a latitudinal gradient.

Why do we see these latitudinal differences, with top-down controls being relatively more important in the tropics? We offer several speculative suggestions. First, the exploitation ecosystem hypothesis posits that greater productivity should favor top-down control because when plant productivity is high, as in the tropics, sufficient resources will be available to allow natural enemies to act as "effective trophic levels" that control herbivore populations (Fretwell, 1977; Oksanen *et al.*, 1981). Second, because tropical climates are more favorable year round, populations of both herbivores and natural enemies do not suffer severe seasonal crashes. This should lead to a more reliable presence of an effective third trophic level in tropical communities. And finally, because natural enemies are predictable due to benign tropical climates, plants have had the evolutionary opportunity to enlist the help of natural enemies in controlling herbivores (Coley and Kursar, in press b). For example, tropical plants more frequently have extrafloral nectaries. They also have twice the levels of tannins and toughness, which slow herbivore growth and increase their susceptibility to natural enemies. Thus, we suggest that the high, year-round productivity of the tropics may be an important factor leading to the observed gradient in trophic controls.

Many aspects of trophic cascades models remain untested in tropical or temperate systems. Most studies have focused on biomass at different trophic levels, and very few studies have examined top-down effects of predators on plant community structure or bottom-up effects of plant resources on animal community structure (Persson 1999). Clearly, more empirical studies are needed to understand the scope of trophic cascades and the conditions under which they occur. Future studies should attempt to test the effects of top-down cascades on plant community structure and bottom-up cascades on consumer community structure, and investigators should utilize creative approaches, such as examining effective trophic levels (Christian and Luczkovich, 1999), to alleviate some of the problems pointed out by critics of trophic cascades theory (Polis and Strong, 1996). These studies will undoubtedly reveal some of the mechanisms driving the strong latitudinal gradient in species diversity.

Acknowledgments

We are grateful to G. Gentry, C. Dodson, S. van Nouhuys, and an anonymous reviewer for comments that improved previous drafts of the manuscript. J. Heitman, A. Schaefer, and C. Squassoni assisted in literature retrieval and data entry for the meta-analysis. Financial support came from Colorado Office of State Colleges (LAD), Mesa State College (LAD), Earthwatch Institute (LAD and G. Gentry) and the National Science Foundation (PDC and LAD). For the tropical parasitism data reported in this chapter, excellent technical assistance was provided by many Earthwatch volunteers.

REFERENCES

Abe, T. and Higashi, M. (1991) Cellulose centered perspective on terrestrial community structure. *Oikos* **60**: 127–133.

Aide, T. M. and Londoño, E. C. (1989) The effects of rapid leaf expansion on the growth and survivorship of a lepidopteran herbivore. *Oikos* **55**: 66–70.

Barone, J. A. (in press a) Herbivory and pathogen damage in a wet and a dry tropical forest: a test of the pest–rainfall hypothesis. *Oikos*.

Barone, J. A. (in press b) Comparison of herbivores and herbivory in the canopy and understory for two tropical tree species. *Biotropica*.

Basset, Y. (1994) Palatability of tree foliage to chewing insects: a comparison between a temperate and a tropical site. *Acta Oecologia* **15**: 181–191.

Becker, P. (1981) Potential physical and chemical defenses of *Shorea* seedling leaves against insects. *Malaysian Forester* **2&3**: 346–356.

Benrey, B. and Denno, R. F. (1997) The slow-growth-high-mortality hypothesis: a test using the cabbage butterfly. *Ecology* **78**: 987–999.

Bernays, E. and Graham, M. (1988) On the evolution of host specificity in phytophagous arthropods. *Ecology* **69**: 886–892.

Bowers, M. D. (1990) Recycling plant natural products for insect defense. In *Insect Defenses: Adaptive Mechanisms and Strategies of Prey and Predators*, ed. D. L. Evans and J. O. Schmidt, pp. 353–386. New York: State University of New York Press.

Brewer, S. W., Rejmanek, M., Johnstone, E. E. and Caro, T. M. (1997) Top-down control in tropical forests. *Biotropica* **29**: 364–367.

Briggs, M. A. (1990) Relation of *Spodoptera eridania* choice to tannins and proteins of *Lotus corniculatus*. *Journal of Chemical Ecology* **16**: 1557–1564.

Bryant, J. P., Chapin III, F. S. and Klein, D. R. (1983) Carbon/nutrient balance of boreal plants in relation to vertebrate herbivory. *Oikos* **40**: 357–368.

Bryant, J. P., Clausen, T. P., Reichardt, P. B., McCarthy, M. C. and Werner, R. A. (1987) Effect of nitrogen fertilization upon the secondary chemistry and nutritional value of quaking aspen (*Populus tremuloides*, Michx.) leaves for the large aspen tortrix (*Choristoneura conflicatana*, Walker). *Oecologia* **73**: 513–517.

Camara, M. D. (1997) Physiological mechanisms underlying the costs of chemical defence in *Junonia coenia* Hubner (Nymphalidae): a gravimetric and quantitative genetic analysis. *Evolutionary Ecology* **11**: 451–469.

Christian, R. R. and Luczkovich, J. J. (1999) Organizing and understanding a winter's seagrass foodweb network through effective trophic levels. *Ecological Modelling* **117**: 99–124.

Coley P. D. (1983) Herbivory and defensive characteristics of tree species in a lowland tropical forest. *Ecological Monographs* **53**: 209–233.

Coley, P. D. (1988) Effects of plant growth rate and leaf lifetime on the amount and type of antiherbivore defense. *Oecologia* **74**: 531–536.

Coley, P. D. and Aide, T. M. (1991) Comparison of herbivory and plant defenses in temperate and tropical broad-leaved forests. In *Plant–Animal Interactions: Evolutionary Ecology in Tropical and Temperate Regions*, ed. P. W. Price, T. M. Lewinsohn, G. W. Fernandes and W. W. Benson, pp. 25–49. New York: John Wiley.

Coley, P. D. and Barone, J. A. (1996) Herbivory and plant defenses in tropical forests. *Annual Review of Ecology and Systematics* **27**: 305–335.

Coley, P. D. and Kursar, T. A. (1996) Anti-herbivore defenses of young tropical leaves: physiological constraints and ecological tradeoffs. In *Tropical Forest Plant Ecophysiology*, ed. S. S. Mulkey, R. Chazdon and A. P. Smith, pp. 305–336. New York: Chapman and Hall.

Coley, P. D. and Kursar, T. A. (in press a) Herbivory, plant defenses and natural enemies in tropical forests. In *Interacciones Quimicas entre Organismos: Aspectos Básicos y Perspectivas de Aplicación*, ed. A. L. Anaya, R. Cruz-Ortega and F. J. Espinosa-García.

Coley, P. D. and Kursar, T. A. (in press b) Herbivory, plant defenses and natural enemies in tropical forests. In *Interacciones Quimicas entre Organismos: Aspectos Básicos y Perspectivas de Aplicación*, ed. A. L. Anaya, R. Cruz-Ortega and F. J. Espinosa-García.

Coley, P. D., Bryant, J. P. and Chapin III, F. S. (1985) Resource availability and plant antiherbivore defense. *Science* **230**: 895–899.

Crankshaw, D. R. and Langenheim, J. H. (1981) Variation in terpenes and phenolics through leaf development in *Hymenaea* and its possible significance to herbivory. *Biochemical Systematics and Ecology* **9**: 115–124.

Crooks, K. R. and Soulé, M. E . (1999) Mesopredator release and avifaunal extinctions in a fragmented system. *Nature* **400**: 563–566.

DeVries, P. J. (1987) *The Butterflies of Costa Rica and their Natural History*, vol. 1. Princeton, NJ: Princeton University Press.

DeVries, P. J. (1997) *The Butterflies of Costa Rica and their Natural History*, vol. 2. Princeton, NJ: Princeton University Press.

Dial, R. and Roughgarden, J. (1995) Experimental removal of insectivores from rain forest canopy: direct and indirect effects. *Ecology* **76**: 1821–1834.

Dobzhansky, T. (1950) Evolution in the tropics. *American Scientist* **38**: 209–221.

Dudt, J. F. and Shure, D. J. (1994) The influence of light and nutrients on foliar phenolics and insect herbivory. *Ecology* **75**: 86–98.

Duffey, S. S. (1980) Sequestration of plant natural products by insects. *Annual Review of Entomology* **25**: 447–477.

Dyer, L. A. (1995) Tasty generalists and nasty specialists? A comparative study of antipredator mechanisms in tropical lepidopteran larvae. *Ecology* **76**: 1483–1496.

Dyer, L. A. (1997) Effectiveness of caterpillar defenses against three species of invertebrate predators. *Journal of Research on the Lepidoptera* **34**: 48–68.

Dyer, L. A. and Floyd, T. (1993) Determinants of predation on phytophagous insects: the importance of diet breadth. *Oecologia* **96**: 575–582.

Dyer, L. A. and Gentry, G. (1999) Larval defensive mechanisms as predictors of successful biological control. *Ecological Applications* **9**: 402–408.

Dyer, L. A. and Letourneau, D. K. (1999a) Relative strengths of top-down and bottom-up forces in a tropical forest community. *Oecologia* **119**: 265–274.

Dyer, L. A. and Letourneau, D. K. (1999b) Trophic cascades in a complex, terrestrial community. *Proceedings of the National Academy of Sciences, USA* **96**: 5072–5076.

Ehrlich, P. R. and Raven, P. H. (1964) Butterflies and plants: a study in coevolution. *Evolution* **18**: 568–608.

Elton, C. S. (1973) The structure of invertebrate populations inside neotropical rain forest. *Journal of Animal Ecology* **42**: 55–104.

Feeny, P. (1976) Plant apparency and chemical defense. *Recent Advances in Phytochemistry* **10**: 1–40.

Fiedler, K. (1998) Diet breadth and host plant diversity of tropical- vs. temperate-zone herbivores: Southeast Asian and West Palaearctic butterflies as a case study. *Ecological Entomology* **23**: 285–297.

Fischer, A. G. (1960) Latitudinal variations in organic diversity. *Evolution* **14**: 64–81.

Floyd, T. (1996) Top-down impacts on creosotebush herbivores in a spatially and temporally complex environment. *Ecology* **77**: 1544–1555.

Fretwell, S. D. (1977) The regulation of plant communities by food chains exploiting them. *Perspectives in Biology and Medicine* **20**: 169–185.

Gauld, I. D. and Gaston, K. J. (1994) The taste of enemy-free space: parasitoids and nasty hosts. In *Parasitoid Community Ecology*, ed. B. A. Hawkins and W. Sheehan, pp. 279–299. New York: Oxford University Press.

Gauld, I. D., Gaston, K. J. and Janzen, D. H. (1992) Plant allelochemicals, tritrophic interactions and the anomalous diversity of tropical parasitoids: the "nasty" host hypothesis. *Oikos* **65**: 353–357.

Grime, J. P. (1979) *Plant Strategies and Vegetation Processes*. Chichester: John Wiley.

Gurevitch, J. and Hedges, L. V. (1993) Meta-analysis: combining the results of independent experiments. In *Design and Analysis of Ecological Experiments*, ed. S. M. Scheiner and J. Gurevitch, pp. 378–398. New York: Chapman and Hall.

Hairston, N. G., Smith, F. E. and Slobodkin, L. B. (1960) Community structure, population control, and competition. *American Naturalist* **94**: 421–424.

Hawkins, B. A. (1994) *Pattern and Process in Host–Parasitoid Interactions*. New York: Cambridge University Press.

Hawkins, B. A., Cornell, H. V. and Hochberg, M. E. (1997) Predators, parasitoids, and pathogens as mortality agents in phytophagous insect populations. *Ecology* **78**: 2145–2152.

Herms, D. A. and Mattson, W. J. (1992) The dilemma of plants: to grow or defend. *Quarterly Review of Biology* **67**: 283–335.

Janzen, D. H. (1970) Herbivores and the number of tree species in tropical forests. *American Naturalist* **104**: 521–528.

Janzen, D. H. (1974) Tropical blackwater rivers, animals and mast fruiting by the Dipterocarpaceae. *Biotropica* **6**: 69–103.

Jeanne, R. L. (1979) A latitudinal gradient in rates of ant predation. *Ecology* **60:** 1211–1224.
Jedrzejewski, W., Jedrzejewski, B., Okarma, H. and Ruprecht, A. L. (1992) Wolf predation and snow cover as mortality factors in the ungulate community of the Bialowieza National Park, Poland. *Oecologia* **90:** 27–36.
Kemp, W. P., Harvey, S. J. and O'Neill, K. M. (1990) Patterns of vegetation and grasshopper community composition. *Oecologia* **83:** 299–308.
Koptur, S. (1991) Extrafloral nectaries of herbs and trees: modelling the interaction with ants and parasitoids. In *Ant–Plant Interactions*, ed. D. Cutler and C. Huxley, pp. 213–230. Oxford: Oxford University Press.
Krieger, R. I., Feeny, P. P. and Wilkinson, C. F. (1971) Detoxification enzymes in the guts of caterpillars: an evolutionary answer to plant defense. *Science* **172:** 579–581.
Kusnezov, N. (1957) Numbers of species of ants in faunae of different latitudes. *Evolution* **11:** 298–299.
Kyto, M., Niemela, P. and Larsson, S. (1996) Insects on trees: population and individual response to fertilization. *Oikos* **75:** 148–159.
Langenheim, J. H., Macedo, C. A., Ross, M. K. and Stubblebine, W. H. (1986) Leaf development in the tropical leguminous tree *Copaifera* in relation to microlepidopteran herbivory. *Biochemical Systematics and Ecology* **14:** 51–59.
Larsson, S., Wiren, A., Lundgren, L. and Ericsson, T. (1986) Effects of light and nutrient stress on leaf phenolic chemistry in *Salix dasyclados* and susceptibility to *Galerucella lineola* (Coleoptera). *Oikos* **47:** 205–210.
Leigh, E. G. (1999) *Tropical Forest Ecology: A View from Barro Colorado Island.* New York: Oxford University Press.
Letourneau, D. K. and Dyer, L. A. (1998a) Experimental manipulations in lowland tropical forest demonstrate top-down cascades through four trophic levels. *Ecology* **79:** 1678–1687.
Letourneau, D. K. and Dyer, L. A. (1998b) Density patterns of *Piper* ant-plants and associated arthropods: top predator cascades in a terrestrial system? *Biotropica* **30:** 162–169.
Levin, D. A. (1976) Alkaloid-bearing plants: an ecogeographic perspective. *American Naturalist* **110:** 261–284.
Levin, D. A. and York, B. M., Jr. (1978) The toxicity of plant alkaloids: an ecogeographic perspective. *Biochemical Systematics and Ecology* **6:** 61–76.
Lindeman, R. L. (1942) The trophic–dynamic aspect of ecology. *Ecology* **23:** 399–418.
Lowman, M. D. (1985) Temporal and spatial variability in insect grazing of the canopies of five Australian rainforest tree species. *Australian Journal of Ecology* **10:** 7–24.
Lowman, M. D. and Box, J. D. (1983) Variation in leaf toughness and phenolic content among five species of Australian rain forest trees. *Australian Journal of Ecology* **8:** 17–25.
MacArthur, R. H. and Wilson, E. O. (1967) *The Theory of Island Biogeography*. Princeton, NJ: Princeton University Press.
Marquis, R. J. and Braker, H. E. (1994) Plant–herbivore interactions: diversity, specificity and impact. In *La Selva: Ecology and Natural History of a Neotropical Rain Forest*, ed. L. M. McDade, K. S. Bawa, G. S. Hartshorn and H. E. Hespenheide, pp. 261–281. Chicago, IL: University of Chicago Press.
Menge, B. A. (1992) Community regulation: under what conditions are bottom-up factors important on rocky shores? *Ecology* **73:** 755–765.
Meserve, P. L., Gutierrez, J. R. and Jaksic, F. M. (1993) Effects of vertebrate predation on a

caviomorph rodent, the degu (*Octodon degus*), in a semiarid thorn scrub community in Chile. *Oecologia* **94:** 153–158.
Mihaliak, C. A. and Lincoln, D. E. (1985) Growth pattern and carbon allocation to volatile leaf terpenes under nitrogen-limiting conditions in *Heterotheca subaxillaris* (Asteraceae). *Oecologia* **66:** 423–426.
Miller, J. C. and Hanson, P. E. (1989) Laboratory feeding tests on the development of gypsy moth larvae with reference to plant taxa and allelochemicals. *Bulletin of the Agricultural Experimental Station, Oregon State University* **674:** 1–63.
Moen, J., Garfjell, H., Oksanen, L., Ericson, L. and Ekerholm, P. (1993) Grazing by food-limited microtine rodents on a productive experimental plant community: does the "green desert" exist? *Oikos* **68:** 401–413.
Moran, M. D., Rooney, T. P. and Hurd, L. E. (1996) Top-down cascade from a bitrophic predator in an old-field community. *Ecology* **77:** 2219–2227.
Murdoch, W. W. (1966) Community structure, population control, and competition. *American Naturalist* **100:** 219–226.
Nichols-Orians, C. M. (1991a) The effects of light on foliar chemistry, growth and susceptibility of seedlings of a canopy tree to an attine ant. *Oecologia* **86:** 552–560.
Nichols-Orians, C. M. (1991b) Environmentally induced differences in plant traits: consequences for susceptibility to a leaf-cutter ant. *Ecology* **72:** 1609–1623.
Nuñez-Farfan, J. and Dirzo, R. (1989) Leaf survival in relation to herbivory in two tropical pioneer species. *Oikos* **55:** 71–74.
Oksanen, L., Fretwell, S. D., Aruda, J. and Niemela, P. (1981) Exploitation of ecosystems in gradients of primary productivity. *American Naturalist* **118:** 240–261.
Pace, M. L., Cole, J. J., Carpenter, S. R. and Kitchell, J. F. (1999) Trophic cascades revealed in diverse ecosystems. *Trends in Ecology and Evolution* **14:** 483–488.
Paine, R. T. (1966) Food web complexity and species diversity. *American Naturalist* **100:** 65–75.
Persson, L. (1999) Trophic cascades: abiding heterogeneity and the trophic level concept at the end of the road. *Oikos* **85:** 385–397.
Pianka, E. R. (1966) Latitudinal gradients in species diversity: a review of concepts. *American Naturalist* **100:** 33–46.
Polis, G. A. (1999) Why are parts of the world green? Multiple factors control productivity and the distribution of biomass. *Oikos* **86:** 3–15.
Polis, G. A. and Strong, D. R. (1996) Food web complexity and community dynamics. *American Naturalist* **147:** 813–846.
Price, P. W. (1991a) Patterns in communities along latitudinal gradients. In *Plant–Animal Interactions: Evolutionary Ecology in Tropical and Temperate Regions*, ed. P. W. Price, T. M. Lewinsohn, G. W. Fernandes and W. W. Benson, pp. 51–69. New York: John Wiley.
Price, P. W. (1991b) The plant vigor hypothesis and herbivore attack. *Oikos* **62:** 244–251.
Price, P. W., Bouton, E. E., Gross, P., McPheron, B. A., Thompson, J. N. and Weis, A. E. (1980) Interactions among three trophic levels: influence of plants on interactions between insect herbivores and natural enemies. *Annual Review of Ecology and Systematics* **11:** 41–65.
Rathcke, B. J. and Price, P. W. (1976) Anomalous diversity of tropical ichneumonid parasitoids: a predation hypothesis. *American Naturalist* **110:** 889–893.
Rohde, K. (1992) Latitudinal gradients in species diversity: the search for the primary cause. *Oikos* **65:** 514–527.

Schaffner, J. V. and Griswold, C. L. (1934) *Macrolepidoptera and their Parasites Reared from Field Collections in the Northeastern Part of the United States*. Washington, DC: US Department of Agriculture.

Scriber, J. M. (1973) Latitudinal gradients in larval feeding specialization of the world Papilionidae (Lepidoptera). *Psyche* **73**: 355–373.

Scriber, J. M. (1984) Larval foodplant utilization by the world Papilionidae (Lepidoptera): latitudinal gradients reappraised. *Tokurana* **6/7**: 1–50.

Scriber, J. M., Tsubaki, Y. and Lederhouse, R. C. (1995) *Swallowtail Butterflies: Their Ecology and Evolutionary Biology*. Orlando, FL: W. B. Saunders.

Sheehan, W. (1991) Host range patterns of hymenopteran parasitoids of exophytic lepidopteran folivores. In *Insect–Plant Interactions*, ed. E. Bernays, pp. 209–247. Boca Raton, FL: CRC Press.

Shure, D. J. and Wilson, L. A. (1993) Patch-size effects on plant phenolics in successional openings of the Southern Appalachians. *Ecology* **74**: 55–67.

Sime, K. R. and Brower, A. V. Z. (1998) Explaining the latitudinal gradient anomaly in ichneumonid species richness: evidence from butterflies. *Journal of Animal Ecology* **67**: 387–399.

Slobodkin, L. B. (1960) Ecological energy relationships at the population level. *American Naturalist* **94**: 213–236.

Spiller, D. A. and Schoener, T. W. (1994) Effects of top and intermediate predators in a terrestrial food web. *Ecology* **75**: 182–196.

Terborgh, J. (1992) Maintenance of diversity in tropical forests. *Biotropica* **24**: 283–292.

Turner, I. M. (1995) Foliar defenses and habitat adversity of three woody plant communitites in Singapore. *Functional Ecology* **9**: 279–284.

Wallace, A. R. (1878) *Tropical Nature and Other Essays*. London: Macmillan.

Waterman, P. G. (1983) Distribution of secondary metabolites in rain forest plants: towards an understanding of cause and effect. In *Tropical Rain Forest: Ecology and Management*, ed. S. L. Sutton, T. C. Whitmore and A. C. Chadwick, pp. 167–179. Oxford: Blackwell Science.

Waterman, P. G., Ross, J. A. M. and McKey, D. B. (1984) Factors affecting levels of some phenolic compounds, digestibility, and nitrogen content of the mature leaves of *Bateria fistulosa* (Passifloraceae). *Journal of Chemical Ecology* **10**: 387–401.

White, T. C. R. (1978) The importance of a relative shortage of food in animal ecology. *Oecologia* **33**: 71–86.

White, T. C. R. (1984) The abundance of invertebrate herbivores in relation to the availability of nitrogen in stressed food plants. *Oecologia* **63**: 90–105.

Whittaker, R. H. and Feeny, P. P. (1971) Allelochemics: chemical interactions between species. *Science* **171**: 757–770.

Wilson, E. O. (1971) *The Insect Societies*. Cambridge, MA: Harvard University Press.

STANLEY H. FAETH AND THOMAS L. BULTMAN

5

Endophytic fungi and interactions among host plants, herbivores, and natural enemies

Introduction

Plant-associated microbes are well known for mediating interactions between plants, herbivores, and natural enemies. Plant pathogens may increase or decrease host resistance to invertebrate herbivores and alter attack by natural enemies of the herbivores (e.g., Hatcher, 1995). Mycorrhizal associations alter plant nutrition and growth and thus indirectly influence herbivores feeding upon host plants (e.g., Gehring and Whitham, 1994; Gehring *et al.*, 1997) as well as their natural enemies. Endophytic fungi (fungi that live asymptomatically within plants, at least for part of their life cycle), however, are the only plant-associated microorganisms that are postulated to directly increase host plant defenses against both vertebrate and invertebrate herbivores (Carroll, 1988; Clay, 1988). Endophytic fungi have been considered as "acquired chemical defenses" (Cheplick and Clay, 1988) in grasses and "inducible defenses" and herbivore "antagonists" in woody plants (Carroll, 1988, 1991).

Endophytic fungi are very abundant and often extremely diverse in both woody (e.g., Carroll, 1991; Faeth and Hammon, 1997 a, b; Stone and Petrini, 1997; Arnold *et al.*, 2000) and grass host plants (Leuchtmann, 1992; Saikkonen *et al.*, 1998; Schulthess and Faeth, 1998). The main mechanism for increased plant resistance to herbivores is the production of mycotoxins. Additionally, endophytes may also alter plant physiology and morphology (Clay, 1990; Bacon, 1993), similar to mycorrhizal associations. Endophytic mycotoxins, notably alkaloids, are now well documented for some systemic endophyte infections in pooid grasses, but far less so for the more diverse and localized infections in grasses and woody plants

(Petrini *et al.* 1992; Siegel and Bush, 1997; Saikkonen *et al.*, 1998). Some endophytes in woody plants have been shown to increase host plant resistance to insect herbivores, especially sedentary insects, such as galling insects (Wilson and Carroll, 1997). However, the vast majority of this diverse group of microorganisms probably are neutral, and even occasionally positive (e.g, Gange, 1996), in their interactions with host plant herbivores (Faeth and Hammon, 1996, 1997a, b; Saikkonen *et al.*, 1998). These endophytes generally form localized infections, are horizontally transmitted via spores, and most have little effect on either the host plant or herbivores (Saikkonen *et al.*, 1998). Consequently, we should not expect widespread and strong effects on the third trophic level, natural enemies (but see Preszler *et al.*, 1996).

In this chapter, we examine the generality of increased host plant resistance via endophytic fungal associations. We focus largely on the systemic, specialized endophytes inhabiting pooid grasses because: (1) these associations have been studied much more than other endophyte–host plant interactions, (2) alkaloidal mycotoxins responsible for increased herbivore resistance are fairly well known, and (3) the interactions of systemic endophytes with their hosts are considered strongly mutualistic. Thus, if plant defenses via endophytes are common in nature, we expect they should be especially evident in these grass–endophyte associations. Further, we predict that any effects of endophytes on the third trophic level should also be most prominent in these systems compared to endophytes that non-systemically infect grasses and to those that infect plants other than grasses.

Grass systemic endophytes

Background

Systemic endophytes of cool season grasses in the Pooideae subfamily are typically members of the ergot family, Clavicipitaceae (Ascomycota). Most of these endophytes are found as the anamorphic (asexual) stage and are transmitted vertically by growing into seeds of maternal plants (Clay, 1988; Schardl *et al.*, 1997). Other clavicipitaceous, systemic endophytes (e.g., *Balansia*) and epiphytes (e.g., *Atkinsonella*) may produce mycotoxins that affect herbivores and natural enemies (Clay, 1989), but they, for the most part, negatively affect their host plants by producing stromata which sterilize the host. Thus, these fungi are not considered strongly mutualistic in terms of increasing host plant resistance to herbivores.

These endophytes and their effects on herbivores have been reviewed elsewhere (e.g., Clay 1988, 1989, 1991, 1998; Breen, 1994; Saikkonen *et al.*, 1998) and we do not consider them further here. Instead, we focus on *Epichloë* and *Neotyphodium*, systemic endophytes of cool-season grasses. There are at least nine known described species of *Epichloë* (Schardl and Leuchtmann, 1999) and at least eight species of *Neotyphodium* (Clay, 1998). However, many others are yet undescribed taxonomically (e.g., White, 1987; White *et al.*, 1996; Miles *et al.*, 1998; Marshall *et al.*, 1999) and still others yet to be discovered and isolated from host grasses (e.g., Leuchtmann, 1992).

Exclusively asexual forms of *Epichloë* have been classified by convention as the genus *Neotyphodium* (formerly *Acremonium* sect. *Albo-lanosa*: Glenn *et al.*, 1996). While *Neotyphodium* is always transmitted vertically (Type III infection: Schardl and Phillips, 1997), strains of *Epichloë* can either be transmitted vertically via seeds or horizontally (Type II infection). *Epichloë* can produce stromata in grass inflorescences that produce disease conditions (choke panicle) and both asexual and sexual spores (Bucheli and Leuchtmann, 1996). Production of stromata depends both on the *Epichloë* strain and environmental conditions (Schardl *et al.*, 1997) which, in turn, can affect the outcome of the interaction. Vertically transmitted *Epichloë* are more mutualistic relative to the host grass than those strictly horizontally transmitted *Epichloë* (Bucheli and Leuchtmann, 1996; Schardl and Clay, 1997; Schardl *et al.*, 1997).

Asexual *Neotyphodium* endophytes, alternatively, are always transmitted from maternal plant to offspring (Type III infection), similar to cytoplasmic organelles (Siegel and Schardl, 1992; Schardl and Tsai, 1992; White *et al.*, 1993a, b; but see White *et al.*, 1996). Evolutionary theory (Law, 1985; Lewis, 1985; Massad, 1987; Ewald, 1988, 1994; Marquis and Alexander, 1992; Frank, 1994) predicts that vertically transmitted strains of *Epichloë* and especially *Neotyphodium* endophytes should exhibit a high degree of mutualistic interaction with the host plant and a high degree of specificity. For the seed-borne endophytes in *Festuca* species of grasses, molecular phylogenic studies support this prediction, showing a high degree of specificity, suggesting long evolutionary relationships with the host plant (An *et al.*, 1992, 1993; Schardl and Tsai, 1992; Schardl *et al.*, 1994, 1997). Molecular phylogenies suggest that *Neotyphodium* endophytes evolved from parasitic, pathogenic strains of *Epichloë* species on multiple occasions (White, 1988; Schardl and Clay, 1997) and may have diversified via hybridization with several *Epichloë* species, at least in perennial rye-grass and tall fescue (e.g., Schardl *et al.*, 1994, 1997; Tsai *et al.*, 1994; Schardl and Phillips, 1997).

Effects on vertebrate herbivores

Undoubtedly, the presence of *Neotyphodium* and *Epichloë* endophytes inhabiting pasture and turf grasses, such as tall fescue and perennial rye-grass, have dramatic biological effects on vertebrate grazers. Endophyte infections in these introduced grasses cause toxicoses to grazing livestock (Reddick and Collins, 1988; Clay, 1989,1990, 1991, 1992; Ball *et al.*, 1993; Hoveland, 1993). In perennial ryegrass (*Lolium perenne*), *Neotyphodium*-linked ergot and indole diterpene-type (e.g., lolitrem B) alkaloids produce staggers, intoxication, and general poor health in sheep and cattle. In tall fescue (*Festuca arundinacea*), pyrrolizidine (lolines) and ergot-type alkaloids cause gangrene of extremities, reduced conception, and general poor health in livestock (see Siegel and Bush, 1996; Bush *et al.*, 1997).

However, while cases of livestock toxicity are well known from tall fescue and perennial ryegrass introduced to North America (Clay, 1988, 1991; Siegel and Schardl, 1991), these grasses appear much less toxic to vertebrates in their native ranges (Siegel and Bush, 1996). Infected tall fescue and perennial ryegrass in native habitats tend to produce far fewer types and lower levels of alkaloids than in the introduced and cultivated varieties of these grasses (e.g. Siegel and Bush, 1997; Saikkonen *et al.*, 1998). Most populations contained only one type of alkaloid (peramine) rather than the three to four types typically found in introduced tall fescue and perennial ryegrass (Saikkonen *et al.*, 1998). Also, tall fescue in native habitats appears to be far less dominant and invasive (Saikkonen, 2000) than its agronomic counterpart in the USA (Clay and Holah, 1999), suggesting reduced competitive advantages related to grazing in native habitats (Saikkonen, 2000).

There are surprisingly few examples of *Epichloë* and *Neotyphodium* endophytes in native grasses with marked biological effects on vertebrate herbivores (Table 5.1) given the widespread occurrence of these endophytes in grasses. Systemic *Epichloë* or *Neotyphodium* endophytes are known from all tribes and most genera in the subfamily Pooideae, including Poeae (Festuceae), Aveneae, Meliceae, Triticeae, Brachypoideae, and Bromeae. Also, if one considers Stipeae as a tribe within the subfamily Arundinoideae (e.g., Barkworth and Everett, 1988), then these endophytes may also occur across grass subfamilies because *Neotyphodium* is found

Table 5.1. *Systemic, seed-borne endophytes of grasses (Type II or III) with reported strong negative effects on vertebrate herbivores*

Grass	Endophyte	Herbivore	Effect	Continent[a]	Reference[b]
Achnatherum (Stipa) robusta (sleepy grass)	*Neotyphodium*	Horses/sheep	Narcotic	NA	1,2,8
Achnatherum inebrians (drunken horse grass)	*Neotyphodium*	Cattle	Narcotic	AS	3,4
Lolium temulentum (Darnel)	*Neotyphodium*	Livestock	Toxicosis	EU, AS	7, 9
Lolium perenne (perennial ryegrass)	*Neotyphodium*	Cattle/sheep	Staggers	NA, NZ, AU	5
Poa huecu	*Neotyphodium*[c]	Livestock	Staggers	SA	7
Melica descumbens (drunk grass)	*Neotyphodium*	Livestock	Staggers/ narcotic	AF	6, 7
Echinopogon ovatus	*Neotyphodium*-like?	Cattle/sheep	Staggers	AU, NZ	7
Festuca argentina	*Neotyphodium*[c]	Livestock	Toxicosis[c]	SA	8
Festuca arundinacea (tall fescue)	*Neotyphodium*	Cattle/sheep	Toxicosis	NA	5
Festuca hieronymi	*Neotyphodium*[c]	Livestock	Toxicosis[c]?	SA	8

Notes:

[a] NA, North America; AS, Asia; EU, Europe; SA, South America; AU, Australia; NZ, New Zealand.

[b] 1, Marsh and Clawson (1929); 2, Kaiser *et al.* (1996); 3, Bruehl *et al.* (1994); 4, Miles *et al.* (1996); 5, Clay (1991); 6, White (1987); 7, Miles *et al.* (1998); 8, Powell and Petroski (1992); 8, Parodi (1950); 9, Moon *et al.* (2000).

[c] Toxicity caused by *Neotyphodium* unconfirmed.

in *Achnatherum* (*Stipa*). Leuchtmann (1992) reported that about 290 grass species are infected by systemic clavicipitaceous endophytes, or about 4% of 8000 known grass species. Leuchtmann (1992), however, considered this as a very conservative estimate, since relatively few grass species have been systematically studied. He estimated that at least 20%–30% of grass species (1600–2400 species) harbor systemic endophytes. Of the 3000+ species of the Pooideae (MacFarlane, 1988), we can expect conservatively 20%–30% are likewise infected with *Epichloë* or *Neotyphodium*. In fact, new discoveries of *Neotyphodium* endophytes appear to be accelerating as more native grasses are tested (e.g., White, 1987; White *et al.*, 1993, 1996; Li *et al.*, 1997; Marlatt *et al.*, 1997; Miles *et al.*, 1998; Marshall *et al.*, 1999; Nan and Li, 2001; Saikkonen *et al.*, 2001).

Neutral or positive effects of grass endophytes on vertebrates are probably underreported (e.g., Carroll, 1991). In contrast, there are probably very few unreported cases of strong negative effects of endophytes on vertebrates. The biological effects of endophyte poisoning on livestock are often striking, and many cool-season grasses are important forage for livestock and wildlife. This contention is borne out by very old reports in botanical, ecological, and agronomic literature of toxic or narcotic grasses, long before the endophytic mechanism of toxicity was known (Hance, 1876; Vogl, 1898; Bailey, 1903; Freeman, 1904; Marsh and Clawson, 1929; White, 1987; Miles *et al.* 1998).

Comprehensive studies of the effects of systemic endophytes in native grasses on vertebrate grazers are scarce, but they suggest that negative effects on vertebrates are uncommon. Contrary to predictions of the antiherbivory hypothesis, our studies of Arizona fescue show no relationship between frequency of endophytes infection and livestock and native ungulate grazing (Saikkonen *et al.*, 1998, 1999, Schulthess and Faeth, 1998). In a recent study of *Elymus canadensis*, wild ryegrass, in North American grasslands, infection by *Neotyphodium* is common and widespread but apparently has little or no effect on grazing mammals (Vinton *et al.*, 2001). In Morocco, *Neotyphodium*-infected *Festuca mairei*, a native grass, apparently does not deter cattle grazing, but does confer drought resistance (Marlatt *et al.*, 1997). Many native grasses in China, although infected by *Neotyphodium*, usually are not toxic to livestock (Nan and Li, 2001).

Even reported cases of strong herbivore resistance to vertebrate grazing (Table 5.1) may be exaggerated. For example, *Achnatherum* (*Stipa*) *robusta*, termed "sleepy grass" for its strong narcotic effects on horses (e.g., Petroski *et al.*, 1992), is toxic in only a few isolated populations in the Sacramento Mountains in New Mexico. Forty-eight other populations from across the southwestern USA harbor high frequencies of endophyte infection but do not exhibit narcotic or toxic effects on livestock (Jones *et al.*, 2000). In another case, Miles *et al.* (1998) reports that toxicity of infected *Echinopogon ovatus* is limited to only certain populations, and other *Echinopogon* species infected with *Neotyphodium* are not toxic to livestock. Finally, two other oft-cited grasses in Table 5.1 causing strong vertebrate toxicity, *Lolium perenne* and *L. temulentum*, are not toxic in many populations within their native range, and in many introduced, agronomic populations (Bor, 1973). Apparently, only the seeds of *L. temulentum* are toxic to vertebrates in populations where vertebrate toxicity is known (Bor, 1973). This evidence supports the hypothesis that alkaloids of seed-borne endophytes should be most effective at seed and seedling stages, rather than adult plant stage (see section "Alternative hypotheses for diversity and maintenance of systemic endophytes," below).

Effects on invertebrate and microherbivores

In addition to deterring vertebrate herbivory, systemic endophytes in grasses are also well known for increasing resistance to invertebrate herbivores and pathogenic microorganisms (Clay, 1987a, 1988, 1989, 1990, 1991; Carroll, 1988; West *et al.*, 1988; Gwinn and Bernard, 1990; Kimmons *et al.*, 1990; Dahlman *et al.*, 1991; Clay *et al.*, 1993; Breen, 1994). Resistance to insect pests in tall fescue appears to result from high levels of peramine and pyrrolizidine (loline) alkaloids produced by the endophyte (Siegel and Bush, 1996, 1997). Clay (1989, 1991: Table 17.1) presents a list of 14 pooid and 10 non-pooid endophyte-infected grass species known to have increased resistance to insect pests. Most of the grasses and invertebrate species tested thus far, however, are non-native, agronomic grasses and generalist pest species introduced to North America. One would expect that native specialists are less affected by alkaloids in their host plants (Saikkonen *et al.*, 1998) so any conclusions regarding the generality of endophytes increasing resistance to invertebrates must be tempered since few native grass/invertebrate systems have been examined.

Even in the agronomic grass/insect pest studies, systemic endophytes do not always increase, and, in some cases, may decrease, resistance to herbivores. Saikkonen *et al.* (1998) summarized studies of agronomic grasses and found only 66% and 71% of bioassays showed negative effects of infected tall fescue and perennial ryegrass, respectively, on invertebrates. These percentages are surprisingly low because endophyte infection in these introduced grasses are thought to be universally detrimental to insect herbivores. Selective breeding, low genetic diversity, and intense grazing selection apparently have selected for endophytes that produce unusually high levels and multiple types of alkaloids relative to most known infected grasses, but even these do not always increase herbivore resistance (Saikkonen *et al.*, 1998).

Relatively few native grass populations and their systemic endophytes have been studied for effects on native invertebrate herbivores. We have extensively tested the role of *Neotyphodium* in Arizona fescue (*Festuca arizonica*), a native southwestern US grass, in resistance to a variety of native and non-native invertebrates. We have found either no increase in resistance or decreases in resistance due to the presence of *Neotyphodium* (Lopez *et al.*, 1995; Saikkonen *et al.*, 1999; Tibbets and Faeth, 1999). Infected Arizona fescue produces one type of alkaloid, peramine, and at highly variable levels within and among populations (L. P. Bush and S. H. Faeth, unpublished data). Since these low and variable levels of alkaloids appear typical of many *Neotyphodium*-infected native grasses (Siegel and Bush, 1997; Saikkonen *et al.*, 1998; Leuchtmann *et al.*, 2001), we predict that resistance to invertebrate herbivores should not be common in *Neotyphodium*-infected native grasses, and certainly less frequent than invertebrate resistance found in agronomic grasses.

Herbivory on grasses: the *raison d'être* for endophytes?

Increased resistance to herbivores has been postulated as the main selective pressure maintaining high frequencies of *Epichloë*, and especially, *Neotyphodium*, endophytes in pooid grasses (Clay, 1988, 1991, 1998; Siegel and Schardl, 1991). This hypothesis is partially supported in that most cases of strong anti-herbivore effects are associated with seed-borne endophytes. Vertically transmitted symbionts are predicted to be more mutualistic with their hosts than horizontally transmitted endophytes by evolutionary theory (Law, 1985; Massad, 1987; Ewald, 1988, 1994; Marquis and Alexander, 1992; Frank, 1994; Schardl *et al.*, 1997; Wilkinson and Schardl, 1997). One expects frequent and strong anti-herbivore mutualisms (e.g., Clay, 1998) since endophyte and host growth and reproduction are closely linked. Further, since most *Epichloë* or *Neotyphodium* endophytes produce some alkaloids (Siegel and Bush, 1997), albeit often at low levels (Saikkonen *et al.*, 1998, 1999), systemic endophytes generally have the basic metabolic pathways in place for alkaloid production. Additionally, variation in alkaloid types and levels exists within and among populations of infected grasses (Siegel and Bush, 1997) and is genetically based (Wilkinson *et al.*, 2000). For example, peramine levels in infected Arizona fescue plants range from zero to >3 ppm within the same population (S. H. Faeth and L. P. Bush, unpublished data). Thus, one would expect that if herbivory exerts strong selective pressure, then high concentrations and numerous types of alkaloids should characterize many more endophyte–grass associations. This begs the question: why aren't there more cases of systemic, vertically transmitted endophytes resulting in strong anti-herbivore effects? Possible explanations include:

1. Herbivory on adult plants may not be a strong selective force in maintaining endophytes that produce high levels of alkaloids. The effects of herbivory on plant fitness are highly variable and often indirect (e.g., Crawley, 1983; Marquis, 1992) and in some cases, absent or even positive (e.g., Paige and Whitham, 1987). For grasses, these effects may even be weaker due to the evolutionary trajectory taken by graminoids to tolerate grazing with adaptations like below- or near-ground meristematic tissue, rather than evolving defense against herbivores (Crawley, 1983; Strauss and Agrawal, 1999). Intensive livestock grazing and even herbivory by insect pest species (Siegel and Bush, 1996) may change competitive abilities of grasses and lead to higher frequencies of infected agronomic grasses (Clay, 1996, 1998; Rambo and Faeth, 1999). However, grazing pressure by vertebrates in natural grasslands is generally less intense and more sporadic than agronomic pastures. Likewise, herbivory by insect pest species is often persistent and severe on agronomic grasses of low genetic diversity grown in near monocultures (Siegel and Bush, 1996). In natural grasslands, increased plant diversity and host plant heterogeneity, as well as increased natural enemy attack (e.g., Barbosa and Schultz, 1987), generally act to reduce herbivore loads relative to agronomic pastures and lawns.
2. The prevalence of mutualisms tends to be greatest under stressful environmental conditions where the partnership ameliorates the stress (e.g., Hacker and Gaines, 1997). Maintenance of high frequencies of endophytes may be more related to continuously stressful abiotic factors, such as low water and nutrient availability, rather than inconsistent and unpredictable herbivory. Indeed, infections increase

resistance to drought (see below) and there is some evidence that endophytes, like mycorrhizal counterparts, may enhance nutrient uptake by altering root structure and releasing phenolic acids into root zones (Malinowski *et al.*, 1998; Malinowski and Belesky, 1999).

3. Alkaloid production by endophytes is costly to the host plant. Synthesis of nitrogen-rich alkaloid compounds by endophytes may compete with host plant requirements for limited nitrogen. Experiments of Cheplick *et al.* (1989) found that infected tall fescue performed worse than uninfected plants under conditions of low soil nitrogen, but the benefits of infection increased with increasing nitrogen. Since most non-native, agronomic grasses are grown under conditions of supplemented soil nutrients, alkaloid levels and, consequently, effects on herbivores may be exaggerated relative to native grasses and soils (Saikkonen *et al.*, 1998). We have found that supplementing soil nutrients to Arizona fescue increases peramine levels twofold. This result suggests that alkaloid production may be limited by available soil nutrients (S. H. Faeth and L. P. Bush, unpublished data), but also suggests that endophyte mutualisms may not be common in nutrient stressful situations (see previous section) unless enhanced nutrient uptake for the hosts outweighs competition for nitrogen by endophyte alkaloid production. In addition to metabolic costs of production, Carroll (1991) suggested that some mycotoxins might directly damage the host plant, and alkaloids, at least those produced by plants, are known to damage plants at high concentrations (Karban and Baldwin, 1997). Finally, alkaloid production by endophytes may also inhibit mycorrhizal colonization and reproduction (Chu-Chou *et al.*, 1992; Goldson *et al.*, 1992) and thus exacerbate nutrient uptake and limitation of the host grass.
4. Plant genotype effects on resistance to herbivores, drought and nutrient stress, and other selective pressures may subsume endophyte effects. Even in agronomic grasses with limited genetic diversity, drought resistance may depend on genotypic variation between the host grass and endophyte (Elbersen and West, 1996; Buck *et al.*, 1997). In Arizona fescue, we have found that plant genotype usually explains more variation in plant growth and reproduction than the presence or absence of the endophyte, under varying nutrients and water regimes (Saikkonen *et al.*, 1999; T. J. Sullivan and S. H. Faeth, unpublished data).

Alternative hypotheses for diversity and maintenance of systemic endophytes

Clavicipitaceous endophytes, especially *Epichloë*, and *Neotyphodium* are widespread and diverse among grass host species. When present in populations, frequencies of endophytes tend to be high (Leuchtmann and Clay, 1997; Saikkonen *et al.*, 1998). Asexual and vertically transmitted *Neotyphodium* symbionts are never gained and can only be lost from plants, either by failure of hyphae to grow into seed heads or tillers (imperfect transmission: Ravel *et al.*, 1997) or loss of hyphal viability in seeds or plants (e.g., Siegel *et al.*, 1984; but see White *et al.*, 1996). Therefore, positive selective pressures must maintain high frequencies in populations, because if neutral or negative, then frequencies should rapidly decline. What other explanations are there for the widespread occurrence of systemic, seed-borne endophytes across grass genera and maintenance of high levels within many species? We suggest several alternative explanations. These are not novel and have been discussed elsewhere (e.g., Siegel and Latch, 1987; Clay, 1988, 1991; Siegel and Schardl, 1992; Bacon, 1993; Breen, 1994; Schardl and Phillips, 1997; Siegel and Bush, 1997). However, these hypotheses have traditionally been relegated as secondary to the herbivore defense hypothesis (Saikkonen *et al.*, 1998). The alternatives include:

Endophytes increase resistance to seed and seedling predators and pathogens

Loss of seeds and seedlings to predators (we group seedling "herbivores" as predators, since herbivory by either vertebrates or invertebrates often results in death) and pathogens directly reduces plant fitness. Endophytes in agricultural grasses may enhance germination success and seedling survival (Clay, 1987b; Bacon, 1993; Clay *et al.*, 1993, 1998), and the seed and seedling stages are usually critical stages influencing plant population dynamics (e.g., Louda, 1983). Production of alkaloids, if primarily effective against seed and seedling predators, should be concentrated in these life stages. Indeed, limited studies have shown that alkaloids tend to be concentrated in seeds and seedlings (Siegel *et al.*, 1990; Bush *et al.*, 1993, Welty *et al.*, 1994). In infected agronomic grasses, seed predation by invertebrates and vertebrates is reduced (Wolock-Madej and Clay, 1991; Knoch *et al.*, 1995). Leuchtmann *et al.* (2000) found that alkaloids were particularly high and diverse in seeds of native European grasses where the endophyte was strictly seed-borne. It is possible that alkaloids are concentrated in seeds simply because hyphae are more dense there, but protection against seed or seedling predators remains a viable hypothesis. If alkaloids reduce herbivory, then we predict their effects should be greatest at the seed and seedling stage. Since alkaloids are costly to produce, accumulation in the seed and seedling stage may be most cost-effective. The presence of alkaloids in adult stages of grasses may be, under most circumstances, ancillary, or even incidental, to seed and seedling production.

Tests of seedling protection against pathogens are rare for native grasses. We have found no difference in seedling mortality from other seed-borne fungi between infected and uninfected Arizona fescue seed (C. E. Hamilton and S. H. Faeth, unpublished data). However, we did find that growth of potential seed pathogens was delayed in *Neotyphodium*-infected seeds compared to uninfected seeds, suggesting an anti-fungal effect of the endophyte.

Endophytes confer drought tolerance

The presence of *N. coenophialum* in introduced tall fescue increases resistance to drought stress (Richardson *et al.*, 1992, 1993; Piper and West, 1993, West *et al.*, 1993, 1995; Elbersen *et al.*, 1994; West, 1994), manifested by higher tiller survival during and improved regrowth after drought (Read and Camp, 1986; Arachevaleta *et al.*, 1992; Bacon, 1993; Elbersen *et al.*, 1994). The mechanisms for increased drought resistance in infected tall fescue appear to involve a combination of factors including: lower leaf conductance and more leaf-rolling during drought periods (Elbersen *et al.*, 1994; West, 1994; Elbersen and West, 1996), changes in hormonal signals (Bacon and White, 1994), higher water-use efficiency (Richardson *et al.*, 1990, in Bacon and White, 1994; Bush *et al.*, 1993), greater capacity for osmotic adjustment and turgor maintenance in leaves (West *et al.*, 1995; Elmi and West, 1995), and accumulation of polyhydroxol alcohols (polyols, such as glycerol: see refs. in Bacon, 1993; Bacon and White 1994) or the amino acid proline (Bacon, 1993).

In SHF's laboratory, studies with Arizona fescue (Lopez *et al.*, 1995; Saikkonen *et al.*, 1999; T. Day and S. H. Faeth, unpublished data) indicate that leaves of infected plants have higher water contents, lose water more slowly, and thus maintain higher turgor pressure, than leaves of uninfected plants. Furthermore, infected plants produce more root biomass than uninfected plants (Saikkonen *et al.*, 1999). Currently, we are conducting long-term, controlled experiments where we: (1) control plant genotype (by fungicidal removal of *Neotyphodium*), and (2) vary water and nutrients. Preliminary results indicate that the presence of the endophyte increases growth rate of plants, but only under conditions of very low moisture and nutrients.

Drought resistance may be particularly important during establishment of seedlings. For example, germination of Arizona fescue seeds occurs during late summer rains, which are followed by a dry period until winter precipitation begins in November–December. Therefore, new seedlings typically experience very low soil moisture. Our preliminary evidence suggests that survival during this period is critical and is related to *Neotyphodium* infection. In a 1999 field experiment, post-germination survival of infected seedlings was significantly greater than that of uninfected seedlings ($G = 4.94$, $df = 1$, $P < 0.05$).

Endophytes increase resistance to other abiotic factors

Many grassland systems have been historically maintained and characterized by fires (e.g., Cooper, 1960). Resistance to periodic fires may explain high frequency of *Neotyphodium* infections (Saikkonen *et al.*, 1998). Our preliminary studies on long-term, prescribed burn plots in Arizona do not support, however, the hypothesis that fire maintains frequency of infection. We know of no other studies that have examined the interactions of fire and endophytic infections, although fire is well known for altering mycorrhizal interactions with host plants (e.g.,Taylor, 1991).

K. Saikkonen and M. Helander (personal communication) report that endophyte infections may also increase overwintering success. Most cool-season, perennial grasses undergo senescence and then regrowth after the winter season. Survival during prolonged and severe cold periods may be related to more extensive roots of infected plants (Saikkonen *et al.*, 1999) or endophyte production of polyols that are also known to function as anti-freeze compounds in cold-adapted organisms (Hochachka and Somero, 1984).

Endophytes increase intra- and interspecific competitive abilities of host grasses

Neotyphodium endophytes, at least in tall fescue and perennial ryegrass, are well known for enhancing growth, and thus competitive abilities, of grasses (Cheplick and Clay, 1988; Cheplick *et al.*, 1989; De Battista *et al.*, 1990; Hill *et al.*, 1990; Kelrick *et al.*, 1990; Marks *et. al.*, 1991; Clay, 1990; Clay *et al.*, 1993; Latch, 1993). Increase in competitive ability is not exclusive of the other aforementioned hypotheses. Increased competitive success is a general phenomenon that may be mechanistically linked to increased drought, fire or winter cold resistance, and nutrient uptake (Malinowski *et al.*, 1998; Malinowski and Belesky, 1999) and mediated by biotic factors such as herbivores and pathogens (Cheplick and Clay, 1988; Arachevaleta *et al.*, 1992; Clay *et al.*, 1993). Endophytes may also enhance interference competition if mycotoxins leach into surrounding soils and inhibit germination or growth of uninfected conspecifics or other plant species (e.g., Clay and Holah, 1999).

We argue that herbivory, often considered the primary factor driving ecology and evolution of endophyte–host plant interactions, is simply one of many explanations. We can predict when herbivory should be important in maintaining infected plants from ecological theory and previous empirical studies. Endophyte-mediated defense against herbivores should be most common when: (1) herbivory is intense and predictable in time and space (e.g., Karban and Baldwin, 1997), (2) the herbivores are generalists that cannot detoxify fungal alkaloids, and (3) the costs of alkaloid production and accumulation are low relative to anti-herbivory benefits (Carroll, 1991). Assuming the major cost of harboring alkaloid-producing endophytes is competition for limiting nitrogen, then we should find endophytes that confer strong anti-herbivore resistance in relatively nitrogen-rich habitats that are consistently grazed upon by invertebrate or vertebrate herbivores. Not coincidentally, most examples of grasses harboring endophytes that confer strong anti-herbivore properties come from agricultural or turfgrass systems, where these conditions are usually fulfilled.

In these agronomic systems, we may expect that effects of endophytes on the third trophic level, natural enemies, should also be more pronounced than most native grass–endophyte interactions in natural communities. In the next section, we review evidence for third trophic level effects in these systems and describe ongoing experiments to test for these effects.

Endophytes and the third trophic level, natural enemies

The incorporation of the third trophic level must be considered to fully understand interactions between herbivores and their host plants (Price *et al.*, 1980). Just as endophytes may influence herbivores, they may also affect natural enemies of those herbivores. Because systemic, seed-borne *Neotyphodium* endophytes have a more consistently, although still quite variable, negative impact on herbivores than do vertically transmitted *Epichloë* species (see sections "Effects on vertebrate herbivores" and "Effects on invertebrate and microherbivores," above), we predict third trophic level effects should be more common in the *Neotyphodium*-infected grasses. Due to the paucity of studies on vertebrate natural enemies, we concentrate our discussion on the effects of endophytes on natural enemies of invertebrate herbivores. At least for agronomic grasses, recent studies suggest that endophytes may indeed alter attack by natural enemies in several different ways.

Endophytes indirectly influence natural enemies by altering herbivore development time

The slow growth–high mortality hypothesis (SG–HM) (Clancy and Price, 1987) states that nutritional quality or allelochemistry can prolong larval development rate of herbivores. Slower development prolongs exposure of herbivores to their predators and parasites (Feeny, 1976; Clancy and Price, 1987). While intuitively pleasing, empirical tests of the SG–HM hypothesis have provided only mixed support (Bouton, 1984; Weseloh, 1984; Clancy and Price, 1987; Damman, 1987; Craig *et al.*, 1990; Loader and Damman, 1991; Johnson and Gould, 1992; Benrey and Denno, 1997). The strongest evidence for the hypothesis appears to come from free-ranging folivores that are more vulnerable to enemy attack rather than herbivores, like miners, gallers, and borers, that are concealed in plant tissue (Benrey and Denno, 1997).

Herbivores feeding on endophyte-infected plants may experience prolonged development perhaps because of alkaloidal mycotoxins. Delayed development while feeding on endophyte-infected grasses has been reported for several herbivores (Clay *et al.*, 1985; Hardy *et al.*, 1985, 1986; Breen, 1994; Popay and Rowan, 1994). Under the SG–HM hypothesis, one might expect greater enemy-caused mortality of insects, particularly free-ranging folivores that are feeding on agronomic grasses (see section "Effects on invertebrate and microherbivores," above). Effects of grass endophytes on herbivore development, however, are not always negative, and are often neutral (Clay *et al.*, 1985; Saikkonen *et al.*, 1998; Tibbets and Faeth, 1999) as well as positive (Clay *et al.*, 1985; Lopez *et al.*, 1995; Bultman and Conard, 1998; Saikkonen *et al.*, 1998; Tibbets and Faeth, 1999). Furthermore, the only study to assess endophyte levels in grasses and parasitism of herbivores showed a negative, rather than positive, relationship between the frequency of parasitism and level of endophyte infection of grasses. Goldson *et al.* (2000) found that levels of *Neotyphodium lolii* (and the peramine alkaloid it produces) in perennial ryegrass were inversely related to rates of parasitism of Argentine stem weevil (*Listronotus bonariensis*) by the parasitoid *Microctonus hyperodae*. The weevil, however, is a stem-borer and therefore might not be expected to follow predictions of the SG–HM hypothesis because of the protection it receives from living within grass stems. Future work needs to assess what effects endophytes have on vulnerability of free-ranging insect herbivores to attack by their predators and parasites.

Compared to systemic grass endophytes, the effects of horizontally transmitted tree endophytes on herbivores are much more variable and probably weaker (see section "Introduction") (Faeth and Wilson, 1996; Faeth and Hammon, 1997a, b; Saikkonen *et al.*, 1998). Nonetheless, prolonged development has been suggested for *Cameraria* leaf miners feeding on Emory oak (Faeth, 1987, 1988, 1991; Faeth and Hammon, 1997b). Further, the only published assessment of the pattern between endophyte infection levels of trees (Emory oak) and parasitism of insect herbivores (*Phyllonorycter* leaf miners) showed a positive relationship (Preszler *et al.*, 1996), which is consistent with the SG–HM hypothesis. While that result is suggestive, more work on tree endophytes is necessary to determine if endophyte infection varies with enemy attack and whether delayed herbivore development is the cause.

Endophytes lower resistance of herbivores, particularly generalists, to pathogens and parasites

Plant allelochemicals can lower herbivore resistance to pathogens. For example, the lepidopteran *Trichoplusia ni* experienced enhanced mortality from the pathogen *Bacillus thuringiensis* as dietary nicotine increased (Barbosa, 1988). In like manner, endophytes and their associated alkaloids may indirectly alter susceptibility of insect herbivores to natural enemies. For example, Japanese beetles feeding on roots of endophyte-infected tall fescue were more susceptible to an entomopathogenic nematode than were beetles feeding on roots of uninfected grass (Grewal *et al.*, 1995). While mycelium of grass endophytes is usually absent in roots, endophyte-produced alkaloids are found there (Siegel *et al.*, 1989). The researchers also found beetles fed an artificial diet containing an ergot alkaloid were more susceptible to nematodes than those fed diets lacking alkaloids. They suggested reduced beetle vigor due to starvation caused the increased susceptibility to the parasite. Similar indirect pathways of interaction have not been reported for endophytes infecting woody plants; however, lack of negative correlation between herbivore pupal mass and endophyte infection (Preszler *et al.*, 1996; Faeth and Hammon, 1997b) suggests increased susceptibility due to herbivore starvation is unlikely in these systems.

Another example of an indirect interaction of endophytes is changes in the herbivore's ability to encapsulate parasitoid eggs. The effectiveness of encapsulation depends upon the physiological condition of the host (Muldrew, 1953; van den Bosch, 1964; Vinson and Barbosa, 1987), which may be reduced by endophytes and their mycotoxins. While this hypothesis has received little attention, one study has offered no support. M. McNeill (personal communcation) found variation in endophyte infection levels in pastures of perennial ryegrass in New Zealand was not related to encapsulation of *M. hyperodae* parasitoid eggs by Argentine stem weevils.

Endophytic alkaloids may alter the behavior of herbivores, which in turn, changes their susceptibility to natural enemies. Although few relative to the number of systemic endophyte–grass associations (Table 5.1, and see section "Herbivory on grasses: the *raison d'être* for endophytes?" above), the cases of strong toxic effects often involve radical changes in behavior, ranging from staggers in cattle to narcoses in horses (Table 5.1). In native grass systems, these behaviors should result in increased susceptibility to predators. Although not well studied, changes in invertebrate behavior due to endophytic alkaloids may also occur, and correspondingly increase their vulnerability to predators and parasites. For example, Tibbets and Faeth (1999) found that leaf-cutting queen foundresses exhibited symptoms resembling cattle staggers when fed infected tall fescue, but not an infected native grass, Arizona fescue.

The effect of endophytes on the resistance of vertebrate herbivores to pathogens and parasites has received little attention. However, a symptom of cattle grazing endophyte-infected grasses is reduced serum prolactin levels (Thompson *et al.*, 1987; Cross, 1997). Because prolactin functions as a cofactor in the regulation of the immune response (Reber, 1993), mammals grazing endophyte-infected grasses may have decreased ability to produce antibodies to a protein antigen. Some work suggests this is the case. For example, mice and rats fed infected tall fescue seeds exhibited impaired immune function (Gay *et al.*, 1990). Antibody response of cattle fed infected tall fescue to immunization with tetanus toxoid was reduced compared to cattle on uninfected pastures (Dawe *et al.*, 1997). Moreover, cattle grazing infected grass exhibited lower basal serum IGF-I (insulin-like growth factor-I) values (Hazlett *et al.*, 1998; Filipov *et al.*, 1999) and even further reduced levels following challenge by an injection of *Escherichia coli* (Filipov *et al.*, 1999). These results suggest vertebrates grazing infected grasses would be more susceptible to pathogens and exhibit reduced growth. Filipov *et al.* (1999) also found that cattle grazing infected grasses and challenged by *E. coli* had heightened levels of tumor necrosis factor alpha (TNF-α), a postinflammatory cytokine, which leads to greater muscle catabolism. Hence, the animals should be more susceptible to subsequent disease and stress.

In summary, while plausible, indirect effects of endophytes on natural enemies through slowed herbivore development or enhanced susceptibility to enemies remain largely unstudied. Based upon limited data so far, there is little support for these interaction pathways.

Endophytes may directly influence predators and parasites

Endophyte-produced alkaloids consumed by herbivores may directly reduce the growth and survival of immature enemies feeding upon toxins accumulated in their host's tissues. Similar interactions involving host plants and their allelochemicals have been documented (Pickett *et al.*, 1991; Barbosa and Benrey, 1998). For example, tobacco allelochemicals consumed by *Cotesia* parasitoids reduce the pupal mass and survival of the parasitoids when feeding on tobacco hornworm (Barbosa *et al.*, 1986, 1991). Evidence of similar interactions involving endophytes comes from *Neotyphodium*-infected grasses. Barker and Addison (1996) found *N. lolii* infecting perennial ryegrass retarded development of the parasitoid *M. hyperodae* attacking weevils. When weevils were fed artificial diet containing endophyte-produced alkaloids, parasitoid survival was reduced. Weevil feeding was depressed by the presence of alkaloids, suggesting host quality was compromised due to starvation. T. L. Bultman and M. McNeill (unpublished data) recently extended this work by testing if several different strains of *Neotyphodium* altered the growth and survival of the parasitoid. This work also complements the accumulating evidence that variation in plant genotype can influence multitrophic interactions (Hare, chapter 2, this volume). All endophyte strains differed from one another with respect to the profile of alkaloids they produce. Plants containing alkaloid-producing strains reduced parasitoid development relative to plants infected with stains that did not produce alkaloids or those that lacked fungal infection (Fig. 5.1).

Bultman *et al.* (1997) tested the effect of grass endophytes on two *Euplectrus* parasitoid species of fall armyworm larvae (*Spodoptera frugiperda*), a lepidopteran, feeding on infected tall fescue. They found the presence of endophyte-infected plants in the diet of fall armyworm had a negative impact on the pupal mass of parasitoids, particularly *E. comstockii*. In con-

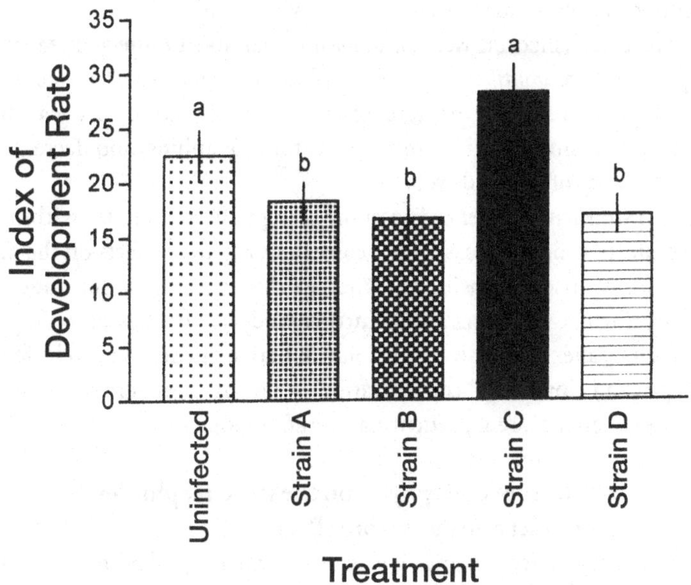

Fig. 5.1. Index of development rate of *Microctonus hyperodae* parasitoids when reared from *Listronotus bonariensis* weevils fed perennial ryegrass containing different endophyte strains. Fungal stain affected development rate ($F_{4,222} = 6.55$, $P < 0.0001$). Development rate was faster when reared from hosts fed plants lacking alkaloids (a) (uninfected and strain C) compared to those containing alkaloid-producing endophytes (b) (orthogonal contrast, $F_{2,222} = 27.3$, $P < 0.0001$). (T. L. Bultman and M. McNeill, unpublished data.)

trast to *M. hyperodae* (above), developmental rate of the parasitoids was not adversely affected (and was in fact accelerated) by the fungal endophyte. They also tested if effects of the endophyte on *E. comstockii* were due to the specific alkaloids, *N*-acetyl and *N*-formyl loline, produced by the fungus. When added to artificial diets of fall armyworm, both lolines caused reduced survival of parasitoids. Additional experiments with aphids and their parasitoids have given similar results. *Rhopalosiphum padi* aphids were fed either *N. coenophialum*-infected or uninfected tall fescue and then parasitized by *Aphelinus asychis*. Mass of adult parasitoids reared from hosts feeding on infected grass was reduced 23% relative to uninfected grass diet (K. C. Tonkel and L. T. Bultman, unpublished data).

The laboratory-based studies showing effects of endophytes on performance and preference of parasitoids predict the fungi should have multitrophic effects within grazing food-webs. This prediction was recently supported by Ormacini *et al.* (2000) who documented the insect food-webs established on *Neotyphodium*-infected and uninfected Italian rye-grass (*Lolium multiflorum*) in field plots in Argentina. They found the endophyte reduced the rate of parasitism by secondary parasitoids (hyperparasitoids and mummy parasitoids of aphids) and decreased the complexity of the food-web.

In summary, evidence is accumulating that endophytes within agronomic, non-native grasses have consistent negative effects on the performance of insect parasitoids. Whether this is also true for endophytes infecting native grasses awaits further study. However, we would predict much weaker effects since alkaloids in native grasses tend to be fewer in type and lower in concentration than in the agronomic grasses (Saikkonen *et al.*, 1998; Leuchtmann *et al.*, 2000).

Do effects of endophytes on the third trophic level counteract anti-herbivore effects?

The negative effects of grass endophytes on parasitoid performance set up the possibility that herbivores may experience some release from their parasitoids that could compromise the defense the endophyte purportedly provides for the plant. If herbivore populations are at least partially regulated by parasitoids, then a reduction in parasitoid performance could lead to more herbivores and greater herbivory on infected plants. Nonetheless, reduced quality of the parasitoids' hosts may not lead to increased plant damage due to the weakening of top-down effects. For example, parasitoids of the Mexican bean beetle were still effective biological control agents although soybeans resistant to the beetle increased development time and reduced survival and reproduction of the beetle's parasitoids (Kauffman and Flanders, 1985). Control was achieved because the reduction in population growth of the parasitoid was less than that of its beetle host. Whether the effectiveness of endophytes in increasing host grass resistance to herbivores is compromised by reducing natural enemy populations is yet untested.

Alternatively, endophytes may enhance efficacy of natural enemies if host preference or host location by enemies is facilitated by endophyte-related chemical changes in the host plant. An extensive literature has accumulated that shows natural enemies, especially parasitoids, use chemical cues of plants to locate their hosts (Vinson, 1976; Nordland *et al.*, 1988; Barbosa and Letourneau, 1988; Whitman, 1988; Turlings *et al.*, 1990, 1991, 1995, and chapter 7, this volume; Dicke, 1994, 1995). Furthermore, herbivore-inflicted plant damage can result in emission of aromatic signals that attract natural enemies (Dicke, 1999), although the consequences of this attraction at the population level are debated, especially in natural systems (Faeth, 1994). Interestingly, recent work has shown endophyte-infected fescue grasses emit volatile compounds that are quantitatively and qualitatively different from those produced by uninfected plants (Wang *et al.*, 1998). In addition, mechanical damage to plants dramatically elevates the levels of compounds emitted (T. Gianfagna, personal communication). These observations suggest endophytes, like many plants, may provide chemical cues that alter host preferences and host location by natural enemies.

Despite the plausibility of endophytes altering host preferences of natural enemies, little work has occurred in this area. In the only published study to date, Barker and Addison (1997) investigated effects of *N. lolii* in perennial ryegrass on preferences of *M. hyperodae* for Argentine stem weevils in New Zealand. They found naïve parasitoids showed no preference for hosts feeding on infected or uninfected plants. Parasitoids with prior experience of weevils on a particular ryegrass diet were subsequently more efficient at parasitizing hosts on the same diet. These results, while limited to one study, are exciting because they suggest grass endophytes may modify host acceptance by parasitoids.

D. Heim and L. T. Bultman (unpublished data) tested the generality of this effect in perennial ryegrass with fall armyworm larvae and its *Euplectrus comstockii* parasitoids. *Euplectrus comstockii* females showed no innate preferences for hosts reared from uninfected or infected perennial ryegrass based on the total number of eggs laid. However, wasps with prior experience of hosts fed infected plants tended to lay more eggs on hosts fed infected compared to uninfected grass (Fig. 5.2). B. L. Bennett and L. T. Bultman (unpublished data) performed similar preference experiments with *Euplectrus* parasitoids, except they used another agronomic grass, tall fescue. Naïve parasitoids tended to prefer hosts fed uninfected plants, and as in the previous study, parasitoids that had prior ovipositional experience with hosts fed infected plants tended to subsequently prefer hosts fed infected grass. These results corroborate those of Barker and Addison (1997) and suggest endophytes may modulate interactions between grasses and insect herbivores in agronomic grasses by altering the likelihood of attack by natural enemies. Presumably, aromatic differences between infected and uninfected plants are stimuli that parasitoids associate with host suitability, just as many parasitoids show

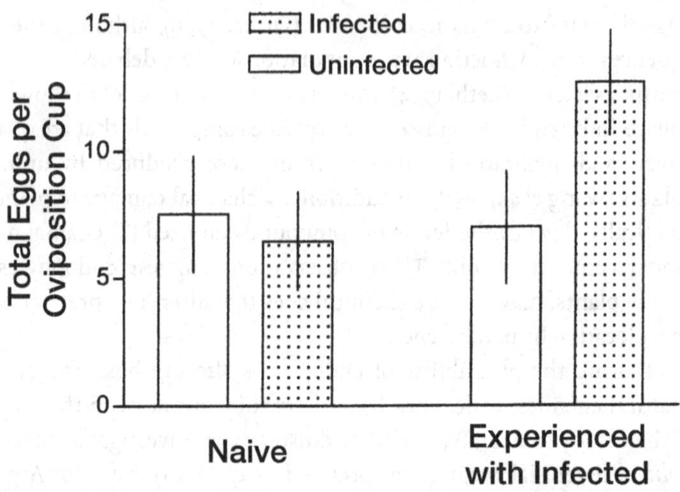

Fig. 5.2. Number of eggs laid by *Euplectrus comstockii* parasitoids that had never before encountered hosts (naïve) or had prior experience with hosts fed fungally infected plants. The experiment was a "no choice" design. Data were analyzed by ANCOVA using mean host weight per oviposition cup as the covariate. Parasitoids with prior experience with hosts fed infected plants tended to subsequently prefer those types of hosts ($F_{2,27} = 4.43$, $P < 0.06$). (D. Heim, unpublished data.)

associative learning between various plant odors and host suitability (Vet and Groenewold, 1990; Turlings *et al.*, chapter 7, this volume). It is not yet clear whether these alterations in parasitoid preference will translate into discernible differences in parasitoid or host insect population dynamics, or if they "trickle down" to affect relative fitness of infected and uninfected agronomic grasses.

The study of plant–endophyte–herbivore interactions has only recently begun to consider the third trophic level. Nonetheless, some patterns are beginning to emerge. Endophytes, at least those infecting agricultural grasses, where alkaloid types and levels are usually very high, often reduce performance of parasitoids, modify their preferences and may even influence trophic structure and diversity (Ormacini *et al.*, 2000). It seems clear that endophyte-produced alkaloids mediate the negative effects on natural enemies of herbivores. Indeed, some researchers (Barker and Addison, 1996; Bultman *et al.*, 1997) have found effects of specific alkaloids on parasitoids. Indirect effects, such as increasing exposure to enemies though prolonged herbivore development or reduced resistance to enemies have little support, but also have received very little attention. Even if endophytes indirectly affect natural enemies by prolonging herbivore development, the net effect of this on the plant is unclear; prolonged development could lead to greater damage to the plant. Further work is needed to determine how the effects of endophytes on natural enemies via the indirect pathway factor back to affect plant performance.

Information on how endophytes affect natural enemy preferences in native grasses is non-existent. However, if these alterations in natural enemy attack are mediated by endophytic alkaloids, then given that native grasses harbor fewer types and lower levels (Siegel *et al.*, 1990; Petrini *et al.*, 1992; Powell and Petroski, 1992; Siegel and Bush, 1996, 1997; Saikkonen *et al.*, 1998), these effects are not expected to be as important. Furthermore, if increased resistance to herbivores does not generally maintain systemic endophytes in grasses (see "Herbivory on grasses: the *raison d'être* for endophytes?" above), then consideration of their effect on the third trophic level becomes moot. At the very least, the effects should be much more variable in more heterogeneous, natural systems.

Likewise, given that endophytic fungal infections in woody plants are probably neutral or at best, weak and indirect, relative to herbivores, we would predict even fewer and weaker effects of endophytes on the third trophic level. Further work with native grasses and woody plants is needed to test this prediction. It is also apparent that the study of third trophic level effects by endophytes has been mostly limited to parasitoids. Predators and pathogens have received almost no attention; future work should determine what effect endophytes have on these natural enemies. Finally, effects of endophytes on natural enemies of vertebrate herbivores have not been generally addressed and await future study.

Summary and future directions

Endophytic fungi are diverse and abundant in woody plants and grasses, and some may produce mycotoxins, especially systemic, vertically transmitted endophytes in pooid grasses. As such, endophytes potentially moderate plant interactions with herbivores and their natural enemies. Evidence suggests that although the influences of endophytes on herbivores and the natural enemies are important in some agronomic grasses, they are weaker and more variable in native grasses, and even more so for horizontally transmitted endophytes in woody plants. Thus, agronomic grasses and their systemic, seed-borne endophytes are at one end of a spectrum of endophyte effects on tritrophic interactions. At the opposite end, are the localized, horizontally transmitted endophytes and their host plants and grasses, with the systemic endophytes of native grasses falling somewhere in between.

Clearly, many questions about the ecology and evolution of endophyte, plants, herbivores, and their natural enemies remain unanswered. Lack of basic ecological information is particularly severe for systemic endophytes in native grasses, where lessons learned from agronomic grasses may not apply. Native grasses and endophytes, mainly because of their variability, are ideal and often tractable systems in which to test evolutionary theories involving persistence of mutualisms, geographic and temporal variation in species interactions, virulence and mode of transmission, symbiosis and adaptive radiation, coevolution, ecological theories of host plant defense, host specialization, and tritrophic level interactions and food-web structure. The first step should be documenting patterns of endophyte diversity and abundances in natural grasslands. Many of these natural grasslands have already disappeared because of agriculture and urbanization, or are threatened by these factors as well as accidental or intentional invasion by introduced species. Therefore the call to understand interactions between grasses, endophytes, herbivores, and natural enemies is an urgent one.

Acknowledgments

We thank C. E. Hamilton, B. Hawkins, A. Leuchtmann, J. J. Rango, T. J. Sullivan, and an anonymous reviewer for their helpful comments on earlier versions of the manuscript. Access to unpublished data was generously provided by L. P. Bush, C. E. Hamilton, C. Miles, and T. J. Sullivan. This research was supported by National Science Foundation grant DEB 97-27020 to SHF and National Science Foundation grants DIR 93-00466 and DEB 95-27600 and US Department of Agriculture grant 1999–02652 to TLB.

REFERENCES

An, Z.-Q., Liu, J.-S., Siegel, M. R., Bunge, G. and Schardl, C. L. (1992) Diversity and origins of endophytic fungal symbionts of the North American grass *Festuca arizonica*. *Theoretical Applied Genetics* **85**: 366–371.

An, Z.-Q., Siegel, M. R., Hollin, W., Tsai, H.-F., Schmidt, D. and Schardl, C. L. (1993) Relationships among non-*Acremonium* endophytes in five grass species. *Applied Environmental Microbiology* **59**: 1540–1548.

Arachevaleta, M., Bacon, C. W., Plattner, R. D., Hoveland, C. S. and Radcliffe, D. E. (1992) Accumulation of ergopeptide alkaloids in symbiotic tall fescue grown under deficits of soil water and nitrogen fertilizer. *Applied Environmental Microbiology* **58**: 857–861.

Arnold, A. E., Maynard, Z., Gilbert, G. S., Coley, P. D. and Kursar, T. A. (2000) Are tropical fungal endophytes hyperdiverse? *Ecology Letters* **3**: 267–274.

Bacon, C. W. (1993) Abiotic stress tolerances (moisture, nutrients) and photosynthesis in endophyte-infected tall fescue. *Agriculture, Ecosystems, Environment* **44**: 123–141.

Bacon, C. W. and White, J. F., Jr. (1994) *Biotechnology of Endophytic Fungi of Grasses*. Boca Raton, FL: CRC Press.

Bailey, V. (1903) Sleepy grass and its effects on horses. *Science* **17**: 392–393.

Ball, D. M., Pedersen, J. F. and Lacefield, G. D. (1993) The tall-fescue endophyte. *American Scientist* **81**: 370–381.

Barbosa, P. (1988) Natural enemies and herbivore–plant interactions: influence of plant allelochemicals and host specificity. In *Novel Aspects of Insect–Plant Interactions*, ed. P. Barbosa and D. Letourneau, pp. 201–210. New York: John Wiley.

Barbosa, P. and Benrey, B. (1998) The influence of plants on insect parasitoids: implications for conservation biological control. In *Conservation Biological Control*, ed. P. Barbosa, pp. 55–82. New York: Academic Press.

Barbosa, P. and Letourneau, D. K. (eds.) (1988) *Novel Aspects of Insect–Plant Interactions*. New York: John Wiley.

Barbosa, P. and Schultz, J. C. (eds.) (1987) *Insect Outbreaks*. New York: Academic Press.

Barbosa, P., Saunders, J. A., Kemper, J., Tumbule, R., Olechno, J. and Martinat, P. (1986) Plant allelochemicals and insect parasitoids: effects of nicotine on *Cotesia congregata* and *Hyposoter annulipes*. *Journal of Chemical Ecology* **12**: 1319–1328.

Barbosa, P., Gross, P. and Kemper, J. (1991) Influence of plant allelochemicals on the tobacco hornworm and its parasitoid, *Cotesia congregatata*. *Ecology* **72**: 1567–1575.

Barker, G. M. and Addison, P. J. (1996) Influence of clavicipitaceous endophyte infection in ryegrass on development of the parasitoid *Microctonus hyperodae* Loan (Hymenoptera: Braconidae) in *Listronotus bonariensis* (Kuschel) (Coleoptera: Curculionidae). *Biological Control* **7**: 281–287.

Barker, G. M. and Addison, P. J. (1997) Clavicipitaceous endophytic infection in ryegrass influences attack rate of the parasitoid *Microctonus hyperodae* (Hymenoptera: Braconidae, Euphorinae) in *Listronotus bonariensis* (Coleoptera: Curculionidae). *Environmental Entomology* **26**: 416–420.

Barkworth, M. E. and Everett, J. (1988) Evolution in the Stipeae: identification and relationships of its monophyletic taxa. In *Grass Systematics and Evolution*, ed. T. R. Soderstorm, K. W. Hilu, C. S. Campbell and M. E. Barkworth, pp. 251–264. Washington, DC: Smithsonian Institution Press.

Benrey, B. and Denno, R. F. (1997) The slow-growth–high-mortality hypothesis: a test using the cabbage butterfly. *Ecology* **78**: 987–999.

Bor, N. L. (1973) *The Grasses of Burma, Ceylon, India, and Pakistan*. International Series of Monographs on Pure and Applied Biology, Koenigstein, Germany: Otto Koeltz Antiquariat.

Bouton, C. E. (1984) Plant defensive traits: translation of their effects on herbivorous insects into reduced plant damage. PhD thesis, University of Illinois at Urbana-Champaign.

Breen, J. P. (1994) *Acremonium* endophyte interactions with enhanced plant resistance to insects. *Annual Review of Entomology* **39**: 401–423.

Bruehl, G. W., Kaiser, W.J. and Klein, R.E. (1994) An endophyte of *Achnatherum inebrians*, an intoxicating grasss of northwest China. *Mycologia* **86**: 773–776.

Bucheli, E. and Leuchtmann, A. (1996) Evidence for genetic differentiation between choke-inducing and asymptomatic strains of the *Epichloë* grass endophyte from *Brachypodium sylvaticum*. *Evolution* **50**: 1879–1887.

Buck, G. W, West C. P. and Elbersen, H. W. (1997) Endophyte effect on drought tolerance in diverse *Festuca* species. In *Proceedings of the 3rd International* Neotyphodium/*Grass Symposium*, Athens, GA, ed. C. W. Bacon, J. H. Bouton and N. S. Hill, pp. 141–144.

Bultman, T. L. and Conard, N. J. (1998) Effects of endophytic fungus, nutrient level, and plant damage on performance of fall armyworm (Lepidoptera: Noctuidae). *Environmental Entomology* **27**: 631–635.

Bultman, T. L., Borowicz, K. L., Schneble, R. M., Coudron, T. A., Crowder, R. J. and Bush, L. P. (1997) Effect of a fungal endophyte and loline alkaloids on the growth and survival of two *Euplectrus* parasitoids. *Oikos* **78**: 170–176.

Bush, L. P. Gay, S. and Burhan, W. (1993) Accumulation of alkaloids during growth of tall fescue. In *Proceedings of the 17th International Grassland Congress*, Palmerston North, New Zealand, pp. 1379–1381.

Bush, L. P., Wilkinson, H. W. and Schardl, C. L. (1997) Bioprotective alkaloids of grass–fungal endophyte symbioses. *Plant Physiology* **114**: 1–7.

Carroll, G. C. (1988) Fungal endophytes in stems and leaves: from latent pathogen to mutualistic symbiont. *Ecology* **69**: 2–9.

Carroll, G. C. (1991) Beyond pest deterrence: alternative strategies and hidden costs of endophytic mutualisms in vascular plants. In *Microbial Ecology of Leaves*, ed. J. H. Andrews and S. S. Monano, pp. 358–378. New York: Springer-Verlag.

Carroll, G. C. (1992) Fungal mutualism. In *The Fungal Community: Its Organization and Role in the Ecosystem*, Mycology Series vol. 9, ed. G. C. Carroll and D. T. Wicklow. New York: M. Dekker.

Cheplick, G. P. and Clay, K. (1988) Acquired chemical defenses of grasses: the role of fungal endophytes. *Oikos* **52**: 309–318.

Cheplick, G. P., Clay, K. and Marks, S. (1989) Interactions between infection by endophytic fungi and nutrient limitation in the grasses *Lolium perenne* and *Festuca arundinacea*. *New Phytologist* **111**: 89–97.

Chu-Chou, M., Guo, B. Z., An, Z.-Q., Hendrix, J. W., Ferriss, R. S., Siegel, M. R., Dougherty, C. T. and Burrus, P. H. (1992) Suppression of mycorrhizal fungi in fescue by *Acremonium coenophialum* endophyte. *Soil Biology and Biochemistry* **24**: 633–637.

Chunlin, W., Gianfagna, T., Chee-Kok, C., Richardson, M. and Meyer, W. (1998) Analysis of volatile compounds from endophyte-infected turfgrasses. *Phytopathology* **88**: S94.

Clancy, K. M. and Price, P. W. (1987) Rapid herbivore growth enhances enemy attack: sublethal plant defenses remain a paradox. *Ecology* **68**: 736–738.

Clay, K. (1987a) The effect of fungi on the interactions between host plants and their herbivores. *Canadian Journal of Plant Pathology* **9**: 380–388.

Clay, K. (1987b) Effects of fungal endophyte on the seed and seedling biology of *Lolium perenne* and *Festuca arundinacea*. *Oecologia* **73**: 358–362.

Clay, K. (1988) Fungal endophytes of grasses: a defensive mutualism between plants and fungi. *Ecology* **69**: 10–16.

Clay, K. (1989) Clavicipitaceous endophytes of grasses: their potential as biocontrol agents. *Mycological Research* **92**: 1–12.

Clay, K. (1990) Fungal endophytes of grasses. *Annual Review of Ecology and Systematics* **21**: 275–297.

Clay, K. (1991) Fungal endophytes, grasses, and herbivores. In *Microbial Mediation of Plant–Herbivore Interactions*, ed. P. Barbosa, V. A. Krischik and C. G. Jones, pp. 199–226. New York: John Wiley.

Clay, K. (1992) Fungal endophytes of plants: biological and chemical diversity. *Natural Toxins* **1**: 147–149.

Clay, K. (1996) Fungal endophytes, herbivores, and the structure of grassland communities. In *Multitrophic Interactions in Terrestrial Systems*, ed. A. C. Gange, pp. 151–169. Oxford: Blackwell Science.

Clay, K. (1998) Fungal endophyte infection and the population biology of grasses. In *The Population Biology of Grasses*, ed. G. P. Cheplick, pp. 255–285. Cambridge: Cambridge University Press.

Clay, K. and Brown, V. K. (1997) Infection of *Holcus lanatus* and *H. mollis* by *Epichloë* in experimental grasslands. *Oikos* **79**: 363–370.

Clay, K. and Holah, J. (1999) Fungal endophyte symbiosis and plant diversity in successional fields. *Science* **285**: 1742–1744.

Clay, K., Hardy T. N. and Hammond, A. M., Jr. (1985) Fungal endophytes of grasses and their effects on an insect herbivore. *Oecologia* **66**: 1–6.

Clay, K., Marks, S. and Cheplick, G. P. (1993) Effects of insect herbivory and fungal endophyte infection on competitive interactions among grasses. *Ecology* **74**: 1767–1777.

Cleland, J. B. (1926) Plants poisonous to stock in Australia. In *Third Report of the Bureau of Microbiology*, pp. 187–217. Sydney, Australia: New South Wales Government.

Coley, A. B., Fribourg, H. A., Pelton, M. R. and Gwinn, K. D. (1995) Effects of tall fescue endophyte infestation on relative abundance of small mammals. *Journal of Environmental Quality* **24**: 472–475.

Cooper, C. F. (1960) Changes in vegatation, structure, and growth of southwestern pine forests since white settlement. *Ecological Monographs* **30**: 129–164.

Craig, T. P., Itomi, J. K. and Price, P. W. (1990) The window of vulnerability of a shoot galling sawfly to attack a parasitoid. *Ecology* **71**: 1471–1482.

Crawley, M. J. (1983) *Herbivory: The Dynamics of Animal–Plant Interactions*. Berkeley, CA: University of California Press.

Cross, D. L. (1997) Fescue toxicosis in horses. In *Proceedings of the 3rd International* Neotyphodium/*Grass Symposium*, Athens, GA, ed. C. W. Bacon, J. H. Bouton and N. S. Hill, pp. 289–309.

Dahlman, D. L., Eichenseer, H. and Siegel, M. R. (1991) Chemical perspectives of endophyte–grass interactions and their implications to insect herbivory. In *Microbial Mediation of Plant–Herbivore Interactions*, ed. P. Barbosa, V. A. Krischik and C. L. Jones, pp. 227–252. New York: John Wiley.

Damman, H. (1987) Leaf quality and enemy avoidance by larvae of a pyralid moth. *Ecology* **68**: 87–97.

Dawe, D. L., Studemann, J. A., Hill, N. S. and Thompson, F. N. (1997) Immunosuppression in cattle with fescue toxicosis. In *Proceedings of the 3rd International* Neotyphodium/*Grass Symposium*, Athens, GA, ed. C. W. Bacon, J. H. Bouton and N. S. Hill, pp. 411–412.

De Battista, J. P., Bouton, J. H., Bacon, C. W. and Siegel, M. R. (1990) Rhizome and herbage production of endophyte-removed tall fescue clones and populations. *Agronomy Journal* **82**: 652–654.

Dicke, M. (1994) Local and systemic production of volatile herbivore-induced terpenoids: their role in plant–carnivore mutualism. *Journal of Plant Physiology* **143**: 465–472.

Dicke, M. (1995) Why do plants "talk"? *Chemoecology* **5/6**: 159–165.

Dicke, M. (1999) Evolution of induced indirect defense of plants. In *The Ecology and Evolution of Inducible Defenses*, ed. R. Tollrian and C. D. Harval, pp.l 62–88. Princeton, NJ: Princeton University Press.

Elbersen, H. W. and West, C. P. (1996) Growth and water relations of field-grown tall fescue as influenced by drought and endophyte. *Grass and Forage Science* **51**: 333–342.

Elbersen, H. W., Buck, G. W., West, C. P. and Joost, R. E. (1994) Water loss from tall fescue leaves is decreased by endophyte. *Arkansas Farm Research* **43**: 8–9.

Elmi, A. A. and West, C. P. (1995) Endophyte infection effects on stomatal conductance, osmotic adjustment and drought recovery of tall fescue. *New Phytologist* **131**: 61–67.

Ewald, P. W. (1988) Cultural vectors, virulence, and the emergence of evolutionary epidemiology. *Oxford Surveys of Evolutionary Biology* **5**: 215–245.

Ewald, P. W. (1994) *Evolution of Infectious Disease*. Oxford: Oxford University Press.

Faeth, S. H. (1987) Community structure and folivorous insect outbreak: the roles of vertical and horizontal interactions. In *Insect Outbreaks*, ed. P. Barbosa and J. C. Schultz, pp. 135–171. New York: Academic Press.

Faeth, S. H. (1988) Plant-mediated interactions between seasonal herbivores: enough for evolution or coevolution? In *Chemical Mediation of Coevolution*, ed. K. C. Spenser, pp. 391–414. New York: Academic Press.

Faeth, S. H. (1991) Variable induced responses: direct and indirect effects on oak folivores. In *Phytochemical Induction by Herbivores*, ed. M. J. Raupp and D. W. Tallamy, New York: John Wiley.

Faeth, S. H. (1994) Induced plant responses: effects on parasitoids and other natural enemies of phytophagous insects. In *Parasitoid Community Ecology*, ed. B. A. Hawkins and W. S. Sheehan, pp. 245–260. Oxford: Oxford University Press.

Faeth, S. H. and Hammon, K. E. (1996) Fungal endophytes and the phytochemistry of oak foliage: determinants of oviposition preference of leafminers? *Oeocologia* **108**: 728–736.

Faeth, S. H. and Hammon, K. E. (1997a) Fungal endophytes in oak trees. I. Long-term patterns of abundances and associations with leafminers. *Ecology* **78**: 810–819.

Faeth, S. H. and Hammon, K. E. (1997b) Fungal endophytes in oak trees II. Experimental analyses of interactions with leafminers. *Ecology* **78**: 820–827.

Faeth, S. H. and Wilson, D. (1996) Induced responses in trees: mediators of interactions among macro- and micro-herbivores? In *Multitrophic Interactions in Terrestrial Systems*, ed. A. C. Gange and V. K. Brown, pp. 201–215. Blackwell Science.

Feeny, P. (1976) Plant apparency and chemical defenses. In *Biochemical Interaction between Plants and Insects*, ed. J. S. Wallace and R. L. Mansell, pp. 1–40. New York: Plenum Press.

Filipov, N. M., Thompson, F. N., Stuedemann, J. A., Elsasser, T. H., Kahl, S., Kahl, R. P., Sharma, R. P., Young, C. R., Stanker, L. H. and Smith, C. K. (1999) Increased responsiveness to intravenous lipopolysaccharide (LPS) challenge in steers grazing endophyte-infected tall fescue compared with steers grazing endophyte-free tall fescue. *Journal of Endocrinology* **163**: 213–220.

Frank, S. A. (1994) Genetics of mutualisms: the evolution of altruism between species. *Journal of Theoretical Biology* **170**: 393–400.

Freeman, E. M. (1904) The seed fungus of *Lolium temulentum* L., the darnel. *Philosophical Transactions of the Royal Society of London (Biology)* **196**: 1–27.

Gange, A. C. (1996) Positive effects of endophytic infections in sycamore aphids. *Oikos* **75**: 500–510.

Gay, N., Dew, R. H., Boissonneault, G. A., Boling, J. A., Cross, R. J. and Cohen, D. A. (1990) Effects of endophyte infected tall fescue on growth and immune functions in the rate, mice and bovine. *Beef Cattle Research Reports* **326**: 44–47.

Gehring, C. A. and Whitham, T. G. (1994) Interactions between above-ground herbivores and the mycorrhizal mutualists of plants. *Trends in Ecology and Evolution* **9**: 251–255.

Gehring, C. A., Cobb, N. S. and Whitham, T.G. (1997) Three-way interactions among ectomycorrhizal mutualists, scale insects, and resistant and susceptible pinyon pines. *American Naturalist* **149**: 824–841.

Glenn, A. E., Bacon, C. W., Price, R. and Hanlin, R. T. (1996) Molecular phylogeny of *Acremonium* and its taxonomic implications. *Mycologia* **88**: 369–383.

Goldson, S. L., Gou, B. Z., Hendrix, J. W., An, Z.-Q. and Ferriss, R. S. (1992) Role of *Acremonium* endophyte of fescue on inhibition of colonization and reproduction of mycorrhizal fungi. *Mycologia*, **84**: 882–885.

Goldson, S. L., Proffitt, J. R., Fletcher, L. R. and Baird, D. B. (2000) Multitrophic interaction between the ryegrass *Lolium perenne*, its endophyte *Neotyphodium lolii*, the weevil pest *Listronotus bonariensis*, and its parasitoid *Microctonus hyperodae*. *New Zealand Journal of Agricultiural Research* **43**: 227–233.

Grewal, S. K., Grewal, P. S. and Gaugler, R. (1995) Endophytes of fescue grasses enhance susceptibility of *Popillia japonica* larvae to an entomophagous nematode. *Entomologia Experimentalis et Applicata* **74**: 219–224.

Gwinn, K. D. and Bernard, E. C. (1990) Nematode reproduction on endophyte-infected and endophyte-free tall fescue. *Plant Disease* **74**: 57–761.

Hacker, S. D. and Gaines, S. D. (1997) Some implications of direct positive interactions for community species diversity. *Ecology* **78**: 1990–2003.

Hance, H. F. (1876) On a Mongolian grass producing intoxication in cattle. *Journal of Botany* **14**: 210–212.

Hardy, T., Clay, K. and Hammond, A. M., Jr. (1985) Fall armyworm (Lepidoptera: Noctuidae): a laboratory bioassy and larval performance study for the fungal endophyte of perennial ryegrass. *Journal of Economic Entomology* **78**: 571–574.

Hardy, T., Clay, K. and Hammond, A. M., Jr. (1986) Leaf age and related factors affecting endophyte-mediated resistance to fall armyworm (Lepidoptera: Noctuidae) in tall fescue. *Environmental Entomology* **15**: 1083–1089.

Hatcher, P. E. (1995) Three-way interactions between plant pathogenic fungi, herbivorous insects and their host plants. *Biological Review* **70**: 639–694.

Hazlett, W. D., Lester, T. L. and Rorie, R. W. (1998) Influence of endophyte-infected fescue on serum and intra-uterine insulin growth factor I and II in beef heifers. *Journal of Animal Science* **76** (Suppl. 1): 230.

Hill, N. S., Stringer, W. C., Rottinghaus, G. E., Belesky, D. P., Parrott, W. A. and Pope, D. D. (1990) Growth, morphological, and chemical component responses of tall fescue to *Acremonium coenophialum*. *Crop Science* **30**: 156–161.

Hochachka, P. W. and Somero, G. N. (1984) *Biochemical Adaptations*. Princeton, NJ: Princeton University Press.

Hoveland, C. S. (1993) Importance and economic significance of the *Acremonium* endophytes to performance of animals and the grass plant. *Agriculture, Ecosystems, Environment* **44**: 3–12.

Johnson, T.M. and Gould, F. (1992) Interaction of genetically engineered host plant resistance and natural enemies of *Heliothus virescens* (Lepidoptera: Noctuidae) in tobacco. *Environmental Entomology* **21**: 586–597.

Jones, T.A., Ralphs, M. H., Gardner, D. R. and Chatterton, N. J. (2000) Cattle prefer endophyte-free robust needlegrass. *Journal of Range Management* **53**: 427–431.

Kaiser, W. J., Breuhl, G. W., Davitt, C. M. and Klein, R. E. (1996) *Acremonium* isolates from *Stipa robusta*. *Mycologia* **88**: 539–547.

Karban, R. and Baldwin, I. T. (1997) *Induced Responses to Herbivory*. Chicago, IL: University of Chicago Press.

Kauffman, W. C. and Flanders, R. V. (1985) Effects of variably resistant soybean and lima bean cultivars on *Pediobius foveolatus* (Hymenoptera: Eulophidae), a parasitoid of the Mexican bean beetle, *Epilachna varivestis* (Coleoptera: Coccinellidae). *Environmental Entomology* **14**: 678–682.

Kelrick, M. I., Kasper, N. A. Bultman, T. L. and Taylor, S. (1990) Direct interactions between infected and uninfected individuals of *Festuca arundinacea*: differential allocation to shoot and root and shoot biomass. In *Proceedings of the International Symposium on* Acremonium/*Grass Interactions*, ed. S. S. Quisenberry, R. E. Kimmons and C. A. Joost, pp. 21–29. Baton Rouge, LA: Louisiana Agricultural Experiment Station.

Kimmons, C. A., Glenn, K. D. and Bernard, E. C. (1990) Nematode reproduction on endophyte-infected and endophyte-free tall fescue. *Plant Disease* **74**: 757–761.

Knoch, T. R., Faeth, S. H. and Arnott, D. L. (1995) Endophytic fungi alter foraging and dispersal by desert seed-harvesting ants. *Oecologia* **95**: 470–475.

Latch, G. C. M. (1993) Physiological interactions of endophytic fungi and their hosts: biotic stress tolerance imparted to grasses by endophytes. *Agriculture, Ecosystems, Environment* **44**: 143–156.

Law, R. (1985) Evolution in a mutualistic environment. In *The Biology of Mutualisms*, ed. D. H. Boucher, pp. 145–170. London: Croom Helm.

Leuchtmann, A. (1992) Systematics, distribution, and host specificity of grass endophytes. *Natural Toxins* **1**: 150–162.

Leuchtmann, A. and Clay, K. (1990) Isozyme variation in the *Acremonium/Epichloë* fungal endophyte complex. *Phytopathology* **80**: 1133–1139.

Leuchtmann, A. and Clay, K. (1997) The population biology of grass endophytes. In *The Mycota*, vol. 5, *Plant Relationships*, part B, ed. G. C. Carroll and P. Tudzynski, pp. 185–204. Berlin, Germany: Springer-Verlag.

Leuchtmann, A., Schmidt, D. and Bush, L. P. (2000) Different levels of protective alkaloids in grasses with stroma-forming and seed-transmitted *Epichloë/Neotyphodium* endophytes. *Journal of Chemical Ecology* **26**: 1025–1036.

Lewis, D. H. (1985) Symbiosis and mututalism: crisp concepts and soggy semantics. In *The Biology of Mutualisms*, ed. D. H. Boucher, pp. 29–39. London: Croom Helm.

Li, B., Zheng, X. and Suichang, S. (1997) Survey of endophytic fungi in some native forage grasses of Northwestern China. In *Proceedings of the 3rd International* Neotyphodium/*Grass Interactions Symposium*, Athens, GA, ed. C. W. Bacon, J. H. Boulton and N. S. Hill, pp. 69–72.

Loader, C. and Damman, H. (1991) Nitrogen content of food plants and vulnerability of *Pieris rapae* to natural enemies. *Ecology* **72**: 1586–1590.

Lopez, J. E., Faeth, S. H. and Miller, M. (1995) The effect of endophytic fungi on herbivory by redlegged grasshoppers (Orthoptera: Acrididae) on Arizona fescue. *Environmental Entomology* **24**: 1576–1580.

Louda, S. M. (1983) Seed predation and seedling mortality in the recruitment of a shrub, *Haplopappus venetus* (Asteraceae) along a climatic gradient. *Ecology* **64**: 511–521.

MacFarlane, T. D. (1988) Poaceae: Subfamily Pooideae. In *Grass Systematics and Evolution*, ed. T. R. Soderstorm, K. W. Hilu, C. S. Campbell and M. E. Barkworth, pp. 263–276. Washington, DC: Smithsonian Institution Press.

Malinowski, D. and Belesky, D. P. (1999) *Neotyphodium coenophialum* infection affects the ability of tall fescue to use sparingly available phosphorous. *Journal of Plant Nutrition* **22**: 835–853.

Malinowski, D., Alloush, G. A. and Belesky, D. P. (1998) Evidence for chemical changes on the root surface of tall fescue in response to infection with the fungal endophyte *Neotyphodium coenophialum*. *Plant and Soil* **205**: 1–12.

Marks, S., Clay, K. and Cheplick, C. P. (1991) Effects of fungal endophytes on interspecific and intraspecific competition in the grasses *Festuca arundinacea* and *Lolium perenne*. *Journal of Applied Ecology* **28**: 194–204.

Marlatt, M. L., West, C. P., McConnell, M. E., Sleper, M. E., Buck, D. A., Correll, S. C. and Saida, S. (1997) Investigations on xeriphytic *Festuca* spp. from Morocco and their associated endophytes. In *Proceedings of the 3rd International* Neotyphodium/*Grass Interactions Symposium*, Athens, GA, ed. C. W. Bacon, J. H. Boulton and N. S. Hill, pp. 73–76.

Marquis, R. J. (1992) The selective impact of herbivores. In *Plant Resistance to Herbivores and Pathogens*, ed. R. S. Fritz and E. L. Simms, pp. 301–325. Chicago, IL: University of Chicago Press.

Marquis, R. J. and Alexander, H. M. (1992) Evolution of resistance and virulence in plant–herbivore and plant–pathogen interactions. *Trends in Ecology and Evolution* **7**: 126–129.

Marsh, C. D. and Clawson, A. B. (1929) *Sleepy Grass* (Stipa vaseyi) *as a Stock-Poisoning Plant*. USDA Technical Bulletin no. 114. Washington, DC: USDA.

Marshall, D., Tunali, B. and Nelson, L. R. (1999) Occurrence of fungal endophytes in species of wild *Triticum*. *Crop Science* **39**: 1507–1512.

Massad, E. (1987) Transmission rates and the evolution of pathogenicity. *Evolution* **41**: 1127–1130.

Miles, C. O., Lane, G. A., Garthwaite, I., Piper, E. L., Ball, O. J.-P., Latch, G. C. M., Allen, J. M., Hunt, M. B., Min, F. K., Fletcher, I. and Harris, P. S. (1996) High levels of ergonovine and lysergic acid amide in toxic *Achnatherum inebrians* accompany infection by an *Acremonium*-like endophytic fungus. *Journal of Agricultural Food Chemistry* **44**: 1285–1290.

Miles. C. O., DiMenna, M. E., Jacons, S. W. L., Garthwaite, I., Lane, G. A., Prestidge, R. A., Marshall, S. L., Wilkinson, H. W., Schardl, C. L., Ball, O. J.-P. and Latch, G. C. M. (1998) Endophytic fungi in indigenous Australasian grasses associate with toxicity to livestock. *Applied and Environmental Microbiology* **64**: 601–606.

Moon, C. D., Scott, B., Schand, C. C. and Christensen, M. J. (2000) Evolutionary origins of *Epichloë* endophytes from annual ryegrasses. *Mycologia* **92**: 1103–1118.

Muldrew, J. A. (1953) The natural immunity of the larch sawfly (*Pristiphora erichsonii*

[Htg.]) to the introduced parasite *Mesoleius tenthredinis* Morley, in Manitoba and Saskatchewan. *Canadian Journal of Zoology* **31**: 313–332.

Nan, Z. B. and Li, C. J. (2001) *Neotyphodium* in native grasses in China and observations on endophyte/host interactions. In *4th International Symposium on* Neotyphodium/*Grass Interactions*, ed. P. Dapprich and V. H. Paul, pp. 41–50.

Nordland, D. A., Lewis, W. J. and Altieri, M. A. (1988) Influences of plant-produced allelochemicals on the host/prey selection behavior of entomophagous insects. In *Novel Aspects of Insect–Plant Interactions*, ed. P. Barbosa and D. K. Letourneau, pp. 65–90. New York: John Wiley.

Ormacini, M., Chaneton, E. J., Ghersa, C. M. and Muller, C. B. (2000) Symbiotic fungal endophytes control insect host–parasite interaction webs. *Nature* **409**: 78–81.

Paige, K. N. and Whitham, T. G. (1987) Overcompensation in response to mammalian herbivory: the advantages of being eaten. *American Naturalist* **129**: 407–416.

Parodi, L. R. (1950) Las gramineas toxicas para el ganado en la republica. *Revista Argentina Agricultura* **17**: 164–227.

Petrini, O., Sieber, T. H., Toti, L. and Viret, O. (1992) Ecology, metabolite production, and substrate utilization in endophytic fungi. *Natural Toxins* **1**: 185–196.

Petroski, R. J., Powell, R. G. and Clay, K. (1992) Alkaloids of *Stipa robusta* (sleepygrass) infected with an *Acremonium* endophyte. *Natural Toxins* **1**: 84–88.

Pickett, J. A., Powell, W., Wadhams, L. J. Woodcock, C. M. and Wright, A. F. (1991) Biochemical interactions between plant–herbivore–parasitoid. *Redia* **74**: 1–14.

Piper, E. L. and West, C. P. (1993) The role of endophytes in tall fescue. *Proceedings of the Arkansas Academy of Science* **47**: 89–92.

Popay, A. J. and Rowan, D. D. (1994) Endophytic fungi as mediators of plant–insect interactions. In *Insect–Plant Interactions*, vol. 5, ed. E. A. Bernays, pp. 83–103. Boca Raton, FL: CRC Press.

Powell, R. G. and Petroski, R. J. (1992) Alkaloid mycotoxins in endophyte-infected grasses. *Natural Toxins* **1**: 163–170.

Preszler, R. W. Gaylord, E. S. and Boecklen, W. J. (1996) Reduced parasitism of a leaf-mining moth on trees with high infection frequencies of an endophytic fungus. *Oecologia* **108**: 159–166.

Price, P. W., Bouton, C. E., Cross, P., McPheron, B. A., Thompson, J. N. and Weis, A. E. (1980) Interactions among three trophic levels: influence of plants on interactions between insect herbivores and natural enemies. *Annual Review of Ecology and Systematics* **11**: 41–65.

Rambo, J. L. and Faeth, S. H. (1999) The effect of vertebrate grazing on plant and insect community structure. *Conservation Biology* **13**: 1047–1054.

Ravel, C., Michalakis, Y. and Charmet, G. (1997) The effect of imperfect transmission on the frequency of mutualistic seed-borne endophytes in natural populations of grasses. *Oikos* **80**: 18–24.

Read, J. C. and Camp, B. J. (1986) The effect of the fungal endophyte *Acremonium coenophialum* in tall fescue on animal performance, toxicity, and stand maintenance. *Agronomy Journal* **78**: 848–850.

Reber, P. M. (1993) Prolactin and immunomodulation. *American Journal of Medicine* **95**: 637–644.

Reddick, B. B. and Collins, M. H. (1988). An improved method for detection of *Acremonium coenophialum* in tall fescue plants. *Phytopathology* **78**: 418–420.

Richardson, M.D., Chapman, G. W., Hoveland, C. S. and Bacon, C. W. (1992) Sugar alcohols in endophyte-infected tall fescue under drought. *Crop Science* **32:** 1060–1061.

Richardson, M.D., Hoveland, C. S. and Bacon, C. W. (1993) Photosynthesis and stomatal conductance of symbiotic and nonsymbiotic tall fescue. *Crop Science* **33:** 145–149.

Saikkonen, K. (2000) Kentucky 31, far from home. *Science* **287:** 1887a.

Saikkonen, K., Faeth, S. H., Helander, M. and Sullivan, T. J. (1998) Fungal endophytes: a continuum of interactions with host plants. *Annual Review of Ecology and Systematics* **29:** 319–343.

Saikkonen, K., Helander, M., Faeth, S. H., Schulthess, F. and Wilson, D. (1999) Endophyte–grass–herbivore interactions: the case of *Neotyphodium* endophytes in Arizona fescue populations. *Oecologia* **121:** 411–420.

Saikkonen, K., Ahlholm, J., Helander, M., Lehtimäki, S. and Niemaläinen, O. (2001) Endophytic fungi in wild and cultivated grasses in Finland. *Ecography* **23:** 360–366.

Schardl, C. L. and Clay, K. (1997) Evolution of mutualistic endophytes from plant pathogens. In *The Mycota*, vol. 5, *Plant Relationships*, part B, ed. G. C. Carroll and P. Tudzynski, pp. 221–238. Berlin, Germany: Springer-Verlag.

Schardl, C. L. and Leuchtmann, A. (1999) Three new species of *Epichloë* symbiotic with North American grasses. *Mycologia* **91:** 95–107.

Schardl, C. L. and Phillips, T. D. (1997) Protective grass endophytes: where are they from and where are they going? *Plant Disease* **81:** 430–438.

Schardl, C. L. and Tsai, H.-F. (1992) Molecular biology and evolution of the grass endophytes. *Natural Toxins* **1:** 171–184.

Schardl, C. L., Leuchtmann, A., Tsai, H.-F., Collett, M. A., Watt, D. M. and Scott, D. B. (1994) Origin of a fungal symbiont of perennial ryegrass by interspecific hybridization of a mutualist with ryegrass choke pathogen, *Epichloë typhina*. *Genetics* **136:** 1307–1317.

Schardl, C. L., Leuchtmann, A., Chung, K.-R., Penny, D. and Siegel, M. R. (1997) Coevolution by common descent of fungal symbionts (*Epichloë* spp.) and grass hosts. *Molecular Biology and Evolution* **14:** 133–143.

Schulthess, F. M. and Faeth, S. H. (1998) Distribution, abundances, and associations of the endophytic fungal community of Arizona fescue (*Festuca arizonica*). *Mycologia* **90:** 569–578.

Siegel, M. R. (1985) *Acremonium* fungal endophytes of tall fescue and perennial ryegrass: significance and control. *Plant Disease* **69:** 179–183.

Siegel, M. R., and Bush, L. P. (1996) Defensive chemicals in grass–fungal endophyte associations. *Recent Advances in Phytochemistry* **30:** 81–118.

Siegel, M. R., and Bush, L. P. (1997) Toxin production in grass/endophyte association. In *The Mycota*, vol. 5, *Plant Relationships*, Part A, ed. G. C. Carroll and P. Tudzynski, pp. 185–208. Berlin, Germany: Springer-Verlag.

Siegel, M. R. and Latch, G. C. M. (1987) Fungal endophytes of grasses. *Annual Review of Phytopathology* **25:** 293–315.

Siegel, M. R. and Schardl, C. L. (1992) Fungal endophytes of grasses: detrimental and beneficial associations. In *Microbial Ecology of Leaves*, ed. J. H. Andrews and S. S. Hirano, pp. 198–221. New York: Springer-Verlag.

Siegel, M. R., Varney, D. R., Johnson, M. C., Nesmith, W. C. Buckner, R. C., Bush, L. P., Burris, P. B. II and Hardison, J. R. (1984) A fungal endophyte of tall fescue: evaluation of control methods. *Phytopathology* **74:** 937–941.

Siegel, M. R., Dahlman, D. L. and Bush, L. P. (1989) The role of endophytic fungi in grasses: new approaches to biological control of pests. In *Integrated Pest Management for Turfgrass and Ornamentals*, ed. A. R. Leslie and R. L. Metcalf, pp. 169–186. Washington, DC: USEPA.

Siegel, M. R., Latch, G. C. M., Bush, L. P., Fannin, F. F., Rowan, D. D., Tapper, B. A., Bacon, C. W. and Johnson, M. C. (1990) Fungal endophyte-infected grasses: alkaloid accumulation an aphid response. *Journal of Chemical Ecology* **16**: 3301–3315.

Stone, J. K. and Petrini, O. (1997) Endophytes of forest trees: a model for fungus–plant interactions. In *The Mycota*, vol. 5, *Plant Relationships*, part B, ed. G. C. Carroll and P. Tudzynski, pp. 129–142. Berlin, Germany: Springer-Verlag.

Strauss, S. Y. and Agrawal, A. A. (1999) The ecology and evolution of plant tolerance to herbivory. *Trends in Ecology and Evolution* **14**: 179–185.

Taylor, R. J. (1991) Plant fungi, and bettongs: a fire-dependent co-evolutionary relationship. *Australian Journal of Ecology* **16**: 409–411.

Thompson, F. N., Stuedemann, J. A., and Sartin, J. L., Jr. (1987) Selected hormonal changes with summer fescue toxicosis. *Journal of Animal Science* **65**: 727–733.

Tibbets, T. M. and Faeth, S. H. (1999) *Neotyphodium* endophytes in grasses: deterrents or promoters of herbivory by leaf-cutting ants? *Oecologia* **118**: 297–305.

Tsai, H.-F., Liu, J. S., Christensen, M. J., Latch, G. C. M., Siegel, M. R. and Schardl, C. L. (1994) Evolutionary diversification of fungal endophytes of tall fescue grass by hybridization with *Epichloë* species. *Proceedings of the National Academy of Sciences, USA* **91**: 2542–2546.

Turlings, T. C. J., Tumlinson, J. H. and Lewis, W. J. (1990) Exploitation of herbivore-induced plant odors by host-seeking parasitic wasps. *Science* **250**: 1251–1253.

Turlings, T. C. J., Tumlinson, J. H., Eller, F. J. and Lewis, W. J. (1991) Larval-damaged plants: source of volatile synomones that guide the parasitoid *Cotesia marginiventris* to the micro-habitat of its hosts. *Entomologia Experimentalis et Applicata* **58**: 78–82.

Turlings, T. C. J., Loughrin, J. H., McCall, P. J., Rose, U. S. R., Lewis, W. J. and Tumlinson, J. H. (1995) How caterpillar-damaged plants protect themselves by attracting parasitic wasps. *Proceedings of the National Academy of Sciences, USA* **92**: 4169–4174.

van den Bosch, R. (1964) Encapsulation of the eggs of *Bathyplectes curculionis* (Thomson) (Hymenoptera; Ichneumonidae) in larvae of *Hypera brunneipennis* (Boheman) and *Hypera postica* (Gyllenhal) (Coleoptera: Curculionidae). *Journal of Insect Pathology* **6**: 343–367.

Vet, L. E. M. and Groenewold, A. W. (1990) Semiochemicals and learning in parasitoids. *Journal of Chemical Ecology* **16**: 3119–3135.

Vinson, S. B. (1976) Host selection by insect parasitoids. *Annual Review of Entomology* **21**: 109–138.

Vinson, S. B. and Barbosa, P. (1987) Interrelationships of nutritional ecology of parasitoids. In *Nutritional Ecology of Insects, Mites, and Spiders and Related Invertebrates*, ed. F. Slansky and J. G. Rodriguez, pp. 673–695. New York: John Wiley.

Vinton, M. A., Kathol, E. S., Vogel, K. P. and Hopkins, A. A. (in press) Endophytic fungi in Canada wild rye: widespread occurrence in natural grasslands in the central United States. *Journal of Range Management*.

Vogl, A. (1898) Mehl und die anderen Mehlprodukte der Cerealien und Leguminosen. *Zeitschrift für Nahrungsmittel Untersuchung. Hygiene Wahrenkunde* **12**: 25–29.

Wang, C., Gianfagna, T., Chin, E., Richardson, M. and Meyer, W. (1998). Analysis of volatile compounds from endophyte-infected turfgrasses. *Phytopathology* **88**(9) Supplement: 594.
Welty, R. E., Craig, A. M. and Azevedo, M. D. (1994) Variability of ergovaline in seeds and straw and endophyte infection in seeds among endophyte-infected genotypes of tall fescue. *Plant Disease* **78**: 845–849.
Weseloh, R. M. (1984) Effects of the feeding inhibitor Plictran and low *Bacillus thuringiensis* Berliner doses on *Lymantria dispar* (L.) (Lepidoptera: Lymantridae): implications for *Cotesia melanoscelus* (Ratzeburg) (Hymenoptera: Braconidae). *Environmental Entomology* **13**: 1371–1376.
West, C. P. (1994) Physiology and drought tolerance of endophyte-infected grasses. In *Biotechnology of Endophytic Fungi of Grasses*, ed. C. W. Bacon and J. F. White, Jr., pp. 87–99. Boca Raton, FL: CRC Press.
West, C. P., Izekor, E., Oosterhuis, D. M. and Robbins, R. T. (1988) The effect of *Acremonium coenophialum* on the growth and nematode infestation of tall fescue. *Plant Soil* **112**: 3–6.
West, C. P., Izekor, E., Turner, K. E. and Elmi, A. A. (1993) Endophyte effects on growth and persistence of tall fescue along a water-supply gradient. *Agronomy Journal* **85**: 264–270.
West, C. P., Elberson, H. W., Elmi, A. A. and Buck, G. W. (1995) *Acremonium* effects on tall fescue growth: parasite or stimulant? In *Proceedings of the 50th Southern Pasture and Forage Crop Improvement Conference*, pp. 102–111.
White, J. F., Jr. (1987) Widespread distribution of endophytes in the Poaceae. *Plant Disease* **71**: 340–342.
White, J. F., Jr. (1988) Endophyte–host associations in forage grasses XI. A proposal concerning origin and evolution. *Mycologia* **80**: 442–446.
White, J. F., Jr., Glenn, A. E. and Chandler, K. F. (1993a) Endophyte–host associations in grasses XVIII. Moisture relations and insect herbivory of the emergent stromal leaf of *Epichloë*. *Mycologia* **85**: 195–202.
White, J. F., Jr., Morgan-Jones, G. and Morrow, A. C. (1993b) Taxonomy, life cycle, reproduction and detection of *Acremonium* endophytes. *Agriculture, Ecosystems, Environment* **44**: 13–37.
White, J. F., Jr. Martin, T. I. and Cabral, D. (1996) Endophyte–host associations in grasses XXII. Conidia formation by *Acremonium* endophytes on the phylloplanes of *Agrostis hiemalis* and *Poa rigidifolia*. *Mycologia* **88**: 174–178.
Whitman, D. W. (1988) Allelochemical interactions among plant, herbivores, and their predators. In *Novel Aspects of Insect–Plant Interactions*, ed. P. Barbosa and D. K. Letourneau, pp. 11–64. New York: John Wiley.
Wilkinson, H. H. and Schardl, C. L. (1997) The evolution of mutualism in grass–endophyte associations. In Neotyphodium/*grass interactions*, ed. C. W. Bacon and N. S. Hill, pp. 13–26. New York: Plenum Press.
Wilkinson, H. H., Siegel, M. R., Blankenship, J. D., Mallory, A. C., Bush, L. P. and Schardl, C. L. (2000) Contribution of fungal loline alkaloids to protection from aphids in an endophyte–grass mutualism. *Molecular Plant–Microbe Interactions* **13**: 1027–1033.
Wilson, D. and Carroll, G. C. (1997) Avoidance of high-endophyte space by gall-forming insects. *Ecology* **78**: 2153–2163.
Wolock-Madej, C. and Clay, K. (1991) Avian seed preference and weight loss experiment: the role of fungal-infected fescue seeds. *Oecologia* **88**: 296–302.

SASKYA VAN NOUHUYS AND ILKKA HANSKI

6

Multitrophic interactions in space: metacommunity dynamics in fragmented landscapes

Introduction

The distribution and abundance of consumers are necessarily limited by the distribution and abundance of their resources. With the exception of obligate mutualisms, a species at a higher trophic level in a food chain will occupy a subset of the locations occupied by species at lower trophic levels (Holt, 1995, 1997). In order to persist, species at higher trophic levels must be able to colonize, at a sufficiently high rate, sites occupied by populations of the lower trophic level species. The interaction may be spatially dynamic in both directions because species at any trophic level may influence the dynamics of one another. Once the interacting species occur as local populations in a shared habitat patch, phenomena traditionally addressed by studies of multitrophic interaction take place. Thus for species living in fragmented landscapes it is critical to keep in mind both processes occurring at large spatial scales and those occurring within a single habitat patch or local population. This chapter is about the interplay between spatial dynamics and multitrophic level interactions.

Species involved in a trophic interaction, such as the interaction between a predator and its prey, are influenced directly and indirectly by the trophic levels above and below them. The indirect effect of a non-adjacent trophic level can be either positive or negative. For example, herbivorous hosts may be concealed from (Weis and Abrahamson, 1985; Hawkins *et al.*, 1990) or exposed to (Price *et al.*, 1980; Walde, 1995a; Turlings *et al.*, 1995, chapter 7, this volume; Thaler, 1999) a foraging parasitoid by attributes of their food plant. Similarly, attributes of a herbivore and/or its food plant may protect a parasitoid from hyperparasitism (Weis and Abrahamson, 1985; Yeargan and Braman, 1989) or alternatively, increase susceptibility of a parasitoid to hyperparasitism (Singh and Srivastava, 1988; van Baarlen *et al.*, 1996; Sullivan and Völk, 1999). Within communities, the indirect effects of the lower and upper trophic levels can either increase or decrease the stability of populations, by moderating the use of a potentially limiting resource or by facilitating the consumption of the resource until it has gone locally extinct.

The primary focus of the study of multitrophic interactions is to analyze and comprehend the attributes of organisms that influence non-adjacent trophic levels either directly or indirectly. One such attribute of an organism is its spatial distribution (dispersion), both within and among habitat patches suitable for occupancy. Distribution is generally viewed as the outcome of many ecological processes, and is influenced by the many ecological factors traditionally labeled as habitat requirements and niche. We can also consider distribution as another attribute of an organism, which potentially affects its trophic interactions as much as, or even more than its chemical makeup, sensory perception, phenology, growth rate, and other such factors. The addition of the spatial structure of the landscape and spatial population dynamics to the study of multitrophic interactions is becoming an increasingly necessary consideration with increasing fragmentation of many natural environments.

Expansion of the study of multitrophic interactions to include space introduces two complementary ecological phenomena. First is the extent to which trophic interactions among individuals taking place at the scale of local populations (e.g. foraging behavior, prey preference, and density-dependent behaviors) might affect the dynamics at the regional or metapopulation scale. Second is how large-scale population dynamics, such as migration among populations and extinction–colonization dynamics, might affect local multitrophic interactions.

In the following section, we briefly outline the theory of multitrophic interactions in fragmented landscapes, which is essentially the theory of metapopulation dynamics (for a review see Hanski, 1999) extended to several interacting species. We then discuss selected empirical findings from the literature that illustrate the range of questions asked by ecologists. The rest of this chapter is devoted to a more detailed analysis of multitrophic interactions in a community of two host plant species, one herbivorous insect, two primary parasitoids and two hyperparasitoids, which occurs in a highly fragmented landscape and which we and others have studied over the past several years as an example of a small metacommunity.

Brief overview of theory

A metapopulation is an assemblage of locally breeding conspecific populations that are connected via migration (Hanski and Gilpin, 1997). The viability of a classical metapopulation, with no extinction-resistant "mainland" populations (Harrison, 1991), depends on the rates of local extinction and colonization, and on the degree of asynchrony in local population dynamics (Hanski, 1998). Interacting species, each with their own spatial population structure, may persist as single populations, patchy populations, or as metapopulations (Harrison and Taylor, 1997) in the same fragmented landscape, primarily depending on the scale of migration and hence the degree of mixing of neighboring populations. For example, a single patchy plant population might support a herbivore metapopulation, which might support a relatively continuous parasitoid population (a mobile species), which in turn might support a hyperparasitoid metapopulation (a more sedentary species). The stability of the entire system would clearly depend on the spatial population structures of each species (Taylor, 1988, 1991).

The natural theoretical framework to consider the spatial dimension of multitrophic interactions is to expand single-species metapopulation models to several interacting species. Robert Holt (1995, 1997) in particular has developed such a theory with simple patch occupancy models, extending the previously studied two-species competition (Levins and Culver, 1971; Slatkin, 1974; Hanski, 1983, 1999; Nee and May, 1992; Nee *et al.*, 1997) and predator–prey models (Taylor, 1991; May, 1994; Harrison and Taylor, 1997; Nee *et al.*, 1997; Hassell, 2000) to three or more species. The key assumptions made by Holt (1997) are that the food chain in a particular habitat patch is built up via sequential colonization, and that the extinction of a prey population automatically leads to the extinction of the predator population (and, naturally, any species at even higher trophic levels). The most noteworthy simplification of the models is the presence–absence description of local populations, common to all patch occupancy models (Hanski, 1999). Structured models, involving a description of local dynamics as well as of metapopulation dynamics, have been constructed and analyzed for two species at most (e.g., Reeve, 1988; Hassell *et al.*, 1991; Rohani *et al.*, 1996).

A basic conclusion emerging from the models is that metapopulation dynamics can constrain the length of specialist food chains in fragmented landscapes, that is, species located at higher trophic levels may not be able to persist in a landscape where the species at lower trophic levels are specialized to an uncommon habitat or have a restricted distribution for other reasons. In certain situations, alternative stable states may occur, such that an intermediate predator can only occur in the presence of the top predator (Holt, 1997).

In heterogeneous fragmented landscapes, with more than one kind of habitat patch present, species may persist either by being specialists on one patch type with low extinction and/or high colonization rate, or by being generalists and thereby having access to a larger number of habitat patches, which by itself facilitates colonization. Different species in a multitrophic interaction may exhibit different degrees of specialization. For instance, a predator may use two alternative prey species each specializing in a different habitat type, and thereby the predator population generates an indirect interaction between the prey species in a mosaic of the two types of habitat (Holt, 1997). It is clear that the complexities that one may build up with such considerations for multitrophic interactions are considerable – not only do we have a web of interspecific interactions but the structure of that web may be critically modified by the web of spatial interactions among the species.

Empirical studies of habitat fragmentation and multitrophic interactions

There have been very few if any studies in which the classical metapopulation processes have been shown clearly to mediate the regional coexistence of interacting species (the most convincing example is Holyoak and Lawler, 1996; for a review see Harrison and Taylor, 1997). This is likely because few studies of multitrophic level interactions are conducted at a large spatial scale over many generations, and because a great deal has to be understood about the habitat and the biology of each species in order to draw conclusions about the relative contributions of large-scale and small-scale factors to the stability of populations of interacting species. In spite of the complexity inherent in simultaneously addressing individual (local) interaction, space, and time, a handful of empirical studies of multitrophic interactions have explicitly considered the large-scale spatial distribution of interacting species (Hopper, 1984; Kareiva, 1987; Roland, 1993; Walde, 1995b; Holyoak and Lawler, 1996; Roland and Taylor, 1997; Tscharntke *et al.*, 1998; Lei and Camara, 1999; Komonen *et al.*, 2000; Kruess and Tscharntke, 2000). These studies compare the potentially critical role of spatial scale, habitat fragmentation, and species' dispersal behavior among species at different trophic levels, or among different species at the same trophic levels.

Several recent studies compare the relative impact of habitat fragmentation for herbivorous insects (or other lower trophic levels) and their natural enemies (Roland and Taylor, 1995; Jones *et al.*, 1996; Lei and Hanski, 1997; Roland, 1998; Tscharntke *et al.*, 1998; Komonen *et al.*, 2000; Kruess and Tscharntke, 2000; case study, this chapter). These studies primarily show, in agreement with theoretical expectation (Holt, 1997), that species at higher trophic levels suffer more than species at lower trophic levels from a decrease in habitat patch size and an increase in patch isolation, the two primary consequences of habitat fragmentation. This is not surprising because in a dynamic system species at each trophic level can only occur in a subset of the locations in which their host is found, so a fragmented habitat is ever more fragmented at higher trophic levels.

The effect of fragmentation naturally depends on the spatial scale relative to the migration range of each species. Entire insect communities associated with unpredictable host plants may be well adapted to fragmented landscapes. Dubbert *et al.* (1998) studied the effect of habitat patch size, host plant density, and isolation from occupied habitat patches on the colonization of the grass *Calamagrostis epigeios* by a community of stem-boring herbivores and their parasitoids. The researchers mowed patches of suitable habitat at several distances from source populations to eliminate the insects, and the regrowth of the grass led to very different shoot densities. After one year there was no effect of isolation or area on colonization, rather the presence of herbivores in a habitat patch was best predicted by local host plant density. Dubbert *et al.* (1998) conclude that the herbivores are adapted to habitat patchiness at the scale of their study (the most isolated habitat patches were 150 m from the source populations), because of the great intrinsic unpredictability in the occurrence of the grass. On the other hand, parasitoids were more likely to colonize less isolated patches, and hence may be considered as less well adapted to the ephemeral occurrence of their host than are the herbivores. It is also important to remember that a parasitoid cannot successfully colonize a habitat patch until a host population has been established, hence it is potentially misleading to compare the dispersal abilities of a parasitoid and a host over a short time interval.

The other focus of recent studies has been on the comparison of the effect of habitat fragmentation on different species at the same trophic level (such as Jones *et al.*, 1996; Roland and Taylor, 1997; Roland, 1998; case study, this chapter). Roland and Taylor (1997) compared the impact of aspen forest fragmentation on the rate of parasitism by four parasitoids of the forest tent caterpillar (*Malacosoma disstria*). Forest structure (level of fragmentation), host population size and the rate of parasitism by each parasitoid was measured at 127 points within a 25×25 km^2 area, and on a smaller scale at 109 points within a 0.8×0.8 km^2 area. The rate of parasitism by three larger species of parasitoid increased with host density and decreased with the degree of fragmentation. Interestingly, the larger-bodied parasitoid species were influenced by fragmentation at the larger spatial scale. The rate of parasitism by the smallest parasitoid increased with decreasing host density and increasing fragmentation. Roland and Taylor (1997) suggest that dense forest probably acts as a dispersal barrier for the small parasitoid but not for the larger parasitoids.

Not surprisingly studies of trophic interactions which include habitat fragmentation generally focus on dispersal behavior. Attributes of species other than dispersal behavior do contribute to their persistence in fragmented landscapes, though for the most part these other attributes are important because they are related to dispersal. For example, within a community of interacting species generalists may be less influenced by habitat fragmentation than specialists if the former are able to use host species living in alternative habitats, reducing the requirement for dispersal to survive (for brief reviews see Harrison and Taylor, 1997; Holt, 1997; for examples see Schoener and Spiller, 1987a, b; Walde, 1994; case study, this chapter). Species with high growth rate or short generation times have more opportunity for dispersal than species that produce few offspring. Species that can reproduce parthenogenetically or are adapted to inbreeding may be less penalized by habitat fragmentation than species lacking these attributes (Godfray, 1994; Hedrick and Gilpin, 1997; Saccheri *et al.*, 1998; Fauvergue *et al.*, 1999; West and Rivero, 2000). The size of a habitat patch can influence many aspects of individual behavior, such as territoriality, with consequences to the response of the species to fragmentation. In a fragmented landscape, habitat patches also differ in ways other than their size and connectivity. These other measures of landscape structure, such as landscape type and complexity, have been shown to influence the relative success of herbivores and their natural enemies, in part through their influence on local interactions (Marino and Landis, 1996; Polis *et al.*, 1998; Tscharntke *et al.*, 1998; Ohsaki and Sato, 1999).

The Glanville fritillary butterfly case study

To illustrate the role of space, habitat fragmentation, and metapopulation dynamics in multitrophic interactions we use results from an extensive research project on the Glanville fritillary butterfly (*Melitaea cinxia*) and its food plants, parasitoids, and hyperparasitoids conducted in the Åland Islands in southwest Finland (Fig. 6.1; for a review see Hanski, 1999). This system is appropriate for the present purpose as it is relatively simple, consisting of just a few species with well-studied local interactions; and because the spatial scale is large (50 × 70 km) and much is known about the spatial dynamics of the species over several years.

Natural history

The Glanville fritillary butterfly uses two host plant species in the Åland Islands, *Plantago lanceolata* (L.) (Plantaginaceae) and *Veronica spicata* (L.) (Scrophulariaceae) (Kuussaari *et al.*, 1995). The plants are patchily distributed over the 50 × 70 km area in which the butterfly is found. *Plantago lanceolata* is common in open areas throughout the Åland Islands, but is suitable as a host to the butterfly only in open dry meadows. *Veronica spicata* is found primarily in the western third of the region, almost exclusively in the dry rocky habitats appropriate for the butterfly. Both plants are perennial and reproduce both vegetatively and from seed (Muenscher, 1955), though *P. lanceolata* grows more often from seed than *V. spicata* (Rusch and van der Maarel, 1992).

Local populations of the butterfly occur in dry meadows where the host plants frequently suffer from summer drought (Rosén, 1995; Kuussaari, 1998) and in many cases from successional replacement by other plant species (see Rusch, 1988 for Öland, a somewhat comparable island in the Baltic). Larvae can completely defoliate plants, and in some dense butterfly populations the leaves of all the suitable plants are consumed as the gregarious larvae move from plant to plant in the spring (Hanski and Kuussaari, 1995; S. van Nouhuys, personal observation). However, both *P. lanceolata* and *V. spicata* are perennial and regenerate well during the same season and in the following year. Individual plants are not often killed by *M. cinxia* larvae, but defoliated plants may produce few or no seeds, and their stored resources are probably depleted, hence it is quite possible that in the course of time herbivory by *M. cinxia* may influence the local abundances of their hosts. There is a spatially and temporally dynamic interaction between the herbivore and its food plants at all

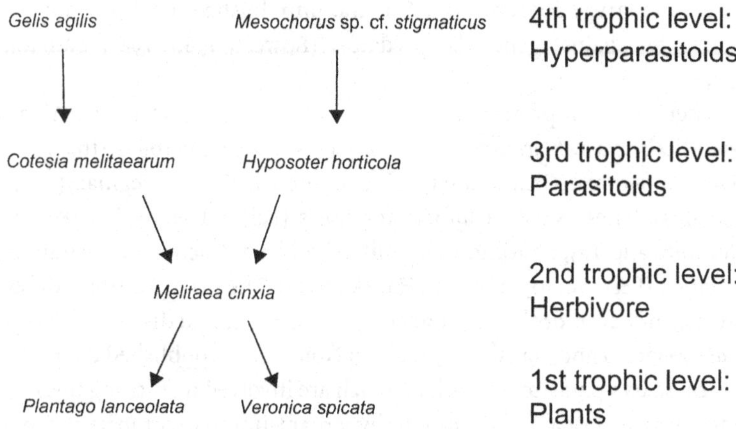

Fig. 6.1. Schematic diagram of the species at each trophic level.

spatial scales, from individual plants to the entire Åland Islands, influenced by the distribution of the two host plant species, spatial variation in genetically based oviposition preference of female butterflies, erratic variation in plant suitability for larval development, and the influence of weather (Hanski, 1999; Kuussaari *et al.*, 2000; I. Hanski and M. Singer, unpublished data; S. van Nouhuys *et al.*, unpublished data).

Each fall the entire study region is surveyed for *M. cinxia* populations. There are some 4000 habitat patches in the study area, of which 300 to 500 are occupied by the butterfly in each year (Kuussaari *et al.*, 1995; Hanski, 1999). Local populations are small, usually made up of a few groups of gregarious larvae, but ranging from one to more than 100 larval groups. Local populations within a cluster of habitat patches (patch networks) comprise classic metapopulations with a high rate of population turnover (Hanski *et al.*, 1995; Hanski, 1997, 1999).

Melitaea cinxia mostly mate once in their natal habitat patch, but substantial migration also occurs, typically to habitat patches within 1 km from the natal patch and especially from small populations in poor-quality habitat (Hanski *et al.*, 1994; Kuussaari *et al.*, 1996; Hanski, 1999). Females lay eggs in clusters of 100 to 200 on the underside of host plant leaves in late June. The larvae hatch and live gregariously in silken nests until their last instar late in the following spring, when they disperse to pupate in the litter. The larval development is interrupted by a seven-month winter diapause. The larvae and their web are visually conspicuous. The host plants contain high concentrations of iridoid glycosides, which the larvae sequester (M. Camara, unpublished data) probably as protection against generalist predators (Bowers, 1980, 1983; Camara, 1997).

There are two primary larval parasitoids of *M. cinxia* in the Åland Islands, *Cotesia melitaearum* (Wilkinson) (Braconidae: Microgastrinae) and *Hyposoter horticola* (Gravenhorst) (Ichneumonidae: Campopleginae) (Lei *et al.*, 1997). These wasps compete for hosts (Lei and Hanski 1998, van Nouhuys and Tay, 2001), and they kill a significant fraction of the butterfly larvae (Lei *et al.*, 1997; Lei and Hanski, 1998). The two parasitoids differ greatly in their morphology, phenology, behaviour and distribution (Lei *et al.*, 1997; Lei and Hanski, 1998; S. van Nouhuys, unpublished data). We use these two parasitoids, both of which are involved in a strong interaction with the host and their specific hyperparasitoids, to compare the role of space and habitat fragmentation for species at the same trophic level, but with dissimilar multitrophic interactions, spatial population structures, and population dynamics (Table 6.1).

Each primary parasitoid has an important secondary parasitoid or hyperparasitoid (Fig. 6.1). *Hyposoter horticola* is parasitized by the mobile solitary larval hyperparasitoid *Mesochorous* sp. cf. *stigmaticus* (Brischke) (Ichneumonidae: Mesochorinae). In contrast, *Cotesia melitaearum* is commonly parasitized by several solitary wingless generalist cocoon parasitoids in the genus *Gelis*, primarily *Gelis agilis* (Fabricius) (Ichneumonidae: Cryptinae) (Table 6.1). While using *C. melitaearum* cocoons *Gelis* are strictly speaking pseudohyperparasitoids, because they lay eggs on the immature parasitoids after they have left the host and made a cocoon (we nonetheless refer to them as hyperparasitoids). In addition to the two primary parasitoids and their two hyperparasitoids, there are four generalist pupal parasitoids of *M. cinxia* about which little is known apart from their names (Lei *et al.*, 1997).

Multitrophic interactions between the plants, the herbivore, and the primary parasitoids

Spatial variation of host plant qualities

Plantago lanceolata and *V. spicata* synthesize and maintain high concentrations of iridoid glycosides which probably deter generalist herbivores and their predators (Bowers, 1991; Stamp, 1992; Stamp and Bowers, 1996; Camara, 1997), but may attract specialists (Bowers, 1983; Oyeyele and Zalucki, 1990) such as *M. cinxia* and its specialist parasitoids. In the Åland Islands the concentrations of aucubin and catalpol, the two main iridoid

Table 6.1.

Species	Generations	Specificity	Dispersal behavior	Regional distribution	Metapopulation structure[a]
Hyperparasitoids					
G. agilis	Several per year	Generalist	Walk, aggregate	Ubiquitous	No
M. stigmaticus	One per year	Specialist?	Strong flier	Wide	No?
Parasitoids					
C. melitaearum	Three per year	Specialist	Weak flier	Narrow	Yes
H. horticola	One per year	Specialist	Strong flier	Wide	No
Herbivore					
M. cinxia	One per year	Specialist	Intermediate flier	Wide	Yes
Plants					
P. lanceolata	Perennial	Widespread	Seed	Wide	No
V. spicata	Perennial	More restricted	Seed	Intermediate	Yes?

Notes:
[a] "Metapopulation structure" means that the spatial occurrence of the species is strongly influenced by the connectivity of habitat patches.

glycosides, vary greatly between plant individuals, and are on average higher in *P. lanceolata* than in *V. spicata* (M. Nieminen and J. Suomi, unpublished data). This variation may contribute to the observed spatial variation in host plant use by the butterfly. Preliminary results indicate that the concentration of aucubin is higher in those *P. lanceolata* individuals on which females have oviposited in comparison with plants on which females have not oviposited (M. Nieminen and J. Suomi, unpublished data).

The non-volatile iridoid glycosides produced by *P. lanceolata* and *V. spicata* are sequestered by *M. cinxia* larvae (M. Camara, unpublished data). Generally, insect larvae that have sequestered iridoid glycosides are unattractive to some predators and parasitoids but attractive to others (Montllor *et al.*, 1991; Stamp, 1992; Theodoratus and Bowers, 1999). *Cotesia melitaearum* females spend a significant amount of time, occasionally even days, attending the web of a particular *M. cinxia* larval group. During this time they touch larvae and groom frequently, but rarely parasitize (Lei and Camara, 1999). Wasps possibly evaluate larvae based on the iridoid glycoside concentration in the larval cuticle, though we do not know whether they would select larvae with high or low levels of iridoid glycosides. While defensive chemicals produced by plants and sequestered by specialist herbivores are likely to be attractive to specialist parasitoids, high levels of compounds such as iridoid glycosides can be detrimental to immature parasitoid development (Campbell and Duffy, 1979; Gauld and Gaston, 1994; Reitz and Trumble, 1996).

Volatile compounds produced by host plants and herbivore-infested host plants are widely known to be attractive to parasitoid wasps (Vet and Dicke, 1992; Turlings *et al.*, 1995, chapter 7, this volume). *Plantago lanceolata* and *V. spicata* produce volatile compounds (Fons *et al.*, 1998), but the behavioral response by herbivores and their natural enemies to these chemicals has not been studied. The role of host plant in the parasitism of herbivores is of course not limited to chemical signals. First, the nutritional quality or toxicity of a host plant may affect the physiological resistance of the herbivore to parasitism, or the length of time it is available to parasitism. Second, the vulnerability of parasitoids to competitors and their own natural enemies may differ between food plant species. *Cotesia melitaearum* compete with *H. horticola* for host larvae, and in the early spring cocoons are subject to extremely high predation and hyperparasitism (Lei and Hanski, 1997, 1998; van Nouhuys and Tay, 2001). If mortality of parasitoids due to competition and natural enemies were to differ between the host plants, or between the habitat patches in which the plants are found, the rate of successful parasitism on the two host plants would also differ. Finally, the small-scale spatial distribution of host plants is likely to affect parasitoid searching efficiency. *Veronica spicata* has a more clumped small-scale occurrence than *P. lanceolata,* which leads to aggregation of host larval groups. If the searching ability of *C. melitaearum* is higher when hosts are aggregated, which seems likely, then larval groups on *V. spicata* would suffer more parasitism than those on *P. lanceolata* (van Nouhuys and Hanski, 1999).

Landscape structure and interactions with the parasitoid *Cotesia melitaearum*

Cotesia melitaearum is a small and rather sedentary species that has two to three generations per year and gregarious larvae (several parasitoid larvae per host individual). *Cotesia melitaearum* is relatively rare both in terms of population sizes and the number of local populations (Lei and Hanski, 1997; van Nouhuys and Tay, 2001). Each spring all known *M. cinxia* populations have been surveyed for *C. melitaearum* cocoons, but this survey remains necessarily somewhat superficial. To obtain a more accurate picture of the occurrence of the parasitoid, each *M. cinxia* population in all patch networks that had ever been occupied by *C. melitaearum* since 1993 was searched thoroughly in 1997–2000 and in some parts of the study area also in the previous years (Fig. 6.2). In this material, the fraction of *M. cinxia* populations occupied by *C. melitaearum* ranged from 9% to 20% and many populations persisted only for a couple of years.

Large well-connected populations of both the butterfly and the parasitoid *C. melitaearum* persist longer than isolated small populations, and population persistence of the parasitoid is associated with large host population size (Hanski, 1999). Within a habitat patch, the oviposition behavior and between-plant movements of both the host butterfly and parasitoids are affected by local host plant distribution. Host plant species affect the butterfly metapopulation dynamics because ovipositing female butterflies have host plant preferences, the plants are not similarly distributed, and there is regional genetic variation in host plant preference (Kuussaari *et al.,* 2000). Additionally, the plants do not respond equally to weather conditions. Host plant species affect the metapopulation dynamics of the primary parasitoid *C. melitaearum* indirectly via the effects on the herbivore population size and distribution, and directly because the parasitoids are more successful where the host larvae feed on *V. spicata*. Local host populations feeding on *V. spicata* are more likely to be

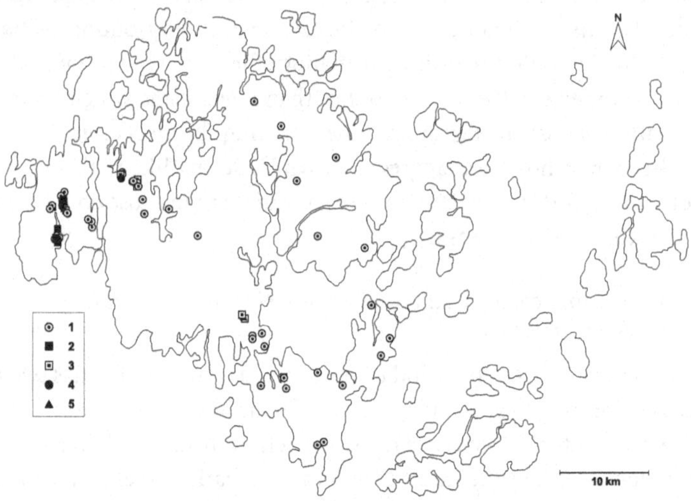

Fig. 6.2. The locations of the known populations of the parasitoid *Cotesia melitaearum* in the Åland Islands from 1997 to 2001. Symbols represent populations that have persisted at least one to five years. Notice that the more permanent parasitoid populations occur in a few clusters. Many of the populations that were observed in only one year are likely to be "remnants" of previously more extensive distributions (the parasitoid has been relatively sparse in recent years).

colonized by the parasitoid, and the parasitoid is less likely to go extinct, than in the case of host populations feeding on *P. lanceolata* (van Nouhuys and Hanski, 1999).

Based on the above-described empirical results, we may infer that if the habitat patches were to become substantially more fragmented, *C. melitaearum* would not be able to persist in the landscape. Similarly, the parasitoid would suffer if the host plant species composition used by the butterfly were to become more *P. lanceolata* dominated. In contrast, if well-connected habitat patches were to become increasingly occupied by the butterfly, or if the host plant use became increasingly *V. spicata* dominated, local populations of *C. melitaearum* would persist longer and would more frequently colonize nearby host populations.

Currently most *C. melitaearum* populations in the Åland Islands are so small that they are unlikely to have a big effect on host population size or host population dynamics (van Nouhuys and Hanski, 1999; van Nouhuys and Tay, 2001), and most host populations currently not used by the parasitoid are unlikely to be quickly colonized because they are out of the range of dispersal by the parasitoid (S. van Nouhuys and I. Hanski, unpublished data). However, in one network of butterfly populations intensely studied in 1993–1996, *C. melitaearum* apparently caused a large decline of many local host populations. In this case the populations of the host butterfly were tightly clustered and some of the populations were initially exceptionally large (Lei and Hanski, 1997). Thus, while the parasitoid currently persists at a very low level, not measurably affecting the population dynamics of its host, the parasitoid could become a more important player in the host dynamics if the host availability were to increase, potentially mediated by the distribution of the host food plants.

Landscape structure and interactions with the parasitoid *Hyposoter horticola*

In contrast to *C. melitaearum*, *H. horticola* is a large, solitary, mobile and abundant parasitoid. Several results suggest that isolation of host populations has little or no effect on the ability of *H. horticola* to colonize host populations. Lei and Hanski (1998) found that in a network of 50 tightly clustered habitat patches there was a negative effect of isolation (distance from possible source populations weighted by the sizes of these populations) on colonization by *C. melitaearum*, but isolation did not have any effect on the colonization and occupancy of habitat patches by *H. horticola*. In the spring of 1999, we sampled *M. cinxia* populations for *H. horticola* throughout the Åland Islands to measure the effect of isolation over a larger spatial scale, within the entire 50 × 70 km study area. We sampled host larvae from 50 populations to cover a range of patch connectivities and population ages. Ten to 60 (mean 26) larvae were sampled from each population, taking haphazardly a few larvae from each larval group. The host population was classified as established "old" ($n = 30$) if it had existed for more than two years, and newly colonized "new" ($n = 17$) if it had been colonized in the previous summer. We calculated the connectivity of each host population using the measure S (Hanski, 1994),

$$S_i = \sum_{j \neq i} \exp(-\alpha d_{ij}) N_j$$

The level of connectivity of patch i is thus calculated by taking into account the distances between the focal patch i and each of the source patches j (d_{ij}), as well as the sizes of the source populations, estimated as the number of host larval groups (N_j). All *M. cinxia* populations were considered to be potential source populations because *H. horticola* is found in the majority of host populations and because the complete distribution of *H. horticola* was unknown. For parameter α we used the value of 1 km^{-1}, which is our rough estimate of the migration range of the parasitoid. The analysis showed no association between the presence of the parasitoid and the age of the host population nor its level of connectivity. We also analyzed whether there was any association between the fraction of host larvae parasitized and the connectivity and the age of the population using analysis of variance. On average, 18% of the larvae were parasitized per sample, and again we found no association between parasitism and connectivity (Fig. 6.3) nor between parasitism and the age of the host population. These data strongly suggest that *H. horticola* can be found in isolated and well-connected populations equally often, and that it does not take the parasitoid long to find newly colonized host populations.

The local population sizes of *H. horticola* are relatively large because 20% to 30% of the larvae in each larval group are parasitized (S. van Nouhuys and I. Hanski, unpublished data). Therefore, *H. horticola* is present basically everywhere in large numbers and it shows no evidence of having a metapopulation structure in the Åland Islands.

The impact of the first trophic level (host food plant) on *H. horticola* is much weaker than the impact of the first trophic level on *C. melitaearum,* as *H. horticola* is not sensitive to plant distribution. Because of its superior dispersal ability, *H. horticola* would likely persist at about the same level if the network of habitat patches were to become substantially more fragmented. However, below we discuss how competition between the parasitoids and their interaction with the hyperparasitoids makes this conclusion more complicated.

Multitrophic interactions between the herbivore, the parasitoids, and the hyperparasitoids

The two abundant hyperparasitoids have very different roles in multitrophic interactions (Table 6.1). *Gelis agilis* (and the other less common *Gelis* species) are flightless generalists that aggregate where *C. melitaearum*

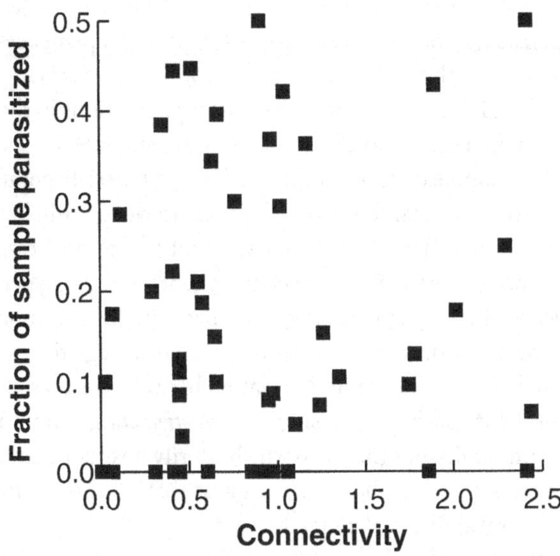

Fig. 6.3. The association between the fraction of *Melitaea cinxia* larvae parasitized by *Hyposoter horticola* in a sample and the connectivity of the habitat patch from which the sample was collected. The fraction of larvae parasitized in samples from isolated populations (low value of connectivity) was not significantly different from the fraction of larvae parasitized in samples from well-connected populations.

density is high and can even cause local extinction of *C. melitaearum* populations, which affects large-scale population dynamics of the host (Lei and Hanski, 1997; van Nouhuys and Hanski, 2000; van Nouhuys and Tay, 2001). Unlike the primary parasitoids, *G. agilis* females are probably not attracted to particular host plant species and they are extreme generalists, using many families of Hymenoptera as well as Lepidoptera, Coleoptera, and even spider egg cases as hosts (Schwarz and Shaw, 1999). In addition, *C. melitaearum* cocoons are not found in particularly close association with the host plants of their host insect. The population dynamics of *Gelis agilis* are likely to be largely disconnected from the dynamics of the primary parasitoid, the butterfly, and the host food plants, even though *G. agilis* is wingless and disperses on foot.

In contrast to *G. agilis*, the hyperparasitoid *Mesochorus* sp. cf. *stigmaticus* probably uses only the host parasitoid *H. horticola* in the Åland Islands (Lei *et al.*, 1997). *Mesochorus stigmaticus* is a true solitary hyperparasitoid that lays eggs into the larvae of *H. horticola* within the host larva on the host food plant. *Mesochorus stigmaticus* may respond to the same host plant cues and is subject to the same herbivore defences as *H. horticola*. *Mesochorus stigmaticus* is a strong flier, and has been found in most host populations in the Åland Islands where it has been sampled (S. van Nouhuys and I. Hanski, unpublished data). The sample of host caterpillars from 50 butterfly populations used to analyze the dispersal ability of *H. horticola* (above) also illustrates the dispersal ability of *M. stigmaticus*. Thirty-seven of the 50 host caterpillar samples contained *H. horticola*, and of them, 23 contained the hyperparasitoid *M. stigmaticus*. In these 23 parasitized populations, the mean fraction of *H. horticola* hyperparasitized by *M. stigmaticus* was 38%. Logistic regression and analysis of variance showed no association between the presence of or the fraction of *H. horticola* hyperparasitized by *M. stigmaticus* and the level of population connectivity nor the age of the butterfly host population. The sample sizes are small, but these data suggest that *M. stigmaticus* is not limited by dispersal ability in the Åland Islands.

At the largest spatial scale, among the different islands in the Åland archipelago, isolation makes a difference. Thus the large island of Kumlinge (*c.* 100 km^2) east of the main Åland Islands (*c.* 30 km isolation, mostly by sea) has a relatively small metapopulation of *Melitaea cinxia*, with some tens of small populations in the past eight years. Of the parasitoids, only *H. horticola* occurs on this island, whereas *C. melitaearum* and the hyperparasitoid *Mesochorus stigmaticus* are absent (M. Nieminen, personal communication).

Because *H. horticola* parasitizes a large but relatively constant fraction of *Melitaea cinxia* larvae, its main effect on *M. cinxia* is to make the local population sizes smaller and more prone to extinction than in the absence of the parasitoid. On the other hand, if *H. horticola* were absent, the butterfly populations would be more vulnerable to parasitism by *C. melitaearum*. The hyperparasitoid *Mesochorus stigmaticus* appears to reduce the numbers of *H. horticola* relatively uniformly over the whole region in the same way as *H. horticola* reduces the population size of the herbivore.

Competition between the primary parasitoids

In order for two parasitoids to share a single host species there must be mechanisms for partitioning the resource or some other specific mechanism of coexistence. *Hyposoter horticola* disperses among host populations much more readily than *C. melitaearum*, and hence the majority of *H. horticola* are in host populations unoccupied by *C. melitaearum*. In contrast, both primary parasitoids are present in the 10%–20% of the butterfly populations occupied by *C. melitaearum*. Because *C. melitaearum* populations are most persistent in well-connected large host populations, and especially where *V. spicata* is the dominant host plant species, it is mostly under these conditions that direct competition between the two primary parasitoids of *Melitaea cinxia* is likely to occur.

Lei and Hanski (1998) showed that *C. melitaearum* is a superior competitor within host populations. They found that when both parasitoids were present in a local population, the fraction of larvae parasitized by *H. horticola* in a larval group also occupied by *C. melitaearum* was low, 18% on average, whereas when *H. horticola* was the only parasitoid in a larval group, the mean fraction of larvae parasitized was 33%. In addition, within a habitat patch the larval groups parasitized by *C. melitaearum* are less isolated from each other than the larval groups without *C. melitaearum*. On the other hand, the competitive interaction between the two parasitoids is complex because there are three generations of *C. melitaearum* during each generation of *H. horticola*. Consequently, immature parasitoids may meet within host larvae under three different competitive conditions. In a laboratory experiment van Nouhuys and Tay (2001) found that when the third generation of *C. melitaearum* parasitize host larvae already occupied by *H. horticola*, the latter suppress the development of *C. melitaearum* larvae. Therefore, though *C. melitaearum* has proven to be the superior competitor at the population level (Lei and Hanski, 1998), it is the inferior competitor in one of its three generations during each *H. horticola* (and host) generation.

Conclusions

In the introduction, we posed the questions to what extent local multitrophic interactions in a fragmented landscape are influenced by regional spatial processes, and to what extent the large-scale spatial processes are influenced by the outcome of local interactions. In a metacommunity with high turnover of local populations, as exemplified by the *M. cinxia* metapopulations and the associated host plants, parasitoids, and hyperparasitoids, the answer to the first question is conclusively affirmative. Not all species are present in every habitat patch, and the interactions among the species that are present are greatly affected by the absence of the remaining species. This is most obvious in the case of the host plant that is not regionally preferred by the ovipositing butterflies (heavily used only if the preferred host plant is absent), and in the case of the competing primary parasitoids.

The answer to the second question – whether local processes influence regional dynamics – is less obvious but also affirmative. There is evidence showing that the "match" between the oviposition preference of migrating butterflies and the host plant composition in the empty habitat patch influences the rate of successful establishment of new populations (I. Hanski and M. Singer, unpublished data), and we have shown how the extinction and colonization rate of the primary parasitoid *C. melitaearum* is influenced by the food plant species of the host insect population (van Nouhuys and Hanski, 1999). It is thus clear that a comprehensive understanding of multitrophic interactions in fragmented landscapes remains incomplete unless the spatial dimension is explicitly considered.

Turning to the theoretical predictions about multitrophic interactions in space, it is evident that the length of the food chain is limited by metapopulation dynamics. We observed that on the island of Kumlinge with a relatively small butterfly metapopulation consisting of some tens of small local populations, of the parasitoids only *H. horticola* is present, whereas its competitor, *C. melitaearum*, and its hyperparasitoid, *Mesochorus stigmaticus*, are absent. This pattern is consistent with the theoretical expectations because *H. horticola* is the better disperser of the two primary parasitoids, and the fourth trophic level (the hyperparasitoid *M. stigmaticus*) is expected to drop out first from the metacommunity in a small patch network. It may also be significant that of the host plant species only *P. lanceolata* occurs in Kumlinge, on which the parasitoid *C. melitaearum* does less well than on *V. spicata*.

A recent study by Komonen *et al.* (2000) has strikingly illustrated how habitat fragmentation is likely to truncate food chains in a different ecological setting. Komonen *et al.* (2000) studied the insect community inhabiting the bracket fungus *Fomitopsis rosea* in continuous and fragmented old-growth forests in Finland. The numerically dominant food chain consisted of the fungus, the tineid moth *Agnathosia mendicella*, and the tachinid parasitoid *Elfia cingulata*. The median number of trophic levels decreased from three in areas of continuous old-growth to one in 10-ha fragments of old-growth that had been isolated for longer than 12–32 years. Such truncation of food chains as a consequence of habitat fragmentation is probably a common occurrence, because species in many metacommunities are not adapted to fragmented or unstable habitats. The mechanisms that lead to the shortening of food chains in newly fragmented habitats involve spatial processes, especially processes limiting dispersal and colonization, but also the traditional multitrophic level interactions.

Acknowledgments

B. Hawkins, M. Singer, T. Tscharntke, and an anonymous reviewer are thanked for helpful comments on the manuscript. M. Siljander helped prepare Fig. 6.2. This work was supported by grants from the Academy of Finland (no. 6400062) and the Finnish Centre of Excellence Programme 2000–2005 (no. 44887).

REFERENCES

Bowers, M. D. (1980) Unpalatability as a defence strategy of *Euphydryas phaeton* (Lepidoptera: Nymphalidae). *Evolution* **35**: 367–375.

Bowers, M. D. (1983) Iridoid glycosides and larval hostplant specificity in checkerspot butterflies (*Euphydryas*, Nymphalidae). *Journal of Chemical Ecology* **9**: 475–493.

Bowers, M. D. (1991) Iridoid glycosides. In *Herbivores: Their Interactions with Plant Secondary Metabolites*, vol. 1, 2nd edn, ed. W. I. Taylor and A. R. Battersby, pp. 297–325. Orlando, FL: Academic Press.

Camara, M. D. (1997) Predator responses to sequestered plant toxins in buckeye caterpillars: are tritrophic interactions locally variable? *Journal of Chemical Ecology* **23**: 2093–2106.

Campbell, B. C. and Duffy, S. (1979) Tomatine and parasitic wasps: potential incompatibility of plant antibiosis with biological control. *Science* **205**: 700–702.

Comins, H. N. and Hassell, M. P. (1996) Persistence of multispecies host–parasitoid interactions in spatially distributed models with local dispersal. *Journal of Theoretical Biology* **183**: 19–28.

Dubbert, M., Tscharntke, T. and Vidal, S. (1998) Stem-boring insects of fragmented *Calamagrostis* habitats: herbivore–parasitoid community structure and the unpredictability of grass shoot abundance. *Ecological Entomology* **23**: 271–280.

Fauvergue, X., Fleury, F., Lemaitre, C. and Allemand, R. (1999) Parasitoid mating structures when hosts are patchily distributed: field and laboratory experiments with *Leptopilina boulardi* and *L. heterotoma*. *Oikos* **86**: 344–356.

Feeny, P. P. (1976) Plant apparancy and chemical defence. *Recent Advances in Phytochemistry* **10**: 1–40.

Fons, F., Rapior, S., Gargadennec, A., Andary, C. and Bessière, J. (1998) Volatile component of *Plantago lanceolata* (Plantaginaceae). *Acta Botanica Gallica* **145**: 265–269.

Gauld, I. D. and Gaston, K. J. (1994) The taste of enemy-free space: parasitoids and nasty hosts. In *Parasitoid Community Ecology*, ed. B. A. Hawkins and W. Sheehan, pp. 278–299. Oxford: Oxford University Press.

Godfray, H. C. J. (1994) *Parasitoids: Behavioural and Evolutionary Ecology*. Princeton, NJ: Princeton University Press.

Hanski, I. (1983) Coexistence of competitors in patchy environments. *Ecology* **64**: 493–500.

Hanski, I. (1994) A practical model of metapopulation dynamics. *Journal of Animal Ecology* **63**: 51–63.

Hanski, I. (1997) Metapopulation dynamics from concepts and observations to predictive models. In *Metapopulation Biology: Ecology, Genetics and Evolution*, ed. I. Hanski and M. E. Gilpin, pp. 69–92. San Diego, CA: Academic Press.

Hanski, I. (1998) Connecting the parameters of local extinction and metapopulation dynamics. *Oikos* **83**: 390–396.

Hanski, I. (1999) *Metapopulation Ecology*. Oxford: Oxford University Press.

Hanski, I. and Gilpin, M. E. (eds.) (1997) *Metapopulation Biology: Ecology, Genetics & Evolution*. San Diego, CA: Academic Press.

Hanski, I. and Kuussaari, M. (1995) Butterfly metapopulation dynamics. In *Population Dynamics: New Approaches and Synthesis*, ed. N. Cappuccino and P. Price, pp. 149–171. San Diego, CA: Academic Press.

Hanski, I., Kuussaari, M. and Nieminen, M. (1994) Metapopulation structure and migration in the butterfly *Melitaea cinxia*. *Ecology* **75**: 747–762.

Hanski, I., Pakkala, T., Kuussaari, M. and Lei, G. C. (1995) Metapopulation persistence of an endangered butterfly in a fragmented landscape. *Oikos* **72**: 21–28.

Harrison, S. (1991) Local extinction in a metapopulation context: an empirical evaluation. *Biological Journal of the Linnean Society* **42**: 73–88.

Harrison, S. and Taylor, A. D. (1997) Empirical evidence for metapopulation dynamics. In *Metapopulation Biology: Ecology, Genetics and Evolution*, ed. I. A. Hanski and M. E. Gilpin, pp. 27–42. San Diego, CA: Academic Press.

Hassell, M. P. (2000) *The Spatial and Temporal Dynamics of Host–Parasitoid Interactions*. Oxford: Oxford University Press.

Hassell, M. P., Comins, H. N. and May, R. M. (1991) Spatial structure and chaos in insect population dynamics. *Nature* **353**: 255–258.

Hawkins, B. A., Askew, R. R. and Shaw, M. R. (1990) Influences of host feeding-niche and food plant type on generalist and specialist parasitoids. *Ecological Entomology* **15**: 275–280.

Hedrick, P. W. and Gilpin, M. E. (1997) Genetic effective size of a metapopulation. In *Metapopulation Biology: Ecology, Genetics and Evolution*, ed. I. A. Hanski and M. E. Gilpin, pp. 166–179. San Diego, CA: Academic Press.

Holt, R. D. (1995) Demographic constraints in evolution: towards unifying the evolutionary theories of senescence and the niche conservatism. *Evolutionary Ecology* **10**: 1–11.

Holt, R. D. (1997) From metapopulation dynamics to community structure: some consequences of spatial heterogeneity. In *Metapopulation Biology: Ecology, Genetics and Evolution*, ed. I. A. Hanski and M. E. Gilpin, pp. 149–165. San Diego, CA: Academic Press.

Holyoak, M. and Lawler, S. P. (1996) The role of dispersal in predator–prey metapopulation dynamics. *Journal of Animal Ecology* **65**: 640–652.

Hopper, K. R. (1984) The effects of host-finding and colonization rates on abundances of parasitoids of a gall midge. *Ecology* **65**: 20–27.

Jones, T. H., Godfray, H. C. J. and Hassell, M. P. (1996) Relative movement patterns of a tephritid fly and its parasitoid wasps. *Oecologia* **106**: 317–324.

Kareiva, P. (1987) Habitat fragmentation and the stability of predator–prey interactions. *Nature* **326**: 388–390.

Komonen, A., Penttilä, R., Lindgren, M. and Hanski, I. (2000) Forest fragmentation truncates a food chain based on an old-growth forest bracket fungus. *Oikos* **90**: 119–126.

Kruess, A. and Tscharntke, T. (2000) Species richness and parasitism in a fragmented landscape: experiments and field studies with insects on *Vicia sepium*. *Oecologia* **122**: 129–137.

Kuussaari, M. (1998) Biology of the Glanville fritillary butterfly (*Melitaea cinxia*). PhD thesis, University of Helsinki, Finland.

Kuussaari, M., Nieminen, M., Pöyry, J. and Hanski, I. (1995) Life history and distribution of the Glanville fritillary *Melitaea cinxia* (Nymphalidae) in Finland. *Baptria* **20**: 167–180.

Kuussaari, M., Nieminen, M. and Hanski, I. (1996) An experimental study of migration in the Glanville fritillary butterfly, *Melitaea cinxia*. *Journal of Animal Ecology* **65**: 791–801.

Kuussaari, M., Singer, M. and Hanski, I. (2000) Local specialization and landscape-level influence on host use in an herbivorous insect. *Ecology* **81**: 2177–2187.

Lei, G. C. and Camara, M. D. (1999) Behaviour of a specialist parasitoid, *Cotesia melitaearum*: from individual behaviour to metapopulation processes. *Ecological Entomology* **24**: 59–72.

Lei, G. C. and Hanski, I. (1997) Metapopulation structure of *Cotesia melitaearum*, a specialist parasitoid of the butterfly *Melitaea cinxia*. *Oikos* **78**: 91–100.

Lei, G. C. and Hanski, I. (1998) Spatial dynamics of two competing specialist parasitoids in a host metapopulation. *Journal of Animal Ecology* **67**: 422–433.

Lei, G. C., Vikberg, V., Nieminen, M. and Kuussaari, M. (1997) The parasitoid complex attacking the Finnish populations of Glanville fritillary *Melitaea cinxia* (Lep: Nymphalidae), an endangered butterfly. *Journal of Natural History* **31**: 635–648.

Levins, R. and Culver, D. (1971) Regional coexistence of species and competition between rare species. *Proceedings of the National Academy of Sciences, USA* **68**: 1246–1248.

Marino, P. C. and Landis, D. A. (1996) Effect of landscape structure on parasitoid diversity and parasitism in agroecosystems *Ecological Applications* **6**: 276–284.

May, R. M. (1994) The effects of spatial scale on ecological questions and answers. In *Large-Scale Ecology and Conservation Biology*, ed. P. J. Edwards, R. M. May and N. R. Webb, pp. 1–18. Oxford: Blackwell Science.

Montllor, C. B., Bernays, E. A. and Cornelius, M. L. (1991) Responses of two hymenopteran predators to surface chemistry of their prey: significance for alkaloid-sequestering caterpillars. *Journal of Chemical Ecology* **17**: 391–399.

Muenscher, W. C. (1955) *Weeds*, 2nd edn. New York: Cornell University Press.

Nee, S. and May, R. M. (1992) Dynamics of metapopulations: habitat destruction and competitive coexistence. *Journal of Animal Ecology* **61**: 37–40.

Nee, S., May, R. M. and Hassell, M. P. (1997) Two-species metapopulation models. In *Metapopulation Biology: Ecology, Genetics and Evolution*, ed. I. A. Hanski and M. E. Gilpin, pp. 123–148. San Diego, CA: Academic Press.

Oyeyele, S. O. and Zalucki, M. P. (1990) Cardiac glycosides and oviposition by *Danaus plexipus* on *Asclepias fruticosa* in South-east Queensland Australia, with notes on the effect of plant nitrogen content. *Ecological Entomology* **15**: 177–186.

Ohsaki, N. and Sato, Y. (1999) The role of parasitoids in the evolution of habitat and larval food plant preference by three *Pieris* butterflies. *Research in Population Ecology* **41**: 107–119.

Price, P. W., Bouton, C. E., Gross, P., McPheron, B. A., Thompson, J. N. and Weis, A. E. (1980) Interaction among three trophic levels: influence of plants on interactions between insect herbivores and natural enemies. *Annual Review of Ecology and Systematics* **11**: 41–65.

Polis, G. A., Hurd, S. D., Jackson, C. T. and Sanchez-Piñero, F. (1998) Multifactor population limitation: variable spatial and temporal control of spiders on gulf of California islands. *Ecology* **79**: 490–502.

Reeve, D. J. (1988) Environmental variability, migration and persistence in host–parasitoid systems. *American Naturalist* **132**: 810–836.

Reitz, S. R. and Trumble, J. T. (1996) Tritrophic interactions among linear furanocoumarins, the herbivore *Trichoplusia ni* (Lepidoptera: Noctuidae), and the polyembryonic parasitoid *Copidosoma floridanum* (Hymenoptera: Encyrtidae). *Environmental Entomology* **25**: 1391–1397.

Rohani, P., May, R. M. and Hassell, M. P. (1996) Metapopulations and local stability: the effects of spatial structure. *Journal of Theoretical Biology* **181**: 107–109.

Roland, J. (1993) Large-scale forest fragmentation increases the duration of a tent caterpillar outbreak. *Oecologia* **93**: 25–30.

Roland, J. (1998) Forest fragmentation and colony performance of forest tent caterpillars. *Ecography* **21**: 383–391.

Roland, J. and Taylor, P. D. (1995) Herbivore–natural enemy interactions in fragmented and continuous forests. In *Population Dynamics: New Approaches and Synthesis*, ed. N. Cappuccino and P. Price, pp. 195–208. San Diego, CA : Academic Press.

Roland, J. and Taylor, P. D. (1997) Insect parasitoid species respond to forest structure at different spatial scales. *Nature* **386**: 710–713.

Rosén, E. (1995) Periodic droughts and long-term dynamics of alvar grassland vegetation on Öland, Sweden. *Folia Geobotanica and Phytotaxonomica* **30**: 131–140.

Rusch, G. (1988) Reproductive regeneration in grazed and ungrazed limestone grassland communities on Öland preliminary results. *Acta Phytogeographica Seucica* **76**: 113–124.

Rusch, G. and van der Maarel, E. (1992) Species turnover and seedling recruitment in limestone grasslands. *Oikos* **63**: 139–146.

Saccheri, I., Kuussaari, M., Kankare, M., Vikman, P., Fortelius, W. and Hanski, I. (1998) Inbreeding and extinction in a butterfly metapopulation. *Nature* **392**: 491–494.

Schoener, T. W. and Spiller, D. A. (1987a) Effects of lizards on spider populations: manipulative reconstruction of a natural experiment. *Science* **236**: 949–952.

Schoener, T. W. and Spiller, D. A. (1987b) High population persistence in a system with high turnover. *Nature* **330**: 470–477.

Schwarz, M. and Shaw, M. R. (1999) Western Palaearctic Cryptinae (Hymenoptera: Ichneumonidae) in the National Museums of Scotland, with nomenclature changes, taxonomic notes, rearing records and special reference to the British check list, part 2, Genus *Gelis* Thunberg (Phygadeuontini: Gelina). *Entomologist's Gazette* **50**: 117–125.

Singh, R. and Srivastava, P. N. (1988) Host acceptance behaviour of *Alloxysta pleuralis*, a cynipoid hyperparasitoid of an aphidiid parasitoid *Trioxys indicus* on aphids. *Entomologia Experimentalis et Applicata* **47**: 89–94.

Slatkin, M. (1974) Competition and regional coexistence. *Ecology* **55**: 128–134.

Stamp, N. E. (1992) Relative susceptibility to predation of two species of caterpillar on plantain. *Oecologia* **92**: 124–129.

Stamp, N. E. and Bowers, D. M. (1996) Consequences for plantain chemistry and growth when herbivores are attacked by predators. *Ecology* **77**: 535–549.

Sullivan, D. J. and Völk, W. (1999) Hyperparasitism: multitrophic ecology and behaviour. *Annual Review of Entomology* **44**: 291–315.

Taylor, A. D. (1988) Large-scale spatial structure and population dynamics in arthropod predator–prey systems. *Annales Zoologici Fennica* **25**: 63–74.

Taylor, A. D. (1991) Studying metapopulation effects in predator–prey systems. *Biological Journal of the Linnean Society* **42**: 305–323.

Thaler, J. (1999) Jasmonite-inducible plant defences cause increased parasitism of herbivores. *Nature* **399**: 686–688.

Theodoratus, D. H. and Bowers, M. D. (1999) Effects of sequestered iridoid glycosides on prey choice of the prairie wolf spider *Lycosa carolinensis*. *Journal of Chemical Ecology* **25**: 283–295.

Tscharntke, T., Gathmann, A. and Steffandewenter, I. (1998) Bioindication using trap-nesting bees and wasps and their natural enemies: community structure and interactions. *Journal of Animal Ecology* **35**: 708–719.

Turlings, T. C. J., Loughrin, J. H., McCall, P. J., Rose, U. S. R., Lewis, W. J. and Tumlinson, J. H. (1995) How caterpillar-damaged plants protect themselves by attracting parasitic wasps. *Proceedings of the National Academy of Sciences, USA* **92**: 4169–4174.

van Baarlen, P., Topping, C. J. and Sunderland, K. D. (1996) Host location by *Gelis festinans*, an egg-sac parasitoid of the linyphiid spider *Erigone atra*. *Entomologia Experimentalis et Applicata* **81**: 155–163.

van Nouhuys, S. and Hanski, I. (1999) Host diet affects extinctions and colonizations in a parasitoid metapopulation. *Journal of Animal Ecology* **68**: 1248–1258.

van Nouhuys, S. and Hanski, I. (2000) Apparent competition between parasitoids mediated by a shared hyperparasitoid. *Ecology Letters* **3**: 82–84.

van Nouhuys, S. and Tay, W. T. (2001) Causes and consequences of small population size for a specialist parasitoid wasp. *Oecologia* **128**: 126–133.

Vet, L. E. M. and Dicke, M. (1992) Ecology of infochemicals used by natural enemies in a tritrophic context. *Annual Review of Entomology* **37**: 141–172.

Walde, S. J. (1994) Immigration and the dynamics of a predator–prey interaction in biological control. *Journal of Animal Ecology* **63**: 337–346.

Walde, S. J. (1995a) How quality of host food plant affects a predator–prey interaction in biological control. *Ecology* **76**: 1206–1219.

Walde, S. J. (1995b) Internal dynamics and metapopulations: experimental tests with predator–prey systems. In *Population Dynamics: New Approaches and Synthesis*, ed. N. Cappuccino and P. Price, pp. 173–193. San Diego, CA: Academic Press.

Weis, A. E. and Abrahamson, W. G. (1985) Potential selective pressures by parasitoids on evolution of plant–herbivore interaction. *Ecology* **66**: 1261–1269.

West, S. A. and Rivero, A. (2000). Using sex ratios to estimate what limits reproduction in parasitoids. *Ecology Letters* **3**: 294–299.

Yeargan, K. and Braman, S. K. (1989) Life history of the hyperparasitoid *Mesochorus discitergus* (Hymenoptera: Ichneumonidae) and tactics used to overcome the defensive behaviour of the green clover worm (Lepidoptera: Noctuidae). *Annals of the Entomological Society of America* **82**: 393–398.

TED C. J. TURLINGS, SANDRINE GOUINGUENÉ, THOMAS DEGEN, AND MARIA ELENA FRITZSCHE-HOBALLAH

7

The chemical ecology of plant–caterpillar–parasitoid interactions

Introduction

Predators and parasitoids can be major mortality factors for many herbivorous insects. As such, these members of the third trophic level may positively affect the fitness of certain plants on which the herbivores feed. Such indirect ecological links between the third and first trophic level appear to have resulted in some spectacular adaptations in plants, as well as in the natural enemies. Most evident are the morphological adaptations that result in a very close relationship between certain plants (e.g., *Cecropia* and *Acacia*) and ants that use these plants as their home and major source of food. These plants may carry structures that allow them to harbor (in hollow thorns or roots) and feed (with extrafloral nectar organs or food bodies) ants. While the ants benefit from lodging and food, the plants benefit from protection against herbivores. It is generally accepted that the plant structures and secretions secure the protection by ants (e.g., Janzen, 1966; McKey, 1988; Oliveira, 1997).

A less obvious plant adaptation are the signals that plants emit when they are damaged by herbivores. These signals, which are emitted in the form of volatile substances, are used by predators and parasitoids to locate potential prey or hosts. Although there are several alternative possible functions for these emissions, evidence for their role in the attraction of the third trophic level is accumulating. This chapter selectively reviews and discusses this evidence. Three aspects relevant to the evolution of herbivore-induced plant signals are emphasized: (1) the factors that lead to and influence the emission of the signals; (2) the specificity and reliability of the signals in terms of the information that they provide to natural enemies of herbivores; and (3) the benefits that plants may derive from attracting parasitoids.

Several aspects of chemically mediated tritrophic level interactions have been extensively reviewed (e.g., Vet and Dicke, 1992; Whitman, 1994; Turlings and Benrey, 1998; Dicke and Vet, 1999; Sabelis *et al.*, 1999). Here we focus on some of the most recent developments in the area. We discuss the usefulness as well as the limitations of plant-provided signals, with an emphasis on results from our own work on cultivated and wild maize plants, and their interactions with caterpillars and parasitoids. With evidence that shows that certain parasitoids can enhance plant fitness, we hope to reinforce the notion that parasitoids may contribute to selective pressures that have helped shape the evolution of induced plant signals. Some future directions of research that may provide more conclusive evidence are discussed (see also chapter 2).

Herbivore-induced plant odor emissions that attract parasitoids

The realization that plants may exhibit an indirect or extrinsic defense in the form of signals that promote the presence and effectiveness of natural enemies has been around for some time (e.g., Price *et al.*, 1980; Price, 1986; Vinson *et al.*, 1987). It has also been known for some time now that plant volatiles play a role in the foraging behavior of natural enemies of herbivores (see review by Nordlund *et al.*, 1988). The first combination of behavioral and chemical evidence showing that plants actively emit signals in response to herbivory was presented by Dicke and Sabelis (1988). They found that the predatory mite *Phytoseiulus persimilis* uses volatile cues emitted by spider-mite-infested plants to locate the spider mite *Tetranychus urticae*, its principle prey. Since this first publication, ample additional evidence emphasizes the importance of herbivore-induced plant volatiles in the host or prey location of many entomophagous arthropods (for reviews see Vet and Dicke, 1992; Dicke, 1994; Turlings and Benrey, 1998; Dicke and Vet, 1999).

We have been studying the host location behavior of larval endoparasitoids that attack Lepidoptera, in particular *Cotesia marginiventris* (Hymenoptera: Braconidae) (Fig. 7.1). This generalist parasitoid mainly uses herbivore-induced plant odors to locate the microhabitat of its hosts (Turlings *et al.*, 1990, 1991a, b). Maize (*Zea mays* L.) and various *Spodoptera* species (Lepidoptera: Noctuidae) have served as the respective plant and herbivore models. Maize and related plants are particularly well suited for such studies because of their rapid and strong response to herbivory (Turlings *et al.*, 1998a; Gouinguené *et al.*, 2001). A healthy maize plant

Fig. 7.1. A female *Cotesia marginiventris* approaching a maize seedling on which a potential host, a *Spodoptera exigua* larva, has been feeding.

releases very small amounts of volatiles, but after an attack by caterpillars, a dramatic change in its odor profile can be detected within hours (Fig. 7.2). The emission is systemic and hence takes place in damaged as well as undamaged leaves of an attacked plant. Mechanical damage inflicted to the leaves of most maize varieties results in a marginal release of volatiles, but when the damaged sites are treated with caterpillar oral secretion, the induction of odor emission is similar to that resulting from caterpillar feeding (Turlings *et al.*, 1990). Specific elicitors in the secretion are responsible for this reaction (see below).

Various other (non-host) insects have been found to elicit the same or similar responses in maize plants (Turlings *et al.*, 1993a, 1998b), thus limiting the information that these odorous plant signals provide on the identity or stage of the herbivore that is present on the plant. Yet, *C. marginiventris* and other hymenopterous parasitoids that attack Lepidoptera larvae appear to rely heavily on these signals for host location.

Plant-provided cues versus host-derived cues

As first pointed out by Vet and Dicke (1992), plant-provided signals have the obvious advantage over cues derived from the host itself in that they are released in much larger quantities and thus easier to detect. The lack of host-derived cues may result from strong selection upon the hosts to be cryptic and difficult to detect by their enemies, while plants may benefit from advertising the presence of herbivores to these enemies (Tumlinson *et al.*, 1992; Vet and Dicke, 1992). The dramatic difference in odor emissions between host and plant is illustrated in Fig. 7.2. An undamaged maize plant (Fig. 7.2A) is virtually odorless compared to plants that have been subjected to feeding by *Spodoptera* larvae (Fig. 7.2B, C). Removing the caterpillars and their feces from a damaged plant does initially not change the odor profile (Fig. 7.2C). Collections of the odor of isolated caterpillars (Fig. 7.2D), or just isolated feces (Fig. 7.2E) confirm that the plant is the principal source of detectable volatiles. The damaged seedlings emit more than a thousandfold the amount of what the other two components of the complete plant host complex emit. The plant-provided signal is highly detectable, but may have the disadvantage that it is not necessarily very reliable in terms of the information it contains (Vet *et al.*, 1991; Vet and Dicke, 1992; Wäckers and Lewis, 1994). The fact that the odors are released only in response to herbivory increases this reliability, but the odors were previously not thought to provide information on the identity of the herbivore or on its suitability as a host for a particular parasitoid. However, as discussed in the next section, recent evidence indicates that plants may actually emit specific signals that do provide such information.

Specificity of the plant signals

Most herbivores will be attacked by multiple natural enemies and some of these natural enemies will be more effective in reducing herbivory than others. It would be useful to a plant to specifically attract the most beneficial natural enemy and not the ones that do not reduce herbivory. However, we have argued in the past that selection will not necessarily favor plants to emit signals that provide very specific information on which herbivore is damaging the plant (Turlings *et al.*, 1995). This argument is based on the notion that the plant cannot control who exploits its signals. Any parasitoid, predator, or even herbivore may adapt its responses to a certain signal if such a signal can guide it to suitable resources. This may be different if there are differences in the detection limits among insect groups. No such evidence exists, but it is possible that there are constraints on what volatile chemicals can be perceived by an insect's chemosensory system and that these constraints vary for

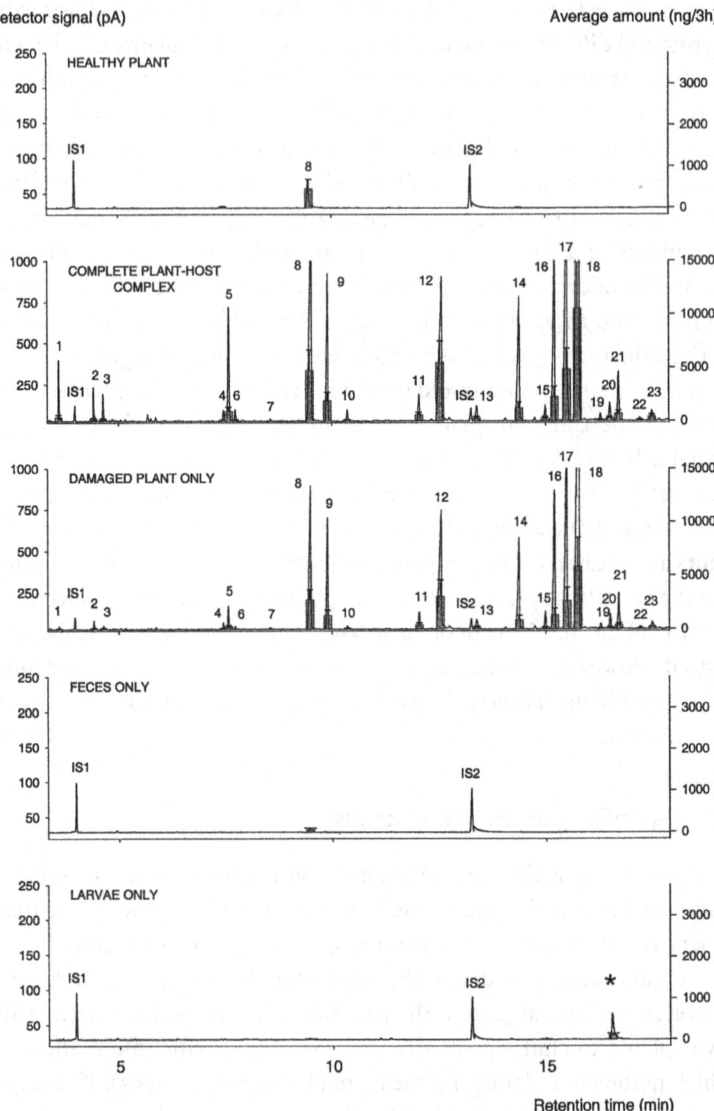

Fig. 7.2. Typical chromatographic profiles overlaid with bar graphs of the average amounts ($n = 5$) of volatiles collected for 2 h from: a healthy, undamaged maize plant (HEALTHY PLANT), a maize plant that has been damaged for 16 h by 10 *Spodoptera littoralis* larvae, with the larvae and their by-products still present (COMPLETE PLANT–HOST COMPLEX), a similarly damaged maize plant, but from which the larvae and by-products were removed (DAMAGED PLANT ONLY); the feces removed from two maize plants damaged each by 10 *S. littoralis* larvae for 16 h (FECES ONLY), and 20 *S.*

different insects. In other words, broadcasting within the chemical range that can only be detected by the insect(s) that the plant "wants" to attract offers an opportunity for specific signals to evolve. But if there are no strict limits to what insects can detect, it seems unlikely that plants can control which natural enemies they attract. Thus, plants that emit a signal in response to the attack by a certain herbivore will do this with the potential consequences of attracting unwanted (e.g., herbivores) or ineffective visitors (e.g., parasitoids that do not reduce herbivory). Attracting natural enemies that cannot successfully attack the herbivores present on a plant is not detrimental to the plant as long as the effective natural enemies are attracted as well and they do not interfere with each other. As a consequence, by lack of any better (more reliable) cues, the natural enemies may be forced to use what the plants offer them, even if this leads them to plants with unsuitable resources.

The suspected absence of specificity of the signals was also based on several observations concerning the behavior of parasitic wasps. For example, *Cotesia marginiventris* and *Microplitis croceipes* are readily attracted to maize plants that were induced to emit volatiles with the regurgitant of grasshoppers, which do not serve as hosts (Turlings *et al.*, 1993a). Similarly, McCall *et al.* (1993) found that *M. croceipes* is just as attracted to plants attacked by *Helicoverpa zea* (a host) as to plants attacked by *Spodoptera exigua* (non-host). However, once these parasitoids have landed on the damaged plants they quickly determine the suitability of the host. They probably use contact kairomones that are present in the hosts' feces or silk (Loke *et al.*, 1983). If they do not detect these innately recognized stimuli, they will quickly depart again (T. C. J. Turlings, personal observations) and search for new cues. It is expected that the wasps use their keen ability to learn through association (Lewis and Tumlinson, 1988; Turlings *et al.*, 1993b; Vet *et al.*, 1995) to eventually pick up subtle differences among

Fig. 7.2. (*cont.*)
littoralis larvae that had been feeding on two maize plants for 16 h (LARVAE ONLY). The labeled peaks are: 1, (*Z*)-3-hexenal; 2, (*E*)-2-hexenal; 3, (*Z*)-3-hexen-1-ol; 4, β-myrcene; 5, (*Z*)-3-hexen-1-yl acetate; 6, 1-hexyl acetate; 7, ocimene; 8, linalool; 9, (3*E*)-4,8-dimethyl-1,3,7-nonatriene; 10, benzyl acetate; 11, phenethyl acetate; 12, indole; 13, methyl anthranilate; 14, geranyl acetate; 15, unknown sesquiterpene; 16, β-caryophyllene; 17, (*E*)-α-bergamotene; 18, (*E*)-β-farnesene; 19, unknown sesquiterpene; 20, β-bisabolene; 21, unknown sesquiterpene; 22, (*E*)-nerolidol; 23, (3*E*, 7*E*)-4,8,12-trimethyl-1,3,7,11-tridecatetraene. The asterisk marks pentadecane, which is emitted from the oral secretion of the larvae (Turlings *et al.*, 1991b). IS1 and IS2 represent the internal standards *n*-octane and nonyl acetate, respectively.

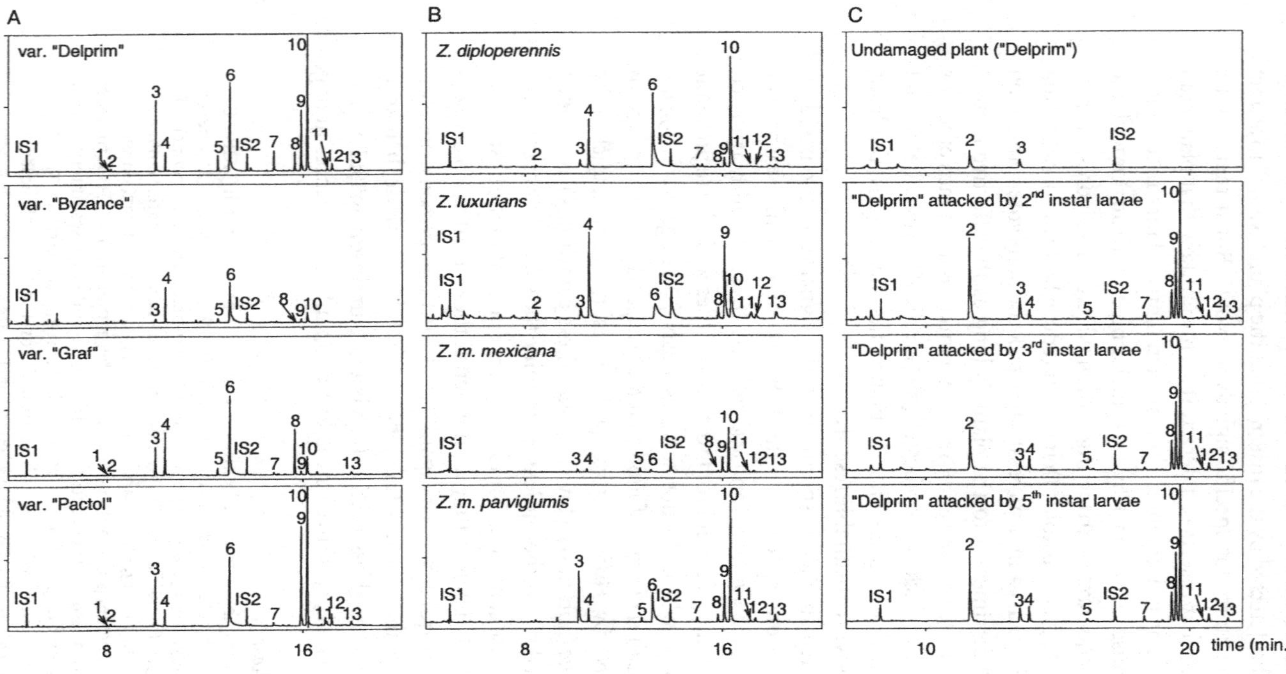

Fig. 7.3. Typical chromatographic profiles of volatiles collected from four different maize genotypes (A), four wild relatives of maize, teosintes (B), and from maize plants of the variety "Delprim" that were left undamaged or incubated in the regurgitant of either second, third, or fifth instar *Spodoptera littoralis* larvae (C). Odors were collected for 2 h about 14 h after a particular treatment was performed or started. For (A) and (B), 2 cm^2 of two leaves of each plant were scratched and 10 μl of *S. littoralis* regurgitant was applied to the damaged sites. The labeled peaks are: 1, β-myrcene; 2, (*Z*)-3–hexen-1–yl acetate; 3, linalool; 4, (3*E*)-4,8–dimethyl-1,3,7–nonatriene; 5, phenethyl acetate; 6, indole; 7, geranyl acetate; 8, β-caryophyllene; 9, (*E*)-α-bergamotene; 10, (*E*)-β-farnesene; 11, β-bisabolene; 12, unknown sesquiterpene; 13, (3*E*, 7*E*)-4,8,12–trimethyl-1,3,7,11–tridecatetraene. IS1 and IS2 represent the internal standards *n*-octane and nonyl acetate, respectively.

host and non-host cues to become more efficient at finding plants that carry hosts. In particular, the use of more than one type of cue, including small amounts of volatiles from host feces (Eller *et al.*, 1988, 1992), as well as visual cues from plant parts and damage patterns (Wäckers and Lewis, 1994), could allow them to detect differences among sites with suitable hosts and sites with non-hosts.

Surprisingly, however, some new studies present strong evidence for high specificity in the plant signals. Dicke (1999) reviews the evidence for and against specificity of plant signals used by parasitoids and predators. One of the most striking examples is the one presented by Takabayashi *et al.* (1995), who showed that maize plants under attack by *Pseudaletia separata* larvae release significant amounts of induced volatiles only when young larvae (first to fourth instar) feed on the plants. Plants that are attacked by larvae of later instars (fifth to sixth) release far less of these signals. In accordance, the parasitoid *Cotesia kariyi*, which can only successfully parasitize the early-instar larvae, is highly attracted to maize plants eaten by early-instar larvae and not by plants eaten by late-instar larvae (Takabayashi *et al.*, 1995). Based on their results, we hypothesized that the regurgitant of younger larvae induces a stronger reaction than that of older larvae. An experiment in which we used the regurgitant of *Spodoptera littoralis* larvae in which we incubated young maize plants does not corroborate this hypothesis (Fig. 7.3). As far as we can judge from the chromatographic profiles, maize plants incubated in the regurgitant of second, third, or fifth instar *S. littoralis* emit very similar blends of odors (Fig. 7.3C). Moreover, caterpillars of the different stages that inflicted equivalent amounts of damage caused comparable odor emissions in individual maize plants and females of the parasitoid *Microplitis rufiventris* did not distinguish among the plants (Gouinguené, 2000). As stressed by Dicke (1999), such assays with natural enemies that exploit the plant-provided signals are ultimately necessary to reveal how specific the signals are.

A second striking example of specific signaling comes from a series of studies on the host location behavior of *Aphidius ervi*, an aphid parasitoid with a limited host range, which includes the pea aphid, *Acyrthosiphon pisum* (Du *et al.*, 1998; Powell *et al.*, 1998). This parasitoid is attracted to pea plants infested by the host aphid and far less so by pea plants infested by a non-host, *Aphis fabae*. Implicated in the specificity of the signal is 6–methyl-5–hepten-2–one, a substance that was only detected in the odor profile of plants infested by *Acyrthosiphon pisum* and not in the profile of *Aphis fabae*-infested plants (Wadhams *et al.*, 1999). Indeed, *Acyrthosiphon ervi* is found to be particularly attracted to 6–methyl-5–hepten-2–one (Du *et al.*, 1998). Interestingly, 6–methyl-5–hepten-2–one is also produced by aphid-infested cereals (Quiroz *et al.*, 1997) on which *A. ervi* finds some of its other hosts (Wadhams *et al.*, 1999). It would be fascinating if 6–methyl-5–hepten-2–one were found to be released only by plants with *A. ervi* hosts, as it would provide this oligophagous parasitoid with a highly reliable signal, which may also have played a role in determining its host range.

Another specificity example comes from a study on *Cardiochiles nigriceps*, a parasitoid that specializes on *Heliothis virescens* (De Moraes *et al.*, 1998). In flight tunnel studies, as well as in field observations, this wasp was able to distinguish between plants that are under attack by its host and plants that are under attack by the closely related noctuid *Helicoverpa zea*, which *C. nigriceps* does not parasitize. De Moraes *et al.* (1998) further showed that some compounds were released in different proportions by maize and tobacco plants, depending on which of the two noctuids fed on the plants. It remains to be determined which of the volatiles provide the wasps with the specific information that allows them to distinguish among plants with suitable hosts and plants with unsuitable larvae.

Reliability of plant-provided signals will also be affected by the variability among plant genotypes (chapter 2). We have screened a large number of maize cultivars and several wild relatives of maize (teosinte) and found considerable differences in the volatile blends that they emit (Fig. 7.3). The cultivars (Fig. 7.3A) as well as the wild species of *Zea* (Fig. 7.3B) differ in the total quantity as well as in the quality of the volatiles that they emit. The variety "Delprim," for instance, released large amounts, while "Byzance" released relatively little. An example of the qualitative differences is the release of β-caryophyllene (peak 8, Fig. 7.3), which was emitted by some cultivars in large amounts, but could not be detected in the blends of others. These results were similar for the wild plants (Fig. 7.3B), showing that this trait has been well conserved despite selection for cultivation. This similarity is contrary to the hypothesis that domestication of plants will result in a decrease of secondary defense substances in favor of increased yield and palatability (Benrey *et al.*, 1998). We had also expected a decrease based on the results from Loughrin *et al.* (1995) who had found much larger quantities of induced volatiles emitted by naturalized cotton than by cultivated cotton. Interestingly, the wild teosinte that is the closest related to cultivated maize, *Zea mays parviglumis* (Wang *et al.*, 1999), was the most odorous of the wild species we tested (Gouinguené *et al.*, 2001).

In contrast to the differences among cultivars, the variation within a particular cultivar is very small. Figure 7.3C shows the odor blends collected from plants of the cultivar "Delprim" after these had been incubated in distilled water or in a solution with the regurgitant of three different larval instars of *S. littoralis*. Plants from the water treatment emitted only few compounds in very small amounts, whereas those that had been incubated in regurgitant showed a dramatic increase in the odor emissions. The emissions were remarkably similar for the different instars (Fig. 7.3C). As mentioned above, this contrasts with the findings of Takabayashi *et al.* (1995), but does certainly not exclude the possibility that certain parasitic wasps are capable of detecting differences among these odor blends. Such differences must be very subtle and are likely to require a particular sensitivity to compounds that are released in small amounts. Given the enormous variation that we find in our screenings, it remains a mystery as to how parasitoids may detect specific signals within this variability. Carefully designed experiments are needed to reveal which of the compounds provide the necessary information. One way by which specificity in signals can be accomplished is if different elicitors operate for different herbivores (Hopke *et al.*, 1994). A good understanding of the mechanisms of induction is therefore essential.

What elicits the odor emissions?

Mechanical damage to leaves triggers the release of various volatile substances, but this release is usually different, or at least weaker than what is emitted in response to herbivore-inflicted damage. For example, Dicke *et al.* (1990a) found that lima bean plants that had been infested by spider mites released several compounds that were not released by plants that had been artificially damaged by rubbing the leaves with carborundum on wet cotton wool. The compounds released only after mite infestation are also the ones that are implicated in the attraction of predatory mites to spider-mite-infested plants (Dicke and Sabelis, 1988; Dicke *et al.*, 1990a, b). Predatory thrips are also capable of making the distinction; they are attracted to spider-mite-infested plants, but not to mechanically damaged plants (Shimoda *et al.*, 1997).

In cabbage plants, the difference between mechanically damaged and herbivore damaged plants is only quantitative (Mattiacci *et al.*, 1994), which may indicate a defense strategy different from that of some other plants. An increased and prolonged release of cabbage volatiles can also be elicited by applying regurgitant from *Pieris brassicae* caterpillar to artificially damaged sites. Mattiacci *et al.* (1995) found that β-glucosidase, which is present in *P. brassicae* regurgitant, triggers this response in cabbage and can render the plants more attractive to *Cotesia glomerata*, a parasitoid of *Pieris* species.

Maize plants that are subjected to mechanical damage release far less of the induced odors than plants that have received a similar amount of damage by *Spodoptera* larvae. However, when regurgitant of these larvae is applied to mechanically damaged sites, a similar reaction takes place as is observed for caterpillar-damaged plants (Turlings *et al.*, 1990). A potent elicitor was isolated from the regurgitant of *S. exigua* (Turlings *et al.*, 2000) and was identified as N-(17–hydroxylinolenoloyl)-L-glutamine (Alborn *et al.*, 1997, 2000). This compound, which was named volicitin, appears to stimulate the octadecanoid pathway in a similar manner as the well-known plant defense signal jasmonic acid.

Work by Paré, Tumlinson, and co-workers has provided additional insight into the processes that lead to the emissions of induced volatiles by plants. In a first study, they demonstrated that in cotton plants induced terpenoids are produced *de novo* and subsequently release several days after a herbivore attack. They passed $^{13}CO_2$ over the plants and found the label back in the induced volatiles, but not in volatiles that are stored and constitutively present in the plants (Paré and Tumlinson, 1997). The *de novo* synthesis may explain why induced volatiles are mainly emitted in the presence of light (Loughrin *et al.*, 1994; Turlings *et al.*, 1995).

Paré *et al.* (1998) further showed that the elicitor volicitin is partially plant-derived. Linolenic acid from the leaves, when ingested by *S. exigua* larvae, binds with a hydroxyl group and glutamine in the caterpillar's oral cavity to form the volicitin molecule. It remains unclear why a compound like volicitin, which is potentially detrimental to the caterpillars, has persisted over evolutionary time. It is likely that volicitin plays an important role in the insect's physiology, perhaps as an emulsifier in digestion. It is clear that the formation of volicitin is partially under the control of the plant.

The structure of volicitin strongly suggests that it is a trigger of the octadecanoid signaling pathway (Alborn *et al.*, 1997). How this further elicits the synthesis and release of the various volatile compounds remains to be fully elucidated. Boland and co-workers (Boland *et al.*, 1999) have made considerable progress in this area and have also demonstrated that unrelated bacterial phytotoxins induce similar volatile emissions. They have raised some doubt about the activity of volicitin and its presence in *S. exigua* regurgitant (Pohnert *et al.*, 1999). However, the fact that they did not detect it in regurgitant that they collected from *S. exigua* may have been the result of a rapid breakdown in crude regurgitant at room temperature. We compared the potency of volicitin with several other known elicitors (Fig. 7.4). For this purpose, maize seedlings (var. "Delprim") were incubated for 16 h with their excised stem in a buffer solution with either 10% *S. littoralis* regurgitant, or solutions of either volicitin, salicylic acid, jasmonic acid, or coronatin at a concentration of 3.16×10^{-7} mol ml^{-1}. Of these known elicitors, coronatin, a pathogen-derived amino acid conjugate (Boland *et al.*, 1995), was by far the most powerful in inducing an emission (Fig. 7.4). For volicitin we used two sources, the Gainesville laboratory (H. Alborn and J. H. Tumlinson) and the Jena laboratory (T. Koch and W. Boland). The tendency for the volicitin from Gainesville to be slightly more active (Fig. 7.4) may have been due to the fact that it was only the active L-configuration (Alborn *et al.*, 1997), while volicitin from the Jena group was a mixture of L- and D- volicitin. The differences between the two were not significant and both versions were just as effective in inducing volatile emissions as caterpillar regurgitant at the chosen concentrations. Jasmonic acid was somewhat less active, while salicylic acid did not induce a response (Fig. 7.4).

How the plant may benefit

Demonstrating that induced plant defenses reduce the damage that herbivores inflict upon the plants to an extent that it increases plant fitness has proven to be difficult. Only recently have some field studies shown the positive effects of induced direct defenses (Agrawal, 1998; Baldwin, 1998). Evidence for the proposed function of induced volatiles in indirect defense has largely been missing (chapter 2). A study by Thaler (1999) made a first step in this direction by showing that parasitization rates in tomato fields that were treated with jasmonic acid (which can induce odor emissions) were higher than in untreated fields. It remained to be shown that the plants actually benefit from increases in parasitism. Lack of such evidence has evoked some skepticism (e.g. Faeth, 1994; van der Meijden and Klinkhamer, 2000). It is clear that our interpretation of results of studies on herbivore-induced changes in plant chemistry is biased by our interest in the effects of the chemicals on parasitoid behavior. It is an intriguing prospect that plants are emitting the volatiles as signal to attract natural enemies of herbivores. Van Loon *et al.* (2000) and Fritzsche-Hoballah and

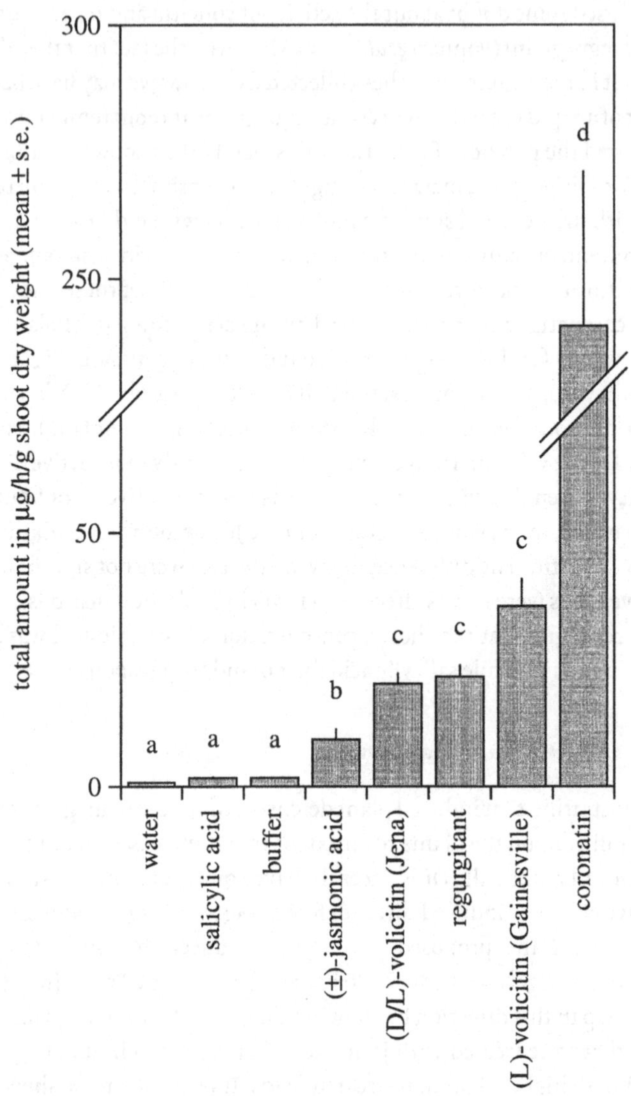

Fig. 7.4. Odor emission of maize seedlings (var. "Delprim") incubated overnight in 0.5 ml of deionized water, buffer (50 mmol Na_2HPO_4 adjusted to pH 8 with citric acid), regurgitant of *Spodoptera littoralis* caterpillars (10× diluted with buffer), and solutions of different elicitors at 3.16×10^{-7} mol ml^{-1} in buffer. Total amount = sum of 13 main compounds in the odor blend (see Fig. 7.3). Treatments with the same letters are not significantly different at the 5% level according to the Tukey–Kramer *post hoc* test (ANOVA on log-transformed values: $F = 131$; df = 7; $P = 0.0001$; $n = 6$ except for regurgitant where $n = 12$).

Turlings (2001) recently showed that the action by parasitoid could indeed increase plant fitness (see below). However, induced plant volatiles may serve various other functions (Turlings *et al.*, 1995).

The main or primary function of plant (induced) volatiles will be very hard to determine, but clearly any benefit that the plant derives from emitting the volatiles will contribute to selective pressures that have shaped their production and release. If the attraction of natural enemies does indeed increase plant fitness, then this effect is likely to have resulted in selective modification of this intriguing plant trait. What are the costs and benefits in this context? The ultimate cost and benefit balance of signaling is extremely difficult to assess because of the complex consequences (Dicke and Sabelis, 1992). Chemicals affect multiple partners within complex food-webs in unpredictable ways. Janssen *et al.* (1998) provide an overview of the various indirect interactions that may take place among the arthropods that inhabit plants and emphasize the need for more studies that consider entire food-webs. Although cost–benefit analyses in which only a few of the interacting partners are considered will not tell us the entire story, they could give us an indication of the use of the signals under particular (assumed) conditions. A first step towards determining the costs of herbivore-induced plant signals is to assess the energy and resource investment.

Secondary compounds are relatively costly in terms of energy and organic matter. Terpenoids, the class of compounds that is emitted in large amounts by many plants species in response to herbivory (Dicke, 1994), are more costly to produce than other induced plant compounds (Gershenzon and Crouteau, 1993). Gershenzon (1994) estimates that for each gram of terpenoids, the equivalent of more than 3 grams of glucose needs to be invested. A large number of reductions and enzymatic conversions are involved in production. Such costs, however, are low in respect to leaf tissue production. Dicke and Sabelis (1989) estimate that induced odor emissions for mite-infested plants is less than 0.001% of leaf production per day. It is generally expected that some cost must be involved otherwise selection may favor plants that continuously produce the defensive substances (Sabelis and de Jong, 1988; Godfray, 1995). Calculations of costs of induced defenses usually lead to underestimates of total physiological costs because they do not consider pleiotropic effects and investments such as enzyme synthesis and avoidance of autotoxicity (Simms, 1992; Zangerl and Bazzaz, 1992). The calculation becomes even more complex when ecological costs are considered.

Ecological costs may come in the form of attracting additional herbivores that exploit the signals. Several beetles are attracted more to plants that are already infested by conspecifics than by uninfested plants. However, other herbivores for which it has been studied tend to be repelled by odors emitted by plants infested by conspecifics (see Turlings and Benrey, 1998 and Dicke and Vet, 1999 for examples).

Ultimately, we need to express the costs and benefits of induced volatile emissions in terms of plant fitness. This is difficult because volatiles may be involved in a multitude of interactions (see also chapters 2, 5, and 6). The attraction of parasitoids and predators is one of those, but other trophic levels will be affected in ways that may be either beneficial or detrimental to the plants. For instance, little is known about how induced plant volatiles affect microorganisms, but some of the terpenoids have antibiotic properties (Langenheim, 1994; Harrewijn *et al.*, 1994/5). It is certainly possible that one of the main (primary) functions of the induced volatiles is to protect the plant against pathogens. In fact, pathogen-derived compounds can elicit plant odor emissions at low concentrations (Boland *et al.*, 1999).

Possible benefits that could be measured within a single plant–herbivore–enemy system are the fitness gains that a plant receives from enhancing the rate at which the specific enemy attacks the herbivore. In the case of predators, an increase in attack would have an immediate and positive effect on plants. The importance of induced plant volatiles was first demonstrated for a plant–mite system where predators are involved (Dicke and Sabelis, 1988). Spider mites tend to overexploit their host plants and, in the absence of predators, the plants may die (Sabelis *et al.*, 1999).

On the other hand, for parasitoids it is not always obvious that they can benefit individual plants. In biological control, introduced parasitoids can reduce herbivory on target plants, but this reduction is accomplished by pest population control and not through immediate protection of individual plants. Most parasitoids for which it has been shown that they are attracted by induced plant odors are koinobiont, which means that they do not immediately kill their hosts, but initially allow them to continue their development. Some hosts of koinobiont parasitoids will actually remain larvae longer and grow larger as a result of parasitization (Rahman, 1970; Parker and Pinnell, 1973; Byers *et al.*, 1993). An individual plant will clearly not benefit from attracting such parasitoids. But what about the parasitoids, such as *C. marginiventris*, that reduce the feeding rate of their hosts (Ashley, 1983; Jalali *et al.*, 1988)?

Will they reduce damage to a plant to an extent that it may increase plant fitness? Experiments with young maize plants clearly suggest that this is the case. We subjected plants either to feeding by a single unparasitized *Spodoptera littoralis* larva, or by a *S. littoralis* larva that was parasitized by *C. marginiventris*. The photograph in Fig. 7.5 shows the consequences for host development. By the time the parasitoid larva emerges from its fourth instar host, the host is far smaller than an unparasitized larva of the same age (Fig. 7.5). This difference is reflected in the dramatic difference in the amount of damage that the larvae caused to young maize plants (Fritzsche-Hoballah and Turlings, 2001). Plants that had been subjected to these different treatments were then transferred to a plot adjacent to a maize field to allow full development. The observed reduction in damage was reflected in a differential seed production, with the plants that had been attacked by a healthy caterpillar yielding less seed than those that had a parasitized caterpillar (Fritzsche-Hoballah and Turlings, 2001). Van Loon *et al.* (2000) obtained similar evidence that certain parasitoids may help increase plant fitness. They show that *Arabidopsis thaliana* plants produce considerably less seed after herbivory by healthy larvae of *Pieris rapae* (Lepidoptera: Pieridae) than after herbivory by larvae that were parasitized by the solitary endoparasitoid *Cotesia rubecula*. Future studies in this area should test for the same effects under natural conditions.

If the attraction of natural enemies is one of the functions of the plant-provided signals, then they should be emitted the most when a plant is in its most vulnerable stage. A small seedling is likely to suffer much more from an attack by a single caterpillar than a fully grown plant. To test if this difference in vulnerability of different plant stages is reflected in the strength of the signal emitted, we compared the odor production for 1-, 2-, 4-, 6-, and 8-week-old maize plants of the variety "Delprim" (Fig. 7.6). Two of the youngest leaves of each plant were damaged by scratching 2×2 cm^2 of the surface, after which 10 μl of *S. littoralis* regurgitant was applied to each damaged site. Plants were covered with a Nalophane cooking bag and volatiles were collected for 2 h about 16 h after treatment as described by Turlings *et al.* (1998b). The very small 1-week-old plants, which carried only two leaves, released relatively small amounts, while the 2-week-old plants (with three fully developed leaves) showed the highest total production of volatiles, even without correction for biomass (Fig. 7.6). The older plants, especially those of 6 and 8 weeks, released much less. Assuming that the very young plants are unlikely to be

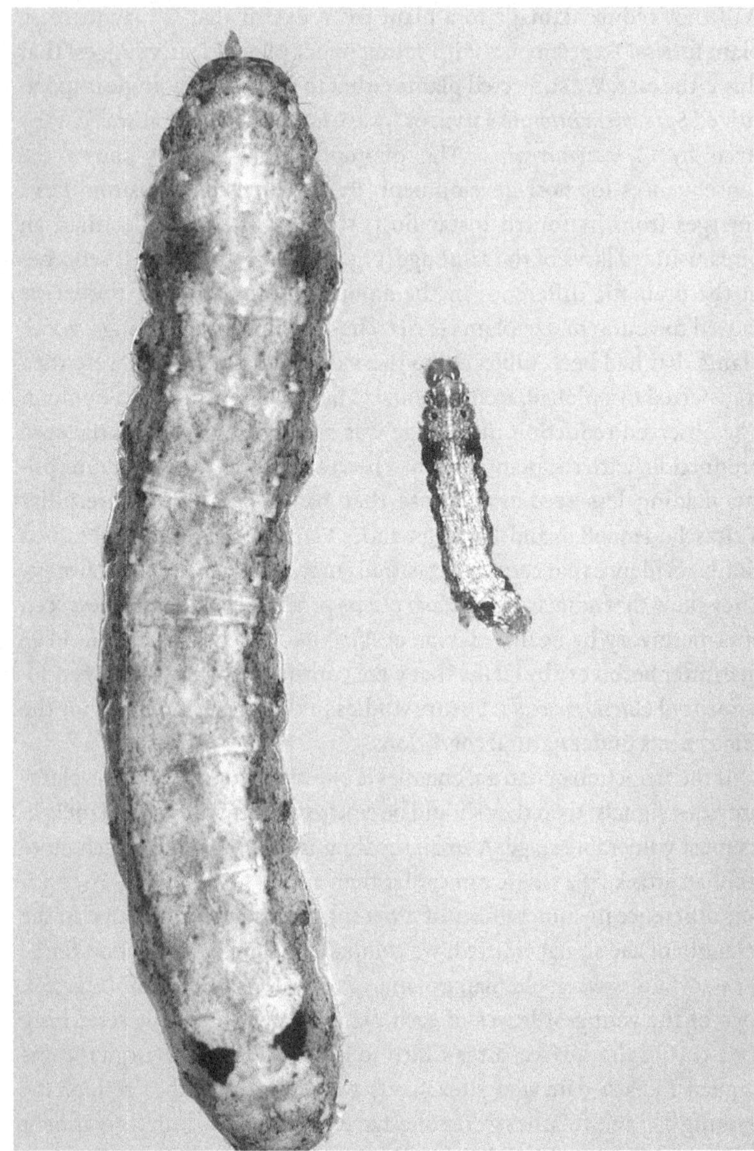

Fig. 7.5. Final size difference between a healthy, unparasitized *S. littoralis* larva (left) and a larva parasitized by *C. marginiventris* (right). (Photo by Yves Borcard.)

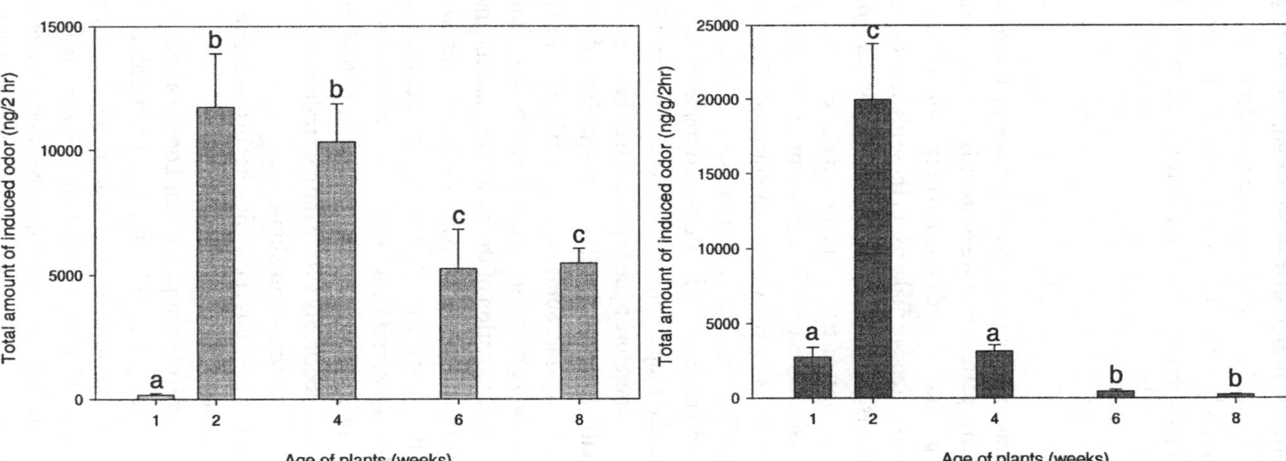

Fig. 7.6. Total of odor emission during 2 h of maize plants of different ages 16 h after being damaged (2 × 2 cm² scratched with a razor blade) and treated with 10 μl of regurgitant of *Spodoptera littoralis* larvae on each damaged site. The left graph presents data without correction for biomass and the right graph presents data after biomass correction (ng/g^{-1} dry weight). Letters above bars indicate significant differences among the different ages (Student–Newman–Keuls *post hoc* test after ANOVA).

attacked, the trend that the next youngest plants released more than the older plants supports the notion that the most vulnerable stage should emit the strongest signal. This, of course, is not evidence for a signaling function, but it fits nicely with the presumptions of the hypothesis that plants may have, in part, evolved this trait to attract the natural enemies of their herbivorous attackers.

Discussion

As ecologists we seek patterns and general rules in the relationships among organisms. One pattern that we observe is that plants respond to herbivory with the production and emission of specific volatiles and that parasitoids of herbivores tend to make use of these plant-provided signals in their search for hosts. But this is perhaps as far as we can go with our generalizations of the interactions. The enormous diversity of plants, herbivores, and parasitoids, each with a different life history, assures an equally immense diversity in the various interactions. Estimates of the number of parasitoid species are as high as several million (Godfray, 1994). Each species will have evolved different strategies to find its hosts. The same is true for the adaptations in the plants that play a key role in these interactions. It is therefore to be expected that the interactions vary in many of their details. One such detail is the specificity and reliability of the signals that plants provide. Some studies clearly indicate specificity, others do not (Dicke, 1999). We argue that availability of a reliable signal in many cases is not an adaptation of the plant, but that the insects take advantage of accidental differences in plant responses. However, with so many variations on the same theme, there are likely to be exceptions. The studies of Powell *et al.* (1998) and De Moraes *et al.* (1998) provide good evidence for specific signals. The next challenge will be to find the mechanisms that allow for this specificity within the constraints of considerable genotypic variation in plant odor emissions.

Equally variable will be the direct benefits that plants derive from parasitoids. We study species that attack the early instars of their hosts and reduce the hosts' lifetime consumption. Van Loon *et al.* (2000) suggest that probably all solitary parasitoids of Lepidoptera reduce food consumption in their host, but others parasitoids do not. Some parasitoids may even increase consumption and thus the damage that is inflicted upon a plant. One might be able to exploit such differences for studies on the evolution of parasitoid–plant interactions by comparing plant species or populations that differ in the degree of direct benefit that they receive from the natural enemies that attack their respective herbivores. An adaptive plant-signaling hypothesis would predict that plant species/populations are more likely to respond with strong odor emissions when the benefits of attracting the natural enemies are high. Long-term selection experiments could be an alternative way to test the same or similar hypotheses if the considerable variation that we observe among maize genotypes is at least partially due to additive genetic variation.

The controversy over the use of transgenic plants in agriculture has put the effects of plant traits on insects, especially on beneficial or cherished insects, in the spotlight (Losey *et al.*, 1999; Schuler *et al.*, 1999). Future research on the consequences of transgenesis will have to address the aspect of plant attractiveness to beneficial insects and may even consider manipulating this trait to enhance the pest control potential of natural enemies. Genetic modifications also have the potential to largely facilitate the testing of the importance of certain substances in attracting parasitoids by allowing for specific alterations in induced odor emissions. Thus, many of the remaining questions addressed in this chapter are likely to be answered in the near future.

Acknowledgments

We thank Philippe Jeanbourquin for conducting the experiments presented in Fig. 7.2. Håkan Häggström and Joannes Turlings provided useful comments on the initial manuscript. Stan Faeth and an anonymous reviewer also provided very constructive suggestions for improvement. We thank Hans Alborn, Jim Tumlinson, Thomas Koch, and Wilhelm Boland for supplying elicitors. Our research was supported by grants from the Swiss National Science Foundation (nos. 31–44459.95 and 31–46237.95) and a grant from the Swiss Agency for Development and Co-operation managed by the Swiss Centre for International Agriculture (ETH Zurich).

REFERENCES

Alborn, H. T., Turlings, T. C. J., Jones, T. H., Stenhagen, G., Loughrin, J. H. and Tumlinson, J. H. (1997) An elicitor of plant volatiles from beet armyworm oral secretion. *Science* **276**: 945–949.

Alborn, H. T., Jones, T. H., Stenhagen, G. and Tumlinson, J. H. (2000) Identification and synthesis of volicitin and related components from beet armyworm oral secretions. *Journal of Chemical Ecology* **26**: 203–220.

Agrawal, A. A. (1998) Induced responses to herbivory and increased plant performance. *Science* **279**: 1201–1202.

Ashley, T. R. (1983) Growth pattern alterations in fall armyworm, *Spodoptera frugiperda*, larvae after parasitization by *Apanteles marginiventris*, *Campoletis grioti*, *Chelonomus insularis*, and *Eiphosoma vitticole*. *Florida Entomologist* **66**: 260–266.

Baldwin, I. T. (1998) Jasmonate-induced responses are costly but benefit plants under attack in native populations. *Proceedings of the National Academy of Sciences, USA* **95**: 8113–8118.

Benrey, B., Callejas, A., Rios, L., Oyama, K. and Denno, R. F. (1998) The effects of domestication of *Brassica* and *Phaseolus* on the interaction between phytophagous insects and parasitoids. *Biological Control* **11**: 130–140.

Boland, W., Hopke, J., Donath, J., Nüske, J. and Bublitz, F. (1995) Jasmonsäure- und Coronatin-induzierte Duftproduktion in Pflanzen. *Angewandte Chemie* **107**: 1715–1717.

Boland, W., Koch, T., Krumm, T., Piel, J. and Jux, A. (1999) Induced biosynthesis of insect semiochemicals in plants. In *Insect–Plant Interactions and Induced Plant Defence*, ed. D. J. Chadwick and J. A. Goode, pp. 110–162. London: Novartis Foundation and Chichester: John Wiley.

Byers, J. R., Yu, D. S. and Jones, J. W. (1993) Parasitism of the army cutworm, *Euxoa auxiliaris* (Grt.) (Lepidoptera: Noctuidae), by *Copidosoma bakeri* (Howard) (Hymenoptera: Encyrtidae) and effect on crop damage. *Canadian Entomologist* **125**: 329–335.

De Moraes, C. M., Lewis, W. J., Paré, P. W. and Tumlinson, J. H. (1998) Herbivore-infested plants selectively attract parasitoids. *Nature* **393**: 570–573.

Dicke, M. (1994) Local and systemic production of volatile herbivore-induced terpenoids: their role in plant–carnivore mutualism. *Journal of Plant Physiology* **143**: 465–472.

Dicke, M. (1999) Are herbivore-induced plant volatiles reliable indicators of herbivore identity to foraging carnivorous arthropods? *Entomologia Experimentalis et Applicata* **91**: 131–142.

Dicke, M. and Sabelis, M. W. (1988) How plants obtain predatory mites as bodyguards. *Netherlands Journal of Zoology* **38**: 148–165.

Dicke, M. and Sabelis, M. W. (1989) Does it pay plants to advertise for bodyguards? Towards a cost–benefit analysis of induced synomone production. In *Causes and Consequences of Variation in Growth Rate and Productivity of Higher Plants*, ed. H. Lambers, M. L. Cambridge, H. Konings and T. L. Pons, pp. 341–358. The Hague, Netherlands: SPB Academic Publishers.

Dicke, M. and Sabelis, M. W. (1992) Costs and benefits of chemical information conveyance: proximate and ultimate factors. In *Insect Chemical Ecology: An Evolutionary Approach*, ed. B. D. Roitberg and M. B. Isman, pp. 122–155. New York: Chapman and Hall.

Dicke, M. and Vet, L. E. M. (1999) Plant–carnivore interactions: evolutionary and ecological consequences for plant, herbivore and carnivore. *In Herbivores: Between Plants and Predators*, ed. H. Olf, V. K. Brown and R. H. Drent, pp. 483–520. Oxford: Blackwell Science.

Dicke, M., Sabelis, M. W., Takabayashi, J., Bruin, J. and Posthumus, M. A. (1990a) Plant strategies of manipulating predator–prey interactions through allelochemicals: prospects for application in pest control. *Journal of Chemical Ecology* **16**: 3091–3118.

Dicke, M., van Beek, T. A., Posthumus, M. A., Ben Dom, N., van Bokhoven, H. and de Groot, Æ. (1990b) Isolation and identification of volatile kairomone that affects acarine predator–prey interactions: involvement of host plant in its production. *Journal of Chemical Ecology* **16**: 381–396.

Du, Y.-J., Poppy, G. M., Powell, W. J., Pickett, J. A., Wadhams, L. J. and Woodcock, C. M. (1998) Identification of semiochemicals released during aphid feeding that attract parasitoid *Aphidius ervi*. *Journal of Chemical Ecology* **24**: 1355–1368.

Eller, F. J., Tumlinson, J. H. and Lewis, W. J. (1988) Beneficial arthropod behavior mediated by airborne semiochemicals: source of volatiles mediating the host-location flight behavior of *Microplitis croceipes* (Cresson) (Hymenoptera: Braconidae), a parasitoid of *Heliothis zea* (Boddie) (Lepidoptera: Noctuidae). *Environmental Entomology* **17**: 745–753.

Eller, F. J., Tumlinson, J. H. and Lewis, W. J. (1992) Effect of host diet and preflight experience on the flight response of *Microplitis croceipes* (Cresson). *Physiological Entomology* **17**: 235–240.

Faeth, S. H. (1994) Induced plant responses: effects on parasitoids and other natural enemies of phytophagous insects. In *Parasitoid Community Ecology*, ed. B. A. Hawkins and W. Sheehan, pp. 245–260. Oxford: Oxford University Press.

Fritzsche-Hoballah, M. E. and Turlings, T. C. J. (2001) Experimental evidence that plants under caterpillar attack may benefit from attracting parasitoids. *Evolutionary Ecology Research* **3**: 553–565.

Gershenzon, J. (1994) Metabolic costs of terpenoid accumulation in higher plants. *Journal of Chemical Ecology* **20**: 1281–1328.

Gershenzon, J. and Croteau, R. B. (1993) Terpenoid biosynthesis: the basic pathway and formation of monoterpenes, sesquiterpenes, and diterpenes. In *Lipid Metabolism in Plants*, ed. T. S. Moore Jr., pp. 339–388. Boca Raton, FL: CRC Press.

Godfray, H. C. J. (1994). *Parasitoids: Behavior and Evolutionary Ecology*. Princeton, NJ: Princeton University Press.

Godfray, H. C. J. (1995) Communication between the first and third trophic levels: an analysis using biological signalling theory. *Oikos* **72**: 367–374.

Gouinguené, S. (2000) Specificity and variability in induced volatile signalling in maize plants. PhD thesis, University of Neuchâtel, Switzerland.

Gouinguené, S., Degen, T. and Turlings, T. C. J. (2001) Variability in herbivore-induced odour emissions among maize cultivars and their wild ancestors (teosinte). *Chemoecology* **11**: 9–16.

Harrewijn, P., Minks, A. K. and Mollema, C. (1994/5) Evolution of plant volatile production in insect–plant relationships. *Chemoecology* **5/6**: 55–73.

Hopke, J., Donath, J., Blechert, S. and Boland, W. (1994) Herbivore-induced volatiles: the emission of acyclic homoterpenes from leaves of *Phaseolus lunatus* and *Zea mays* can be triggered by a β-glucosidase and jasmonic acid. *FEBS Letters* **352**: 146–150.

Jalali, S. K., Singh, S. P. and Ballal, C. R. (1988) Effect of parasitism by *Cotesia marginiventris* on consumption and utilization of artificial diet by larvae of *Spodoptera litura* (Lepidoptera: Noctuidae). *Indian Journal of Agricultural Sciences* **58**: 529–532.

Janssen, A., Pallini, A., Venzon, M. and Sabelis, M. W. (1998) Behaviour and indirect interactions in food webs of plant-inhabiting arthropods. *Experimental and Applied Acarology* **22**: 497–521.

Janzen, D. Z. (1966) Coevolution of mutualism between ants and acacias in Central America. *Evolution* **20**: 249–275.

Langenheim, J. H. (1994) Higher plant terpenoids: a phytocentric overview of their ecological roles. *Journal of Chemical Ecology* **20**: 1223–1280.

Lewis, W. J. and Tumlinson, J. H. (1988) Host detection by chemically mediated associative learning in a parasitic wasp. *Nature* **331**: 257–259.

Loke, W. H., Ashley, T. R. and Sailer, R. I. (1983) Influence of fall armyworm, *Spodoptera frugiperda* (Lepidoptera: Noctuidae) larvae and corn plant damage on host finding in *Apanteles marginiventris* (Hymenoptera: Braconidae). *Environmental Entomology* **12**: 911–915.

Losey, J., Rayor, L. and Carter, M. (1999) Transgenic pollen harms monarch larvae. *Nature* **399**: 214.

Loughrin, J. H., Manukian, A., Heath, R. R., Turlings, T. C. J. and Tumlinson, J. H. (1994) Diurnal cycle of emission of induced volatile terpenoids by herbivore-injured cotton. *Proceedings of the National Academy of Sciences, USA* **91**: 11836–11840.

Loughrin, J. H., Manukian, A., Heath, R. R. and Tumlinson, J. H. (1995) Volatiles emitted by different cotton varieties damaged by feeding beet armyworm larvae. *Journal of Chemical Ecology* **21**: 1217–1227.

Mattiacci, L., Dicke, M. and Posthumus, M. A. (1994) Induction of parasitoid attracting synomone in Brussels sprouts plants by feeding of *Pieris brassicae* larvae: role of mechanical damage and herbivore elicitor. *Journal of Chemical Ecology* **20**: 2229–2247.

Mattiacci, L., Dicke, M. and Posthumus, M. A. (1995) β-Glucosidase: an elicitor of herbivore-induced plant odor that attracts host-searching parasitic wasps. *Proceedings of the National Academy of Sciences, USA* **92**: 2036–2040.

McCall, P. J., Turlings, T. C. J. ,Lewis W. J. and Tumlinson, J. H. (1993) Role of plant volatiles in host location by the specialist parasitoid *Microplitis croceipes* Cresson (Braconidae: Hymenoptera). *Journal of Insect Behavior* **6**: 625–639.

McKey, D. (1988) Promising new directions in the study of ant–plant mutualisms. In *Proceedings of the 14th International Botanical Congress*, ed. W. Greuter and B. Zimmer Koeltz, pp. 335–355. Konigstein, Germany: Taunus.

Nordlund, D. A., Lewis, W. J. and Altieri, M. A. (1988) Influences of plant-produced allelochemicals on the host/prey selection behavior of entomophagous insects. In *Novel Aspects of Insect–Plant Interactions*, ed. P. Barbosa and D. K. Letourneau, pp. 65–90. New York: John Wiley.

Oliveira, P. S. (1997) The ecological function of extrafloral nectaries: herbivore deterrence by visiting ants and reproductive output in *Caryocar brasiliense* (Caryocaraceae). *Functional Ecology* **11**: 323–330.

Paré, P. W. and Tumlinson, J. H. (1997) *De novo* biosynthesis of volatiles induced by insect herbivory in cotton plants. *Plant Physiology* **114**: 1161–1167.

Paré, P. W., Alborn, H. T. and Tumlinson, J. H. (1998) Concerted biosynthesis of an insect elicitor of plant volatiles. *Proceedings of the National Academy of Sciences, USA* **95**: 13971–13975.

Parker, F. D. and Pinnell, R. E. (1973) Effect of food consumption of the imported cabbageworm when parasitized by two species of *Apanteles*. *Environmental Entomology* **2**: 216–219.

Pohnert, G., Jung, V., Haukioja, E., Lempa, K. and Boland, W. (1999) New fatty acid amides from regurgitant of Lepidoptera (Noctuidae, Geometridae) caterpillars. *Tetrahedron* **55**: 11275–11280.

Price, P. W. (1986) Ecological aspects of host plant resistance and biological control:

interactions among three trophic levels. In *Interactions of Plant Resistance and Parasitoids and Predators of Insects*, ed. D. J. Boethel and R. D. Eikenbary, pp. 11–30. New York: John Wiley.

Price, P. W., Bouton, C. E., Gross, P., McPheron, B. A., Thompson, J. N. and Weis, A. E. (1980) Interactions among three trophic levels: influence of plant on interactions between insect herbivores and natural enemies. *Annual Review of Ecology and Systematics* **11**: 41–65.

Powell, W., Pennacchio, F., Poppy, G. M. and Tremblay, E. (1998) Strategies involved in the host location of hosts by the parasitoid *Aphidius ervi* Haliday (Hymenoptera: Braconidae: Aphidiinae). *Biological Control* **11**: 104–112.

Quiroz, A., Pettersson, J., Pickett, J. A., Wadhams, L. J. and Niemyer, H. M. (1997) Semiochemicals mediating spacing behavior of bird cherry-oat aphid, *Rhopalosiphum padi*, feeding on cereals. *Journal of Chemical Ecology* **23**: 2599–2607.

Rahman, M. (1970) Effect of parasitism on food consumption of *Pieris rapae* larvae. *Journal of Economic Entomology* **63**: 820–821.

Sabelis, M. W. and de Jong, M. C. M. (1988) Should all plants recruit bodyguards? Conditions for a polymorphic ESS of synomone production in plants. *Oikos* **53**: 247–252.

Sabelis, M. W., van Baalen, M., Bakker, F. M., Bruin, J., Drukker, B., Egas, M., Janssen, A. R. M., Lesna, I. K., Pels, B., van Rijn, P. C. J. and Scutareau, P. (1999) The evolution of direct and indirect plant defence against herbivorous arthropods. In *Herbivores: Between Plants and Predators*, ed. H. Olf, V. K. Brown and R. H. Drent, pp. 109–166. Oxford: Blackwell Science.

Shimoda, T., Takabayashi, J., Ashirara, W. and Takafuji, A. (1997) Response of predatory insect *Scolothrips takahashii* toward herbivore-induced plant volatiles under laboratory and field conditions. *Journal of Chemical Ecology* **23**: 2033–2048.

Schuler, T., Poppy, G., Kerry, B. and Denholm, I. (1999) Potential side effects of insect-resistant transgenic plants on arthropod natural enemies. *Tibtech* **17**: 210–216.

Simms, E. L. (1992) Costs of plant resistance to herbivory. In *Plant Resistance to Herbivores and Pathogens: Ecology, Evolution, and Genetics*, ed. R. S. Fritz and E. L. Simms, pp. 392–425. Chicago, IL: University of Chicago Press.

Takabayashi, J., Takahashi, S., Dicke, M. and Posthumus, M. A. (1995) Developmental stage of herbivore *Pseudaletia separata* affects production of herbivore-induced synomone by corn plants. *Journal of Chemical Ecology* **3**: 273–287.

Thaler, J. S. (1999) Jasmonate-inducible plant defences cause increased parasitism of herbivores. *Nature* **399**: 686–688.

Tumlinson, J. H., Turlings, T. C. J. and Lewis, W. J. (1992) The semiochemical complexes that mediate insect parasitoid foraging. *Agricultural Zoological Review* **5**: 221–252.

Turlings, T. C. J. and Benrey, B. (1998) Effects of plant metabolites on the behavior and development of parasitic wasps. *Ecoscience* **5**: 321–333.

Turlings, T. C. J., Tumlinson, J. H. and Lewis, W. J. (1990) Exploitation of herbivore-induced plant odors by host-seeking parasitic wasps. *Science* **250**: 1251–1253.

Turlings, T. C. J., Tumlinson, J. H., Eller F. J. and Lewis, W. J. (1991a) Larval-damaged plants: source of volatile synomones that guide the parasitoid *Cotesia marginiventris* to the micro-habitat of its hosts. *Entomologia Experimentalis et Applicata* **58**: 75–82.

Turlings, T. C. J., Tumlinson, J. H., Heath, R. R., Proveaux, A. T. and Doolittle, R. E. (1991b) Isolation and identification of allelochemicals that attract the larval parasitoid *Cotesia marginiventris* (Cresson) to the micro-habitat of one of its hosts. *Journal of Chemical Ecology* **17**: 2235–2251.

Turlings, T. C. J., McCall, P. J., Alborn, H. T. and Tumlinson, J. H. (1993a) An elicitor in caterpillar oral secretions that induces corn seedlings to emit chemical signals attractive to parasitic wasps. *Journal of Chemical Ecology* **19**: 411–425.

Turlings, T. C. J., Wäckers, F., Vet, L. E. M., Lewis, W. J. and Tumlinson, J. H. (1993b) Learning of host-finding cues by hymenopterous parasitoids. In *Insect Learning: Ecological and Evolutionary Perspectives*, ed. D. R. Papaj and A. Lewis, pp. 51–78. New York: Chapman and Hall.

Turlings, T. C. J., Loughrin, J. H., Röse, U., McCall, P. J., Lewis, W. J. and Tumlinson, J. H. (1995) How caterpillar-damaged plants protect themselves by attracting parasitic wasps. *Proceedings of the National Academy of Sciences, USA* **95**: 4169–4174.

Turlings, T. C. J., Lengwiler, U. B., Bernasconi, M. L. and Wechsler, D. (1998a) Timing of induced volatile emissions in maize seedlings. *Planta* **207**: 146–152.

Turlings, T. C. J., Bernasconi, M., Bertossa, R., Caloz, G., Bigler, F. and Dorn, S. (1998b) The induction of volatile emissions in maize by three herbivore species with different feeding habits: possible consequences for their natural enemies. *Biological Control* **11**: 122–129.

Turlings, T. C. J., Alborn, H. T., Loughrin, J. H. and Tumlinson, J. H. (2000) Volicitin, an elicitor of maize volatiles in the oral secretion of *Spodoptera exigua*: isolation and bio-activity. *Journal of Chemical Ecology* **26**: 189–202.

van der Meijden, E., and Klinkhamer, P. G. L. (2000) Conflicting interests of plants and the natural enemies of herbivores. *Oikos* **89**: 202–208.

van Loon, J. J. A., de Boer, G. and Dicke, M. (2000) Parasitoid–plant mutualism: parasitoid attack of herbivore increases plant reproduction. *Entomologia Experimentalis et Applicata* **97**: 219–227.

Vet, L.E.M. and Dicke, M. (1992) Ecology of infochemical use by natural enemies in a tritrophic context. *Annual Review of Entomology* **37**: 141–172.

Vet, L. E. M., Wäckers, F. L. and Dicke, M. (1991) How to hunt for hiding hosts: the reliability–detectability problem in foraging parasitoids. *Netherlands Journal of Zoology* **41**: 202–213.

Vet, L. E. M., Lewis, W. J. and Cardé, R. T. (1995) Parasitoid foraging and learning. In *Chemical Ecology of Insects*, vol. 2, ed. R. T. Cardé and W. J. Bell, pp. 65–101. New York: Chapman and Hall.

Vinson, S. B., Elzen, G. W. and Williams, H. J. (1987) The influence of volatile plant allelochemicals on the third trophic level (parasitoids) and their herbivorous hosts. In *Insects–Plants*, ed. V. Labeyerie, G. Fabres and D. Lachaise, pp. 109–114. Dordrecht, Netherlands: Junk Publishers.

Wadhams, L. J., Birkett, M. A., Powell, W. and Woodcock, C. M. (1999) Aphids, predators, and parasitoids. In *Insect–Plant Interactions and Induced Plant Defence*, ed. D. J. Chadwick and J. A. Goode, pp. 60–67. London: Novartis Foundation and Chichester: John Wiley.

Wäckers, F. and Lewis, W. J. (1994) Olfactory and visual learning and their combined influence on host site location by the parasitoid *Microplitis croceipes* (Cresson). *Biological Control* **4**: 105–112.

Wang, R.-L., Stec, A., Hey, J., Lukens, L. and Doebley, J. (1999) The limits of selection during maize domestication. *Nature* **398**: 236–239.

Whitman, D. W. (1994) Plant bodyguards: mutualistic interactions between plants and the third trophic level. In *Functional Dynamics of Phytophagous Insects*, ed. T. N. Ananthakrishnan, pp. 207–248. New Delhi: Oxford and IBH Publishing Co.

Zangerl, A. R. and Bazzaz, F. A. (1992) Theory and pattern in plant defense allocation. In *Plant Resistance to Herbivores and Pathogens: Ecology, Evolution, and Genetics*, ed. R. S. Fritz and E. L. Simms, pp. 363–391. Chicago, IL: University of Chicago Press.

JÉRÔME CASAS AND IMEN DJEMAI

8

Canopy architecture and multitrophic interactions

Introduction

Predator–prey, parasitoid–host, and other arthropod interactions do not occur in a vacuum, nor in a featureless world, but in a highly structured and complex environment. This basic observation has triggered numerous theoretical and empirical studies at the population level. Many are centered on the dynamics of populations occupying different patches (summarized in Hassell, 2000). A metapopulation framework implies a spatial arrangement of patches and movement of predators between them. However, once in a patch, a homogeneous spatial situation is again assumed, and predators search at random. In fact, we know of very few examples of arthropod predator–prey or host–parasitoid studies which do incorporate the geometry of the environment at a smaller scale than a patch. In particular, we do not know any study that satisfactorily quantifies the architecture of the plant canopies and its influences on the outcomes of the interactions. This is surprising given that a great majority of predator–prey and parasitoid–host interactions occur in vegetation. Filling this gap is the thrust of this chapter.

The disregard for the architecture of the environment, in particular plant architecture, has two explanations. First, concepts and methods for mapping and modeling plant architecture have been developed only recently, i.e., mainly from the 1980s. Plant architecture, in particular tree architecture, has been the subject of intense research for quite some time (see for example Halle and Oldeman, 1970; Halle *et al.*, 1978), but this work was of a qualitative nature. Thus, the knowledge of how to measure and model plant architecture is too recent to have penetrated all fields of ecology. Plant canopies are highly complex modular structures that can be described both in topological and geometrical terms. Topological information specifies the physical relationships between the different components of the structure while geometrical information specifies, *inter alia*, the shape, size, and spatial location of the components (Godin *et al.*, 1999). The architecture of a plant is an emergent property of its morphogenetic rules. Several ways to model morphogenetic rules have recently been proposed and the field is very active (see Room *et al.*, 1996; Michalewicz, 1999; Pearcy and Valladares, 1999; Gauthier *et al.*, 2000; Parker and Brown, 2000). Second, modeling interactions without paying attention to the fine-grained structure of the environment enables us to use variants of diffusion equations, for which a large body of knowledge is available (see Tilman and Kareiva, 1997; Shigesada and Kawasaki, 1997; Turchin, 1998). As comfortable as these assumptions may be, we are left with a large number of interactions for which a consideration of the geometry of the environment seems mandatory. The following example illustrates the kind of situations we envisage and the type of problems we would like to solve.

Imagine a ladybird beetle moving through vegetation, searching for prey of low abundance. Most of the stems and leaves and other structures the beetle explores are void of prey. Except for a few locations with hosts, and a few more with "hints" to the predator such as honeydew drops, the animal is moving in an empty "maze." The "maze" constrains its movement by determining which routes are possible and which are not and is characterized by having components of both order and disorder. The animal itself makes different behavioral decisions in seemly similar conditions, i.e., it also displays some degree of "randomness," real or not, in its movement rules. Finally, the distribution of prey is most likely clumped, implying that the travel time between clusters of hosts will be very long. These long journeys will be spent finding a way through the maze.

The above description calls for a thorough understanding of at least three components of the multitrophic interaction: (1) the architecture of the environment, (2) the distribution of prey in the environment, and (3) the intrinsic movement rules of the predator. It is only after we have all three components that we can answer the following questions:

- What is the relative impact of plant architecture and prey distribution on predator searching efficiency?
- How is the risk of predation distributed among prey?
- Are the basic laws of diffusion equations valid, such as the linear increase mean square displacement?

- If yes, under which conditions can we disregard the architecture of the environment and use a "mean field" approximation?
- If no, what are the consequences of the anomalous diffusion in terms of individual and population parameters?
- Finally, how much biological realism must be sacrificed to construct a robust model for plant architecture and animal movement?

The above questions would be best answered by blending harmoniously the two themes of movement processes and geometrical systems. In practical terms, however, it is much easier to emphasize one of the themes and simplify the other. Thus, this chapter has been written from the perspective of a predator foraging for stationary prey in plant canopies of given architecture. Hence, we do not deal with modeling plant architecture *per se* and point the interested reader to the above entry points in the literature. The framework advocated here could easily be extended to nectar foraging and pollinator movement (see for example Pyke, 1978; Ganeshaiah and Veena, 1988), but these are not multitrophic interactions as understood here. A treatment of these interactions along the lines described in this chapter has not been attempted so far. Also untouched is an aquatic perspective on these issues, as vegetation does act in a very similar way on predator–prey interactions in aquatic environments (see for example Russo, 1987).

The organization of the chapter is as follows. We first conduct a stocktaking of the published works on arthropod interactions in which plant architecture has been studied. While we consistently use predator–prey systems for simplicity, parasitoid–host interactions can be analyzed in the same way. We will see that some of the ideas can even be applied to phytophagous insects, as they must also solve the problem of resource location in a highly heterogeneous environment. Second, we explore the impact of plant architecture on the efficiency of the predator. Then, we turn to the population level and analyze the impact of plant architecture on predation rate. The dozen or so studies provide highly useful information on several aspects, but concomitantly give a somewhat fragmented perspective. The need for a synthesis is obvious, and we present a summary of ideas emerging from the study of random walk in a random medium. We advocate this framework as the best available to examine multitrophic interactions in plant canopies, as simply observing predators moving in plant canopies is not sufficient for tackling the above questions. We end the chapter by calling attention to several other fascinating and unexplored aspects of multitrophic interactions in complex environments.

Canopy geometry, prey distribution, and predator movement: a stocktaking

The aim of this section is to review the few publications demonstrating how the complex geometry of the plant leads to a heterogeneous distribution of prey and a heterogeneous distribution of predator effort. Andow and Prokrym (1990) suggested that there are three components of plant architecture relevant to foraging predators and parasitoids: (1) the plant size and surface area, (2) the structural heterogeneity among plant parts such as flower heads and stems, and (3) the connectivity of the plant parts. Their experiments approximated the structural complexity of plants by providing paper panels of different geometries to the egg parasitoid *Trichogramma nubilalis*. The number of parasitized egg masses of the host-moth *Ostrinia nubilalis* was analyzed as a function of the complexity of the paper panel. They made the important distinction between two different mechanisms acting on parasitism rates. On the one hand, a parasitoid may find the hosts more or less easily due to the structure of the environment, given the same searching intensity. On the other hand, parasitoids may forage with different intensities, irrespective of the presence of hosts, but as a function of the complexity of the environment. Structural complexity caused a threefold decrease in parasitism rate between the simple and complex environments. Part of the decrease in parasitism was due to the fact that *Trichogramma* searched simple surfaces devoid of hosts more intensively than complex ones. The major implication of this work is that decision rules such as giving-up time are influenced by the structural complexity of the environment *per se*. The results obtained by Andow and Prokrym (1990) were later corroborated by similar results by Lukianchuk and Smith (1997) using a different *Trichogramma* species and greater surface complexity. These important results have yet to be incorporated into works dealing with foraging in realistic complex environments.

Given the highly structured environment of plant canopies, prey will not be randomly or evenly distributed in the canopy. Nor do predators forage randomly or evenly. They tend to follow the structure of the canopy, but sometimes only partially. For example, the aphid parasitoid *Aphidius rhopalosiphi* spends most of its time on the leaves and little on the ear of wheat, the preferred feeding site of its aphid host *Sitobion avenae* (Gardner and Dixon, 1985). The parasitoids were reluctant to move on to the ear and normally spent little time there. Another parasitoid of aphids, *Aphidius funebris*, attacks its host in a typical body posture that requires it to attack from leaves adjacent to the host colony (Weisser, 1995). A similar problem of prey accessibility was observed by Grevstad and Klepetka (1992), who found that aphids on *Brassica oleracea caulorapa* were mainly located on the middle of the underside of leaves, an area ladybird beetles could not get at because they could not grip to the undersurface. Consequently, the beetles tended to follow leaf edges and stems rather than the flat surface. Leaf edge is also the preferred route taken by the predator *Anthocoris confusus* during its search for its aphid prey (Evans, 1976). Predators moving in plant canopies composed of needle-like structures rather than leaves encounter similar problems. As described by Vohland (1996), needle density is higher in the upper and outer sectors of pine trees. This strongly influences the time spent by the older stages of the coccinellid *Scymnus nigrinus*, which spend most of their time there. This is also where prey densities are highest. The one-dimensional geometry of the needles "guides" the predator to its prey, and small larvae were very reluctant to cross over the shaft between bark and needle, where the prey feeds. Finally, using normal versus leafless peas, Kareiva and Sahakian (1990) demonstrated that the importance of plant morphological variation to herbivores sometimes becomes apparent only in a multitrophic framework. They demonstrated that different species of ladybirds were less effective in the normal peas, as they fell off the plants more often than in the leafless canopies. In contrast, whereas plant canopy architecture can impede predators by making the "maze" complex, it can also influence the aggregation of prey and predators, as nicely demonstrated by Kaiser (1983) using artificial arenas. He showed that borders influence both prey and predator spider mites in such a way that both stay more often along borders. The shorter the total length of borders, the higher the probability of contacts. As a consequence, different leaf forms of the same area can lead to different predation rates.

These examples show that the spatial coincidence of prey distributions and predator foraging effort is a key aspect of the interaction. They also demonstrate that descriptions of predator movement and prey distributions anchored in a framework based on homogeneous environment are not realistic. Obviously, it will often be difficult to quantify the effect of plant architecture on predators, because prey may change their behavior and location in the presence of predators, and vice versa.

Canopy geometry and predator movement: implications for multitrophic interactions

The effects of the complex geometry of plant canopies operate at both the individual and population levels. We first focus on the efficiency of predators, a behavioral trait, as function of plant architectural complexity.

The efficiency of a moving predator can be defined in many different ways that reflect the influence of plant architecture. Isenhour and Yeargan (1981) defined a measure of efficiency of explored regions as the distance traveled per encounter with a prey. They found very large differences for the bug *Orius insidiosus* attacking thrips on soybeans. The efficiency was at least an order of magnitude higher on the petiolus junction and on the midrib than on the leaf periphery. Efficiency can be also defined as the speed at which an animal travels a given distance. By varying the degree of bean-plant leaf overlap, Kareiva and Perry (1989) created two scenarios for the ladybird *Hippodamia convergens*: a highway and a gap situation. The highway situation enabled the ladybird to travel four times further (net displacement) per minute. The difference was mainly due to the high frequency of reversals in the gap situation. This fascinating study needs confirmation, as advocated by the authors themselves. This example is highly reminiscent of percolation theory, in which the probability of reaching a given point in space or the probability of crossing the whole medium is a function of the connectance between elements (Stauffer, 1985). Efficiency can be also defined as time allocation to given tasks. Suverkropp (1997) found that decreasing the time allocated to *Trichogramma* searching in a single plant *increased* the attack rate. While a longer search on a plant does of course increase the likelihood of finding an egg mass on that plant, the time spent is better allocated to checking other plants, given that hosts are randomly distributed among plants. This is an interesting way to avoid the complex architecture of canopies: instead of getting lost in complex structures, abandon them quickly and move somewhere else. Finally, efficiency can be synonymous with attack rate, which is the number of aphids killed per unit time. Grevstad and Klepetka (1992) found that the attack rate of four different ladybird beetles on aphids was much more influenced by the plant on which they foraged than by the ladybird species. This was due to differences in encounter rates, which were again function of plant species rather than beetle-specific. The rate at which beetles fell off the plant was also a function of the plant species, in particular the slipperiness of the plant surface.

We now turn to the population level and analyze predation rates as function of plant architecture. The interplay between host density and plant architecture in determining parasitism rate at the population level has been worked out for an apple leaf miner by mapping both the architecture of trees over three years old as well as the position of unparasitized and hosts parasitized by *Cirrospilus vittatus* (Casas, 1991). The visit of a female parasitoid to a tree results in the parasitism of one or more hosts. While the first host is assumed to be chosen at random within the tree, further hosts can be parasitized, at random, within a spherical radius of 40 cm. The center of the moving sphere is the host currently under attack. In a young tree, parasitism resulting from a single visit made by a foraging female is usually restricted to hosts on the same branch, as most of the neighbor leaves within the sphere are on the same branch. Older trees have more branches and hence a more complex architecture. Attacks by a parasitoid then include different branches because they often intermix. In these cases, females no longer follow branches individually. As the number of hosts parasitized per attack is low due to the low fecundity of this species, one can expect inversely density-dependent parasitism per visit. In fact, parasitism rates at the branch level will be lower on older trees, as the attacks resulting from a single visit will be spread over several branches (Fig. 8.1). The relationship between the movement of the individual parasitoid (dimensions of the sphere of activity) and the tree architecture (dimensions and relative location of the branches) is of prime importance.

Using artificial plants of varying architecture, Geitzenauer and Bernays (1996) found that paper wasps attacked tobacco budworms at higher rates in architecturally less complex canopies. The mechanisms were behavioral, as it took them less time to locate hosts in the simpler canopies. The giving-up time was also higher in those plants. Also using artificial models, Frazer and McGregor (1994) showed major differences in giving-up time of a coccinellid beetle as a function of the form and angles of attachment of stem and leaf models. These differences are expected to result in major differences in the density of predators in a given crop and major differences in attack rates. All architectural characterizations of plant size, height, leaf number, leaf surface area, and branch number were negatively correlated with the attack rate of *Leptomastix dactylopii*, a parasitoid of the citrus mealybug (Cloyd and Sadof, 2000). The form of the func-

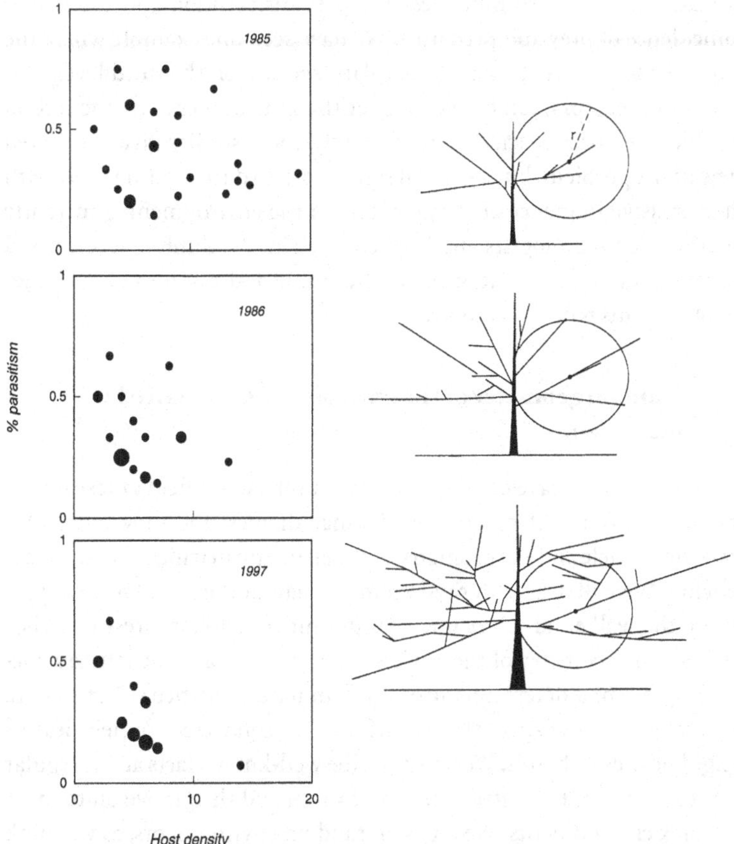

Fig. 8.1. Parasitism at the branch level per year in a tree (left) and canopy architecture of the tree during three successive years (right). Only branches bearing hosts are represented. The radius determining the volume in which a host can be attacked from a given location is also given.

tional response may also change as function of plant architecture. This may have a profound effect on the stability of predator–prey relationships. Messina and Hanks (1998) described a shift from a functional response type II to type III of a ladybird beetle foraging for aphids on two different plants. The shift was due to a density-dependent change in the proportion of aphids in refuges, such as rolled leaves, on one of the plants.

The general message from these studies is that plant architecture has major influences at different levels of multitrophic interactions: at the behavioral decisions of predators, on their own efficiencies, and on the predation rate. It also influences the prey distribution and the spatial coincidence of prey and predators. We have seen one example where the influence of the first trophic level (plant species) on the third level (efficiency of predators) can be even bigger than the influence on the second trophic level (prey species). Unfortunately, these studies give a scattered view of the problem, but a general framework to deal quantitatively with the pervasive influence of the geometry of the environment is currently unavailable for ecologists and is sorely needed. We think that one based on random walks in random geometries could address most of the issues involved. This is our next topic.

Random geometry of the environment and particle movement

We momentarily leave ecology and enter a very active field of research in statistical physics. It covers two distinct themes: the movement of a walking particle and the geometry of the structure in which the particle is moving. We will deal with random movement of the random walk type, where the walker advances one step in unit time to a nearest neighbor site. For the geometry of the environment, we will assume lattice structures, either of a deterministic or random nature. Lattice structures are discrete versions of space that consist of sites connected to their nearest neighbor sites by bonds. We first describe well-known facts about regular diffusion in a regular lattice and continuum and then move on to more complex environments. We focus on random environments, as we think they better represent the architecture experienced by real insects. Movement in a tree-like structure, the comb, is used as an example to illustrate the effect of randomness in the environment's geometry and the effect of bias on the diffusion properties of the particle. We highlight the breakdown of many assumptions underlying the diffusion equation approximation.

Regular diffusion

Let us consider the most simple random walk in an homogeneous environment, the line $(\ldots -4, -3, -2, -1, 0, +1, +2, +3, +4, \ldots)$. The particle hops with steps of $+/-1$. The particle starts at 0 and moves towards the left with probability p and towards the right with probability q. For a small number of steps, it is possible to calculate the exact location of the particle after n steps using the binomial distribution. As an example, assume $p = 0.6$, $q = 0.8$, and $s^2 = 4pq = 0.96$. After 40 steps, the probability of being no further than 10 steps from the origin is 0.68. After 1 million steps, the particle's position will lie almost certainly within a mere 4000 units of its starting-point. In other words, the particle does not move very far from the origin, even after many steps. Working out these probabilities becomes tedious when the number of steps becomes large, and we can make use of the central limit theorem and related theorems of probability theory to find continuum limits of random walks. By using limiting arguments, it is indeed possible to produce differential equations describing the continuum limit of this walk in both space and time. In other words, we give up the lattice structure and enter into a continuum, another homogeneous environment. For simplicity, we use a one-dimensional environment. The position of the particle X(t) becomes then the Gaussian limiting distribution, N (at D^2t), which is the outcome of the diffusion equation, or Brownian motion with drift a and variance D^2 (Fig. 8.2).

Large deviations are rare occurrences in the Gaussian distribution. Indeed, a Gaussian variable with fluctuations σ diverges from the mean by more than 2 σ in only 5% of cases. Fluctuations of more than 10 σ are almost impossible, with a probability of 20^{-23}. The main advantage and the importance of the central limit theorem, which is the basis for the Gaussian limiting distribution, is that only very few quantities are retained from the observed dispersion process. Its detailed structure is lost in the tails that vanish with time. Among the statistics of interest that can be easily obtained are, for example, the diffusion coefficient D and the mean square displacement. It is a convenient means for measuring dispersion of the particle from the origin and increases linearly with time for diffusive processes (i.e., regular diffusion). The number of different sites visited, called the range, is also easily calculated.

We just showed above that discrete random walk processes can be used to generate continuous time processes by taking a continuum limit in both space and time. It is also possible to generate a continuous-time process on a lattice structure, i.e., keeping space discrete, by the use of linear rate equations (Weiss, 1994). Finally, one can also apply a very useful approach, the continuous-time random walk model. It has the advantage of having well-defined steps taking place at well-spaced times. Specifically, one can use a lattice structure and a particle moving between sites where it remains for a given sojourn time t, following some prescribed distribution (Weiss, 1994). Sojourn time is defined as the time

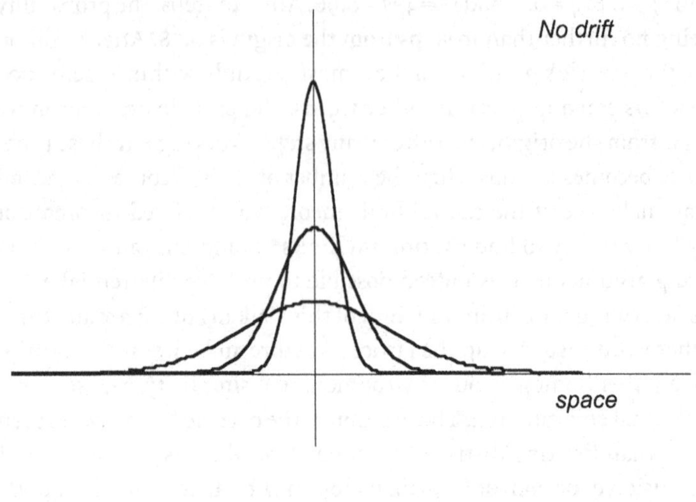

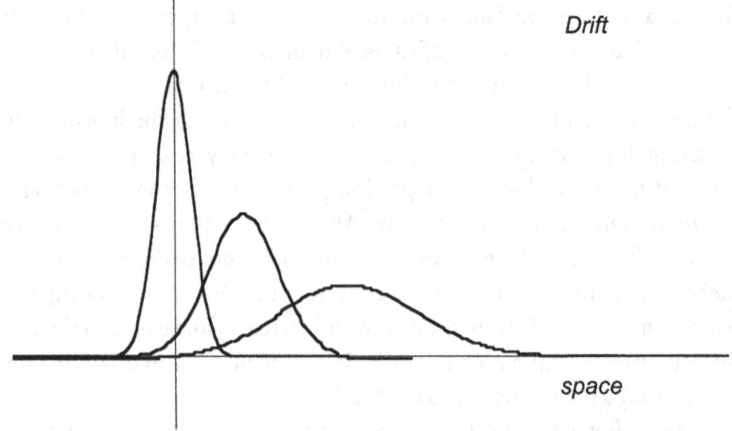

Fig. 8.2. Spread over time of a particle moving according to Brownian motion, without (top) and with (bottom) drift.

spent in one state before moving to another state. Random walks in a homogeneous environment in which the step-length distribution has a fat tail (i.e., long displacements occur relatively often) can change the basic rules of diffusion. In those distributions, the very long displacements do determine the overall dynamics of the system (probability of reaching a location, mean square displacement, etc.). The process becomes superdiffusive, meaning that the particle moves quicker than in the normal diffusion process. This is shown by considering the mean square displacement that increases with the power of time t^{γ}, with $\gamma>1$ (for so-called Lévy walks, see Drysdale and Robinson, 1998). These distributions have been found to be a valid approximation for several species of animals, including insects (Viswanathan *et al.*, 2000). In biological terms, such distributions imply that the likelihood of finding a predator far away from its starting location in a relatively short time is relatively high.

Random environments

Heterogeneity in the environment can be modeled in different ways. One way is to use deterministic models, such as those developed for fractals, percolation, etc. This will not be pursued here, but we refer to Halvin and Ben-Avraham (1987) for an in-depth treatment. Another approach, random environment modeling, is to consider one sample of the environment (a plant) as a single realization of an ensemble (a population of plants). The local properties of the realization, such as the location of gaps in the canopy, are determined following some stochastic process. The position of a particle, and all the statistics associated with it, depends on the history of the particle in the given environment and on the environment itself. Let us denote one realization of such an environment, a single plant, with ω. This environment will remain, for the sake of simplicity, constant through all the walks by the particle. One then has to distinguish two different ways of calculating averages: one over the environment ω and another one over the ensemble of possible environments Ω, the population of plants. This averaging gives the average behavior of the particle in an averaged environment, probably the description nearest to the heart of ecologists. It is only after the second averaging that one can appreciate the general features of the system. In practical terms, it requires the ecologist to map several canopies and predator paths and to come up with a probabilistic model describing both canopy geometry and predator movement.

The environment does not evolve with time in so-called quenched environments. By contrast, in annealed environments, a particle will never experience the same environment. For predators tracking prey in the vegetation, either model can be used, but we focus here on the simplest, the quenched environment. We will come back to annealed environments in the discussion. The randomness in the environment can have two kinds of effects on diffusion (Bouchaud and Georges, 1990):

- It may affect the value of the transport coefficients (velocity, diffusion coefficient, etc).
- It may affect the law of the diffusion process. For example, the mean square displacement may no longer increase linearly in time over long times. Anomalous diffusion, being super- or subdiffusive, corresponds to this kind of movement. In the superdiffusive case, the mean and mean squared displacement increase more quickly than linearly. The subdiffusive behavior leads to a sublinear function of time for both the mean and the mean square displacement.

In order to illustrate our ideas, we use below a specific model of a particle moving in a comb structure. But let us first contrast in general terms the propagation of two packets of predators, both released at the same end ($x = 0$) of a one-dimensional space, and experiencing a drift in the same direction. Assume also that movement is made of hops, or steps of short distance. In the first case, the environment is homogeneous, leading to regular diffusion. In this case, the packet of predators moves as a whole, the location of the mode being the same as the location of the mean (Fig. 8.2). In the second case, the random geometry of the environment leads to long sojourn time at some locations, which then act as temporary traps. A large percentage of the predators experience usual displacement, similar to the predators experiencing regular diffusion. However, the longer the experiment, the greater the likelihood that all predators become trapped at some stage, i.e., hit a relatively rare but quite long sojourn time. In this subdiffusive case, the mode stays at $x = 0$, and the mean position continues to increase, but at a decelerating rate (hence the name subdiffusive). This behavior is in contrast to the regular diffusion in which the mean progresses at a constant rate.

The random comb as an example

The comb structure bears strong similarities to real plant canopies as experienced by insects. It is a simple structure, made of a backbone and branches (Fig. 8.3). Framed in our topic, we ask for example how quickly insects move in a field given that they move up and down in the vegetation. Hence, one may envisage the vertical components of vegetation as acting as "traps" when considering movement in the horizontal plane. The problem is to characterize the movement along the backbone as function of the movement in the branches. Whenever a particle reaches a point on the x-axis it either makes a step along the x-axis with probability p or a step in the y-direction, along a branch, with probability $1 - p$. The particle

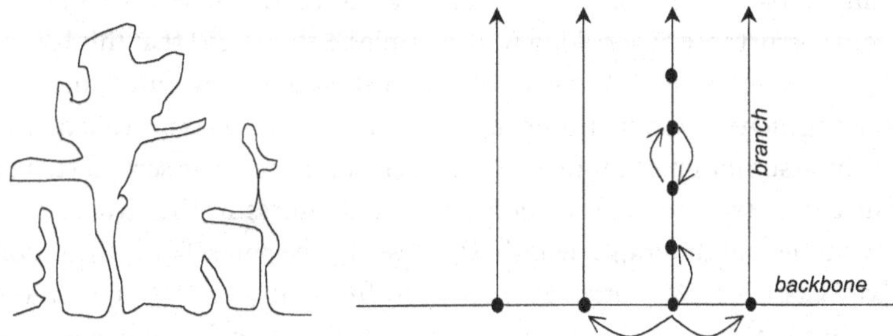

Fig. 8.3. The horizontal displacement of a predator moving up and down in vegetation (left) can be compared to a particle moving along a backbone and spending time in the vertical branches of a comb (right).

will then move within a branch following given jumping rates for the vertical movement. Once back on a point on the x-axis, the process repeats itself. The construction of this example is in three steps of increasing complexity:

1. We first assume no randomness in the branch length, which we set at infinity, and study the movement of the particle along the backbone. The observed movement cannot be modeled by the diffusion equation.
2. Then, we let the length of the branch vary and observe diffusive and subdiffusive behavior as function of the distribution used for modeling branch length.
3. Finally, using the branch distribution that would best correspond to real situations experienced by insects, we add vertical bias, i.e., the tendency for many insects to move up rather than down. A complete breakdown of the diffusive behavior is again observed.

We now demonstrate these three steps in detail.

Let first assume that there is no randomness in the structure and that the branch length is infinite. This biologically unrealistic assumption will be dropped later. The probability of return to the x-axis (τ_n) of a particle moving in one branch is the probability of return to the origin in a one-dimensional random walk for the first time at step $n, \tau_n \sim n^{-3/2}$. This probability distribution is somewhat special in the sense that its mean is infinite. The total time spent in branches is simply the sum of the N sojourn times spent in the different branches. As it is a sum of independent random variables, we can apply the central limit theorem and obtain the mean square displacement $<r^2(n)> \sim n^{1/2}$, which is characteristic of anomalous diffusion. The anomalous transport is due to the average infinite sojourn time of the particle in the branches. This leads to the occasional occurrence of very long waiting times. Let us recall that this breakdown of the diffusive behavior is obtained without any randomness or heterogeneity in the environment. However, this model is unrealistic due to the assumption of infinite length of the branches. The second step in our demonstration consists of adding randomness in the structure by assuming that the branch length (x) is given by the power law distribution $f(x) \sim \gamma x^{-(1+\gamma)}$. We obtain anomalous diffusion if $\gamma \leqslant 1$. If $\gamma > 1$, then the average branch length is finite and the diffusion is again regular at large times. A distribution with finite average branch length seems a priori the best analogy to situations encountered in nature by insects. The final "improvement" of our model is the addition of bias in the particle movement. Bias in random environments has two opposite effects (Halvin and Ben-Avraham, 1987). On the one hand, the particle is following the direction of the field, giving rise to a drift velocity. On the other hand, dead ends act as temporary traps from which particles escape by going against the flow. As a concrete example, let us consider the above case of regular diffusion on a comb by assuming that the branch length is given by an exponential distribution. Then assume that the bias is in the vertical direction. Thus, the particle has a higher probability of going upwards than downwards. Hence, one can ascribe to each branch a delay associated with the branch length. Long branches determine the overall behavior of the particle, as it is "pushed" towards their tip. One can show that the distribution of delays follows a power law distribution and that diffusion is again anomalous. This scenario corresponds in our multitrophic context to situations in which predators move preferentially upwards and may miss prey located on their way up.

Coda

The random geometrical structures in some dimension(s) of the environment cause delays in the movement of the particle in other dimensions. In the comb example, vertical movement in the branches delays progression along the backbone. The delay is generally characterized by a long tail of sojourn times that leads to anomalous transport along other dimension(s).

The study of processes characterized by time distributions with fat tails also brings to the forefront an important problem of scale. Some sojourn times are of the same order of magnitude as the total time of observation. This leads to yet another breakdown of the regular diffusion approach, and its basis, the central limit theorem. In other words, there may not be enough time or steps in the process to attain an equilibrium distribution. By contrast, in the regular diffusion framework, the time-scale is defined by the mean value of the sojourn time distribution, while the physically relevant scale is defined by the variance of the step length distribution. There are no such scales in anomalous diffusion, as those moments diverge (Paul and Baschnagel, 1999).

These very general results, albeit borrowed from statistical physics, are bound to be true for predators moving in plant canopies. The implications are twofold. First, the geometry of the environment will determine the risk of predation of individual prey. Indeed, the probability that prey i located at X_i will be attacked by a predator j located at X_j within some time interval is obviously a function of their respective locations and the possible paths between them. The role and form of risk heterogeneity between prey in population dynamics is a major topic today as it determines the stability of the interaction (Gross and Ives, 1999; Olson *et al.*, 2000). Hence, the estimation of the probability distribution of risk among prey requires at some stage an estimate of accessibility of the prey in a given environment. Second, our understanding of spatial predator–prey population dynamics is built around the advection–diffusion framework championed in ecology by Kareiva and Odell (1987). They and others showed that the predicted spatial patterns, for example waves or uniform distributions, between prey and predators are the result of a delicate interplay between parameters describing random movement of the predator and its tendency to move towards prey (Wollkind *et al.*, 1991; Grünbaum, 1998, 1999; Cantrell and Cosner, 1999). Regular diffusion is often an unstated assumption of this approach. For example, one assumes that predators make many small steps in a relatively short time and that the distance covered is a small fraction of the available space. While these studies show how to incorporate microscopic details about the behavior and movement of predators into a macroscopic image of their distribution, they still lack proper model testing, as acknowledged by the authors themselves and others (Haefner, 1996). Hence, we do not know if these models are adequate, whether the spatial heterogeneity produced by the geometry of the environment is important, and how much a fuller treatment would increase our understanding of the mechanisms leading to spatial stability and our capacity to predict the spatial patterns.

Thus, we conclude that simple random walks in homogeneous environments and the regular diffusion approximation may be poor guides for understanding search strategies of predators and prey location in plant canopies. They are best replaced by a framework built around the concept of random walks in randomly or deterministically determined geometrically structured environments. Once such models are built and tested, simpler approximations can then be tested and the role of the fine-grained geometry of the environment determined.

Application of the framework

The only study we are aware of that follows the approach described above deals with movement of fruit flies foraging in apple trees (Casas and Aluja, 1997), a system similar to a multitrophic interaction as envisaged above. In our study, apple trees lacked fruits, and the framework provided the null hypothesis for inferring the influence of external stimuli, such as fruit color and odor, on the paths of foraging flies. We mapped three trees in cells, or sites, released preconditioned flies, and recorded their behavior and location. For modeling purposes, we discarded cubes devoid of vegetation and concentrated on cubes that could be used as landing points for the flies. This structure is an incomplete lattice structure, because anything within the cube is considered to lie on its lattice point and because empty cubes cannot serve as landing points. Since we were interested in the geometry of the path made by a foraging fly, we discarded both the time spent in the cube and any movement within the cubes. A move or step was defined as a change of cubes. Flies moved mainly to the nearest neighbor cells, but displacement within almost the entire range of possible values was observed. The model closest to the observations was a random walk with a position-dependent bias in the vertical component of movement. The movement rules, i.e., the probability of moving downward, upward, or horizontally, as well as the move distance, were estimated using foraging paths observed in one tree. The model was then applied to a second tree. Five models were built, spanning a range of simplifications in the rules determining the vertical component of movement.

We observed that flies, which generally enter the tree from the lowest half, move quickly upward into the bulk of vegetation. There are two complementary explanations for this behavior in terms of efficiency of movement. First, the presence of a bias not only increases the speed at which flies move away from the starting location, but it also increases the number of sites visited, which is one way of describing the efficiency of a foraging path. As there is no point revisiting previously visited sites, a fly should avoid self-crossing, and the observed number of sites visited was indeed very near the maximum possible, indicating a high searching efficiency. Another interpretation for this behavior comes from the study of the diffusing properties of a set of random walkers (see Yuste and Acedo, 1999 and references therein). When the number of walkers starting at the same time from the same location is large, every possible site is visited in the neighborhood of the starting location within a very short time. But after a very long time, the walkers are so scattered that their paths hardly overlap, and the number of sites visited is simply the number of sites visited by one walker multiplied by the number of walkers. Such a mechanism is postulated as an explanation for the upward bias observed in the apple fruit fly. As flies tend to enter trees at the same height level, a bias in movement would thereby help a foraging fly avoid self-crossings and crossing areas already visited by other flies.

While quite sufficient in two trees, the model failed to reproduce observed movement patterns in a third tree. Testing models of movement in trees different from those in which they were developed is an acid test: models may fail because they are tree-specific. However, if they pass the test, we learn a lot about movement in plant canopies in general. The influence of canopy-to-canopy variation in geometry is best explained using the mean value of the range. The mean value of the range in tree A is dependent on the configuration of tree A (spatial arrangement of gaps in the canopy, geometry of branching, etc.). That mean value is found by observing or simulating many flies in tree A. However, our aim is to characterize movement in apple trees in general, and not just in tree A. Hence, our final interest lies in estimating the mean value over all apple trees, as explained in more general terms above. Thus, the failure of our model to predict movement in a third, geometrically rather different tree is the proof that our model was not robust enough. This is a strong case for developing a stochastic model for the plant canopy that produces an ensemble of canopies, out of which we could select specific realizations that would vary slightly from each other. In parallel, one needs to develop models of movement that are a genuine function of canopy geometry, rather than extract the rules from one environment and apply them as such to another environment.

Outlook

This chapter has been written from the perspective of a predator searching for immobile prey in a quenched environment. Strictly speaking, the description of our problem is valid only for a walking predator. Indeed, passing through connecting locations when moving from one location to another is unavoidable for a walking animal. A flying animal can however reach any point in a single step with some probability. The framework can easily be extended to flying insects once they are near or in a plant canopy. Many flying insects do not make long flights in this environment and follow the structure of the plant to some degree (Casas, 1990). Our approach can accommodate this behavior by using probabilities of moving from one location to another. This chapter has also paid scant attention to the third player in the game, the prey. Prey choose where they are going to end up on the plant, and their locations set the stage for predator movement. While we saw examples where prey are located in places which are difficult for predators to reach, we do not know of any study comparing predator movement foraging in plant canopies for naturally distributed prey versus artificially distributed prey. Coll *et al.* (1997) went a long way along these lines by distinguishing between the direct and indirect effects of plant architecture on predators. They found varying degrees of spatial overlap within plant canopies between prey and predators.

We end by touching on some effects of relaxing the assumption of quenched environments. First, we can allow the canopy architecture to change over time (annealed environments). Suverkropp (1997) calculated the dynamics of the probability of encounter of *Trichogramma* and egg masses of its host as a function of the growing maize canopy over a season. The leaf area was measured while the encounter probability was predicted using a data-rich model. While eggs of *Ostrinia nubilalis* are present, the plant changes from having three or four leaves to having fifteen. This represents an increase in the area to search of more than tenfold. The encounter probability, defined as the probability of a single female encountering an egg mass over a 24–h period, decreases from *c.* 0.3 to *c.* 0.05 during a season (Fig. 8.4). This observation suggests that most females will end the day without finding any hosts in the fully grown maize. This may be even true for their entire lifetime, given that they live for less than 12 days. At the time-scale of a fruiting season, prey may become highly susceptible to predation due to the fruit ontogeny, as

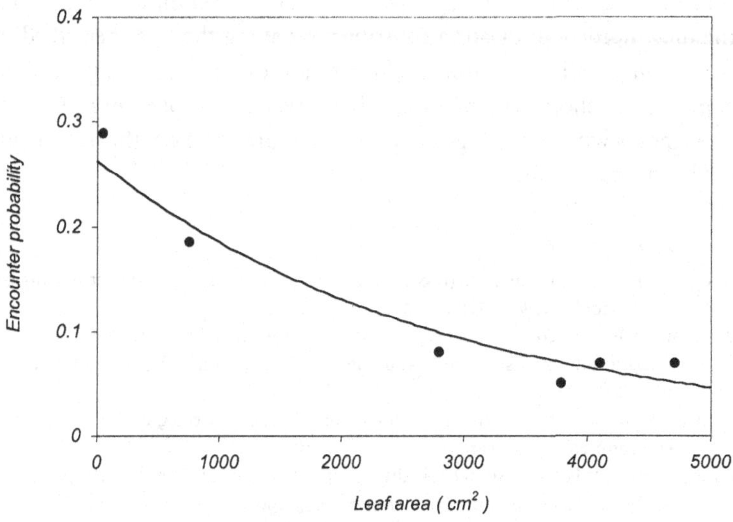

Fig. 8.4. Probability of finding a host by *Trichogramma* as function of the leaf area of maize growing over a season. (After Suverkropp, 1997.)

observed by Udayagiri and Welter (2000) for a mirid bug attacking strawberries. Fruit development resulted in a change in the fruits' structure and hence an increased accessibility of the eggs to its parasitoid. Changes in plant architecture over time-scales of years also influence interspecific interactions. C. R. Fonseca and W. W. Benson (unpublished data) describe ant succession and interspecific relationships during the ontogeny of Amazonian ant trees (tachigali). The canopy of a tree changes from an architecturally simple plant with a couple of leaves to a huge, highly complex canopy of thousands of leaves. More than half a dozen ant species colonize the plant and later disappear during this ontogenic succession. Such studies show that an increase in the complexity of canopy geometry fosters an equivalent increase in complexity of biotic interactions. Ontogeny of the insects suggests that the scale of an individual's range may also vary within its lifespan, as shown by Yang (2000) for a pentatomid predator.

A second possibility is to let the animal itself change the canopy's geometry. Many herbivorous insects are known to eat or tie leaves in very specific ways in order to avoid being eaten by predators (Djemai *et al.*, 2000 and references therein). Except for a few studies, modification of the canopy architecture to avoid predation and parasitism has hardly been considered and rarely quantitatively measured. Overall, we believe that the most needed and lasting contributions along the lines described in this chapter will come from an integration of carefully designed field experiments encompassing detailed observations of prey and predator movements with modeling canopy architecture. To date, this is a virgin field of investigation.

REFERENCES

Andow, D. A. and Prokrym, D. R. (1990) Plant structural complexity and host-finding by a parasitoid. *Oecologia* **82**: 162–165.

Bouchaud, J. P. and Georges, A. (1990) Anomalous diffusion in disordered media: statistical mechanisms, models and physical applications. *Physics Reports* **195**: 127–293.

Cantrell, R. S. and Cosner, C. (1999) A comparison of foraging strategies in a patchy environment. *Mathematical Biosciences* **160**: 25–46.

Casas, J. (1990) Multidimensional host distribution and non-random parasitism: a case study and a stochastic model. *Ecology* **71**: 1893–1903.

Casas, J. (1991) Density dependent parasitism and plant architecture. 4th European Workshop on Insect Parasitoids. *Redia* **74**: 217–222.

Casas, J. and Aluja, M. (1997) The geometry of search movements of insects in plant canopies. *Behavioral Ecology* **8**: 37–45.

Cloyd, R. A. and Sadof, C. S. (2000) Effects of plant architecture on the attack rate of *Leptomastix dactylopii* (Hymenoptera: Encyrtidae), a parasitoid of the citrus mealybug (Homoptera: Pseudococcidae). *Environmental Entomology* **29**: 535–541.

Coll, M., Smith, L. A. and Ridgway, R. L. (1997) Effects of plant on the searching efficency of a generalist predator: the importance of predator–prey spatial association. *Entomologia Experimentalis et Applicata* **83**: 1–10.

Djemai, I., Meyhöfer, R. and Casas, J. (2000) Geometrical games between a host and a parasitoid. *American Naturalist* **156**: 257–266.

Drysdale, P. M. and Robinson, P. A. (1998) Lévy random walks in finite systems. *Physical Review E* **58**: 5382–5394.

Evans, H. F. (1976) The searching behaviour of *Anthocoris confusus* (Reuter) in relation to prey density and plant surface topography. *Ecological Entomology* **1**: 163–169.

Frazer, B. D. and McGregor, R. R. (1994) Searching behaviour of adult female coccinellidae (Coleoptera) on stems and leaf models. *Canadian Entomologist* **126**: 389–399.

Ganeshaiah, K. K. and Veena, T. (1988) Plant design and non random foraging of ants in *Croton bonplandianum* Beill. *Animal Behaviour* **36**: 1683–1690.

Gardner, S. M. and Dixon, A. F. G. (1985) Plant structure and the foraging success of *Aphidius rhopalosiphi* (Hymenoptera: Aphidiidae). *Ecological Entomology* **10**: 171–179.

Gauthier, H., Mech, R., Prusinkiewicz, P. and Varlet-Grancher, C. (2000) 3D architectural modelling of aerial photomorphogenesis in white clover (*Trifolium repens* L.) using L-systems. *Annals of Botany* **85**: 359–370.

Geitzenauer, H. L. and Bernays, E. A. (1996) Plant effects on prey choice in a vespid wasp, *Polistes arizonensis*. *Ecological Entomology* **21**: 227–234.

Godin, C., Costes, E. and Sinoquet, H. (1999) A method for describing plant architecture which integrates topology and geometry. *Annals of Botany* **84**: 343–357.

Grevstad, F. S. and Klepetka, B. W. (1992) The influence of plant architecture on the foraging efficiencies of a suite of ladybird beetles feeding on aphids. *Oecologia* **92**: 399–404.

Gross, K. and Ives, A. R. (1999) Inferring host–parasitoid stability from patterns of parasitism among patches. *American Naturalist* **154**: 489–496.

Grünbaum, D. (1998). Using spatially explicit models to characterize foraging performance in heterogeneous landscapes. *American Naturalist* **151**: 97–115.

Grünbaum, D. (1999) Advection–diffusion equations for generalized tactic searching behaviors. *Journal of Mathematical Biology* **38**: 169–194.

Gupta, H. M. and Campanha, J. R. (2000) The gradually truncated Lévy flight: stochastic process for complex systems. *Physica A* **275**: 531–543.

Haefner, J. W. (1996) *Modeling Biological Systems: Principles and Applications*. New York: Chapman and Hall.

Halle, F. and Oldeman, R. A. A. (1970). *Essai sur l'Architecture et la Dynamique de Croissance des Arbres Tropicaux*. Paris, France: Masson.

Halle, F., Oldeman, R. A. A. and Tolminson, P. B. (1978) *Tropical Trees and Forests: An Architectural Analysis*. Berlin, Germany: Springer Verlag.

Halvin, S. and Ben-Avraham, D. (1987) Diffusion in disordered media. *Advances in Physics* **36**: 695–789.

Hassell, M. P. (2000) *The Spatial and Temporal Dynamics of Host–Parasitoid Interactions*. Oxford: Oxford University Press.

Isenhour, D. J. and Yeargan, K. V. (1981) Interactive behavior of *Orius insidiosus* [Hem.: Anthocoridae] and *Sericothris variabilis* [Thys.: Thripidae]: predator searching strategies and prey escape tactics. *Entomophaga* **26**: 213–220.

Kaiser, H. (1983) Small-scale spatial heterogeneity influences predation success in an unexpected way: model experiments on the functional response of predatory mites (Acarina). *Oecologia* **56**: 249–256.

Kareiva, P. and Odell, G. (1987) Swarms of predators exhibit "preytaxis" if individual predators use area-restricted search. *American Naturalist* **130**: 233–270.

Kareiva, P. and Perry, R. (1989) Leaf overlap and the ability of ladybird beetles to search among plants. *Ecological Entomology* **14**: 127–129.

Kareiva, P. and Sahakian, R. (1990) Tritrophic effects of a simple architectural mutation in pea plants. *Nature* **345**: 433–434.

Lukianchuk, J. L. and Smith, S. M. (1997) Influence of plant structural complexity on the foraging success of *Trichogramma minutum*: a comparison of search on artifical and foliage models. *Entomologia Experimentalis et Applicata* **84**: 221–228.

Messina, F. J. and Hanks, J. B. (1998) Host plant alters the shape of the functional response of an aphid predator. *Environmental Entomology* **27**: 1196–1202.

Michalewicz, M. (ed.) (1999) *Advances in Computational Life Sciences*, vol. 1, *Plants to Ecosystems*. Canberra, Australia: CSIRO Publishing.

Olson, A. C., Ives, A. C. and Gross, K. (2000) Spatially aggregated parasitism on pea aphids, *Acyrthosiphon pisum*, caused by random foraging behaviour of the parasitoid *Aphidius ervi*. *Oikos* **91**: 66–76.

Parker, G. G. and Brown, M. J. (2000) Forest canopy stratification: is it useful? *American Naturalist* **155**: 473–484.

Paul, W. and Baschnagel, J. (1999) *Stochastic Processes: From Physics to Finance*. Berlin, Germany: Springer-Verlag.

Pearcy, R. W. and Valladares, F. (1999) Resource acquisition by plants: the role of crown architecture. In *Physiological Plant Ecology*, ed. J. D. Scholes and M. G. Barker, pp. 45–66. Oxford: Blackwell Science.

Pyke, G. H. (1978) Optimal foraging movement patterns of bumble bees between inflorescences. *Theoretical Population Biology* **13**: 72–98.

Room, P., Hanan, J. and Prusinkiewicz, P. (1996) Virtual plants: new perspectives for ecologists, pathologists and agricultural scientists. *Trends in Plant Sciences* **1**: 33–38.

Russo, A. R. (1987) Role of habitat complexity in mediating predation by the gray damselfish *Abudefduf sordidus* on epiphytal amphipods. *Marine Ecology, Progress Series* **36**: 101–105.

Shigesada, N. and Kawasaki, K. (1997) *Biological Invasions: Theory and Practice*. Oxford: Oxford University Press.

Stauffer, D. (1985) *Introduction to Percolation Theory*. London: Taylor and Francis.

Suverkropp, B. (1997) Host-finding behaviour of *Trichogramma brassicae* in maize. PhD thesis, Wageningen Agricultural University, Netherlands.

Tilman, D. and Kareiva, P. (1997) *Spatial Ecology: The Role of Space in Population Dynamics and Interspecific Interactions*. Princeton, NJ: Princeton University Press.

Turchin, P. (1998) *Quantitative Analysis of Movement*. Sunderland, MA: Sinauer Associates.

Udayagiri, S. and Welter, S. C. (2000) Escape of *Lygus hesperus* (Heteroptera: Miridae) eggs from parasitism by *Anaphes iole* (Hymenoptera: Mymaridae) in strawberries: plant structure effects. *Biological Control* **17**: 234–242.

Viswanathan, G. M., Afanasyev, V. Buldyrev, S. V., Halvin, S., da Luz, M. G. E., Raposo, E. P. and Stanley, H. E. (2000) Lévy flights in random searches. *Physica A* **282**: 1–12.

Vohland, K. (1996) The influence of plant structure on searching behaviour in the ladybird *Scymnus nigrinus* (Coleoptera: Coccinellidae). *European Journal of Entomology* **93**: 151–160.

Weiss, G. H. (1994) *Aspects and Applications of the Random Walk*. Amsterdam, Netherlands: Elsevier.

Weisser, W. W. (1995) Within-patch foraging behaviour of the aphid parasitoid *Aphidius funebris*: plant architecture, host behaviour, and individual variation. *Entomologia Experimentalis et Applicata* **76**: 133–141.

Wollkind, D. J., Collings, J. B. and Barba, M. C. B (1991) Diffusive instabilities in a one-dimensional temperature-dependent model system for a mite predator–prey interaction on fruit trees: dispersal mobility and aggregative preytaxis effects. *Journal of Mathematical Biology* **29**: 339–362.

Yang, L. H. (2000) Effects of body size and plant structure on the movement ability of a predaceous stinkbug, *Podisus maculiventris* (Heteroptera: Pentatomidae). *Oecologia* **125**: 85–90.

Yuste, S. B. and Acedo, L. (1999) Territory covered by N random walkers. *Physical Review E* **60**: R3459–R3462.

VALERIE K. BROWN AND ALAN C. GANGE

9

Tritrophic below- and above-ground interactions in succession

Introduction

Ecological succession is a pivotal process in ecology, since it occurs in all dynamic systems. It is therefore hardly surprising that its study is still a major preoccupation of ecologists, though as one combs the past scientific literature there are clear fashions in areas of interest and the approach adopted. Thus, from early descriptions of specific successional patterns of the vegetation, attention turned to a consideration of the mechanisms underpinning succession and ways in which successional trajectories could be modeled.

The study of succession has traditionally been dominated by plant ecologists. Even though plant–animal and, to a lesser extent, plant–microbial interactions have been in vogue for much of the time span of successional studies, relatively few workers have considered these in the context of succession. Of course, one notable exception is in the practical management of plant succession by the larger herbivores, which has also attracted scientific rigor (e.g., Gibson and Brown, 1992). Successional interactions with other less conspicuous organisms, namely invertebrates and microorganisms, have been given far lower priority by ecologists. Once these interactions become more complex, by involving other organisms or trophic levels, priority has fallen even further. In the few studies that do exist, interest has focused on the interactions that can be seen, namely those between organisms associated with above-ground plant structures. Include soil organisms and there is a gaping void in our knowledge!

This chapter seeks to probe the knowledge base that exists on plant–invertebrate–fungal trends during succession and to present some new data and ideas. It focuses on those complex biotic interactions below ground that may modify the successional dynamics of the vegetation that we see above ground and, indeed, have come to take for granted. We aim to show that tritrophic interactions, based on plant roots as the template, are worthy of study and deserve far more attention than they have yet received. We draw on results from experimental manipulations of these interactions in the laboratory and field and confine ourselves exclusively to terrestrial systems. In particular, we focus on insect–mycorrhizal interactions. Our choice of system is based on the desire to complement trophic interactions described in other chapters of this book, our own area of expertise, and, perhaps most significantly, the potential importance of this particular interaction in terms of plant successional dynamics. In so doing, we aim to whet the appetite and encourage more research in the study of soil-based tritrophic interactions.

Successional attributes and patterns

Terrestrial plant successions, regardless of the soil type on which they occur, are typically characterized by a progression of species of plant and associated organisms. The nature of the plant community over time, in terms of species composition, diversity, and structure provides the template for other organisms (Brown and Southwood, 1987). Successional change in the plant community is invoked by a wide range of abiotic and biotic factors. Abiotic factors, such as soil nutrient and water availability, can have major effects on the rate of succession (Prach *et al.*, 1993) and their interactions with biotic factors have occasionally been documented (e.g., Prach and Pyšek, 1999). Those biotic factors most commonly cited include the seed bank, the basis of the initial floristic composition ideas of Egler (1954), the relative competitive abilities of plant species (e.g., Wilson and Shure, 1993), and plant life-history characteristics (Brown and Southwood, 1987; Brown, 1990). Indeed, the life-strategy concepts of Grime *et al.* (1988) may be considered a further reflection of the latter and certainly change during succession. Recently, Prach and Pyšek (1999) assessed a wide range of species attributes, including those mentioned above, in relation to plant succession and dominance in 15 different seres in the Czech Republic. This review of species traits is rare, since it includes attributes of other trophic levels, namely pollination mode, dispersal, and mycorrhizal status, which were found to relate directly to the nature of the succession. However, it does not include plant species' palat-

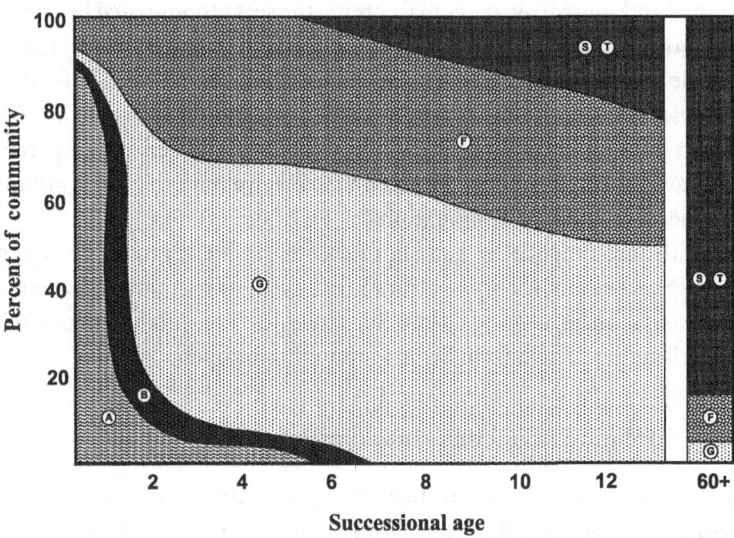

Fig. 9.1. Model to show the transition of plant life-history groups as succession proceeds in a temperate grassland succession (in southern Britain). (A) annual forbs and grasses, (B) monocarpic perennial forbs, (G) perennial grasses, (F) perennial forbs, (S) and (T) woody perennials (shrubs and trees, respectively).

ability to herbivores. It is known that the rate and direction of successional change can be strongly influenced by vertebrate grazing. One only has to consider the effects of introduced herbivores, such as rabbits or goats, to appreciate the dramatic effects these herbivores can have on the succession of native vegetation. Indeed, there is now a prolific literature on the effects of vertebrate herbivory on vegetation dynamics (e.g., Watt, 1981; Noy-Meir *et al.*, 1989; Gibson and Brown, 1992). More recently, there is increasing evidence that the smaller invertebrate herbivores also have an effect on plant succession, sometimes even approaching that of vertebrates (Gibson *et al.*, 1987; Carson and Root, 1999).

From the standpoint of tritrophic interactions, plant life history is of key importance, since it reflects the longevity, phenology, architecture, and chemistry of the components of the plant community. Brown and Southwood (1987) provided a simple model, based on the secondary succession on acidic sandy soil in southern Britain, and the one that we shall return to later in this chapter. Such a model (Fig. 9.1) is generally applicable to many secondary successions, even though the timing of the phases will vary as will the extent of overlap between life-history types. Like most successions, after the initial flush of colonization and growth of annual and short-lived perennial species, the sere is dominated by perennial grasses and forbs, before woody perennials (shrubs and trees) start to invade. It is the balance between perennial grasses and forbs that is often a characteristic feature of a particular type of succession and varies according to soil type and management. Organisms such as herbivores, pathogens and mutualists, which impact on these key plant life-history groups are therefore likely to be significant drivers of succession.

With the composition and abundance of different groups changing during succession, the scene is set for complex interactions between the organisms themselves, vegetation dynamics and plant succession. Surprisingly, the field is challenging though wide open.

Potential interactions in succession

A recent symposium of the British Ecological Society (Gange and Brown, 1997) reviewed the full spectrum of multispecies interactions. In the majority of the interactions described, the plant provides the template with organisms either associated with or feeding on it. Many of the interactions described entail a separation of "the players" either in time, space, or both (e.g., Masters and Brown, 1997). Temporal separation involves the exploitation of the plant at different times, typically with the early feeder influencing the later one through physical or chemical changes in the host plant (Haukioja, 1980; Hunter, 1987). Spatial separation between the organisms is also common and can be simple involving, for example, insects feeding on different plant structures or even on different surfaces of a leaf (e.g., Kidd *et al.*, 1985).

Not surprisingly, most attention has focused on those interactions occurring above ground and most chapters in this volume are a clear indication of this trend (but see chapter 10). For the research scientist, the study of these associations is attractive, since they are immediately apparent and can be experimentally manipulated to understand the underpinning mechanisms. Indeed, one of the most fully explored tritrophic interaction must be that between plant–herbivore–parasite or predator (see chapters 6 and 7), whereas studies on plant–herbivore–fungal pathogen or symbiont interactions are becoming more frequent (e.g., chapter 5). Far fewer studies explore the black box of the soil (chapter 10), though the food-web studies of Moore and De Ruiter (1991, 1997) are an example of the overwhelming biotic complexity that exists. Teasing apart the taxonomic and functional complexity in the soil is a challenge that is less commonly addressed, though there are an increasing number of research groups that champion an understanding of interactions between specific groups of organisms below ground. Gange and Brown (1997) again review some of these.

In this chapter, we focus on a different and in many ways more complex interaction, namely that occurring between organisms that live exclusively either above or below ground. Again, the common denominator in any such interaction is the host plant with which the organisms are associated. We know of no parasites or predators occurring in both domains that could drive such an interaction. We are interested in the potential for herbivores or symbionts, active above or below ground, to modify the structure or physiology of the plant in such a way that the performance of the spatially separated organism is affected. Such an influence may well in turn translate into a difference in growth and fitness of the plant species. Differential effects on individual plant species or life-history groups, that lead to a modification in the competitive balance between species, provide the tools necessary for successional change. Couple this with the difference in the species composition of the vegetation and associated organisms as succession proceeds and we have a potentially significant, though grossly under-appreciated, driver of succession.

Two-way interactions affecting succession

To date, several studies have been published which describe the effects of one particular group of animals or microorganisms on the successional process. Many of these are summarized by Davidson (1993), and here we concentrate on the effects of phytophagous insects and arbuscular mycorrhizal fungi (AMF). The majority of experiments have examined the consequences of excluding one of these groups from developing plant communities, although there are also some studies in which two groups have been manipulated.

It is known that foliar insect herbivores show clear successional patterns, both in terms of species composition and degree of specialization to the host plant (e.g., Edwards-Jones and Brown, 1993) and in life-history traits (Brown, 1986). As succession proceeds, generation time, dispersal ability, and reproductive potential of foliar-feeding insects decline, while host specificity increases (Brown, 1990). The guild structure of foliar insects changes from one dominated by sap-feeding insects in ruderal communities to a more evenly structured community in mid succession

(Brown and Southwood, 1987). Little is known of the successional trends in rhizophagous insect life-history traits, but there are clear changes in their abundance, from low numbers per unit area in early seres to high numbers in mid successional seres (Brown and Gange, 1992). Foliar-feeding insects also increase in abundance through early succession, with their feeding concentrated upon the perennial grasses. The effect is great enough to be easily detectable, if insecticide is applied (Brown and Gange, 1999), leading to a domination of this plant group. In this respect, these insects have a similar effect to that of vertebrate grazers, as by restricting grass growth they tend to reduce the rate of plant succession.

However, far less is known about the succession of soil-dwelling insects, though Clements *et al.* (1987) demonstrated an increase in the number of herbivores in the early stages of grassland succession, before a decline in response to pests and pathogens began. Brown and Gange (1992) also found that populations of subterranean insects were low in early succession, and built up more slowly than those of foliar-feeding species. This has been attributed to the relatively long life cycles and poor dispersal abilities of root-feeding insects, compared to their above-ground counterparts (Brown and Gange, 1990). Despite the fact that populations of rhizophagous insects may be low in early succession, excluding them by the careful application of insecticide to the soil results in significant changes to the plant community. In contrast to foliar feeders, root-feeding insects have their major impact on the perennial forbs, which greatly increase in abundance when insects are absent. Thus, the reduction in forb growth translates to an acceleration of the succession to a grass-dominated sward (Brown and Gange, 1992). Brown and Gange (1989a, 1992) examined the effects of excluding both root- and foliar-feeding insects singly and in combination on an early successional community. Reducing foliar feeders resulted in a lowering of plant species richness, as the community became dominated by perennial grasses. Meanwhile, exclusion of root-feeding insects resulted in a great increase in species richness, caused by the enhanced establishment of a number of perennial forb species. The effects were found to be entirely additive, with no interactions between the treatments.

AMF are also known to change in abundance and diversity through succession (Johnson *et al.*, 1991). Although populations are generally low in early succession, exclusion of these fungi with a soil fungicide has resulted in dramatic effects on the composition of the plant community (Gange *et al.*, 1990). Furthermore, it has been shown that the presence of these fungi greatly aids the establishment of many perennial forb species and that their exclusion leads to decreases in plant species richness (Gange *et al.*, 1993). These relatively crude exclusion experiments do provide good comparisons with the experiments of van der Heijden *et al.* (1998), in which it was found that increasing mycorrhizal species diversity lead to a concomitant increase in vascular plant species diversity in mesocosms.

However, it is a fact that many experiments with mycorrhizal fungi often fail to demonstrate an effect in field situations, when laboratory experiments have suggested that there should be one. One reason for this is likely to be the presence of insects which are ubiquitous in field experiments, but absent in laboratory trials. For example, Gange and Brown (1992) showed that when both root-feeding insects and AMF were experimentally reduced in an early successional plant community, the effect of the mycorrhiza was most clearly seen when insects were absent. It is therefore likely that disruption of the mycorrhizal mycelium occurs, either through mycophagy (by invertebrates such as Collembola or mites) (see chapter 10) or as an occupational hazard of root feeding by macroinvertebrates. Given that densities of both can be high in mid succession, we cannot reasonably consider the effects of the fungi in isolation from the insects or vice versa. However, it is not only subterranean insects that may disrupt mycorrhizal functioning. As outlined below, foliage removal may result in a reduction in carbon translocated below ground, which may adversely affect colonization by the mycorrhiza. In turn, AMF colonization of roots may affect the performance of foliar-feeding insects too. Clearly, in order to fully understand the roles of our three groups in affecting plant succession, we need to detail the potential three-way (multitrophic) interactions that may occur between them.

Examples of multitrophic interactions

Foliar-feeding and root-feeding insects

It has been known for some time that defoliation of vegetation by vertebrate grazers can have variable effects on the soil fauna (Ingham and Detling, 1984; Seastedt *et al.*, 1988). More recent work by Mawdsley and Bardgett (1997) has shown that the nature of the response of soil microbes to defoliation may depend on the plant species defoliated. Such a response for other parts of the soil food-web is currently being explored by Mikola *et al.* (2001).

The effects of foliar-feeding insects on soil insects have only been explored in a few studies. Moran and Whitham (1990) investigated the interactions between root aphids (*Pemphigus betae*) and a foliar gall-forming aphid, *Hayhurstia atriplicis*. While root feeding had no measurable effect on the foliar feeder, the latter decreased the root aphid populations by as much as 91% on susceptible genotypes. Similarly, T. Hunt and B. Blossey (personal communication) showed increasing levels of foliar defoliation resulted in reduced larval survival of *Hylobius transversovittatus* below ground, though larval biomass was not altered. In contrast, Gange and Brown (1989) assessed the effects of the scarab beetle larva, *Phyllopertha horticola*, on the performance of the black bean aphid (*Aphis fabae*) mediated via a common annual host plant, *Capsella bursa pastoris*. They found a positive effect of root feeding on the growth and performance of the foliar feeder, but that this effect was mitigated when soil moisture was plentiful. In a subsequent study, Masters and Brown (1992) found insect herbivory above ground, by the leaf miner *Chromatomyia syngenesiae* on *Sonchus oleraceus*, resulted in a reduced growth rate in the chafer larva. Clearly, results are equivocal and likely reflect the feeding behaviour of the herbivores. It is certainly an area where further research is needed. Partly to encourage debate and experimentation, Masters *et al.* (1993) put forward a simple conceptual model to explain the interactions they had found. With experimental work and plant chemical analyses to support their contention, they suggested that root feeding limits the plant's ability to take up water and nutrients, and leads to a reduced relative water content of the foliage and increased levels of soluble nitrogen (especially amino acids) and carbohydrates. The better-quality food resource leads to increased insect performance and population size in multivoltine foliar-feeding insects. Foliar herbivory, on the other hand, resulted in a reduced root biomass, limiting food availability for subterranean herbivores and causing poorer performance. Field trials supported their model, with experimental plots subjected to reduced below-ground insect herbivory, by the judicious application of soil insecticide, having lower levels of foliar-feeding sucking insects (aphids and Heteroptera) than those with natural or elevated levels of root herbivores (Masters, 1995). A recent study by Masters *et al.* (in press) has shown that tephritid flies, galling the flower heads of *Cirsium palustre*, showed increased abundance and performance as a result of root herbivory and that seed predators and parasitism were also increased, indicating an effect on higher trophic levels.

Arbuscular mycorrhizal fungi and foliar-feeding insects

Given the wealth of studies dealing with the effects of AMF colonization or foliage removal by insects on plant growth, it is perhaps surprising that few studies have investigated the interactions between AMF and insects. It is unfortunate that entomologists tend to ignore fungi when performing experiments, in the same way that mycologists ignore insects. However, given the effects that these organisms can have on their host plants, it is reasonable to expect that multitrophic links between soil microorganisms and above-ground invertebrates are common and important aspects of the functioning of communities.

There are many ways in which the presence of AMF in roots may alter host plant acceptability to an insect. First, it is well known that AMF can increase photosynthetic rates in leaves (e.g., Staddon *et al.*, 1999); such changes may lead to increases in plant growth and ultimately size. A plant that is actively growing and vigorous may be more acceptable to insect herbivores, which perform better upon it (Price, 1991). However, it has also been suggested that insect herbivore performance may be negatively correlated with plant vigor (White, 1984) and if this is so then AMF colonization could lead to decreases in herbivore performance. AMF do not always increase the growth of plants and there are many examples of colonization being apparently antagonistic, leading to plant size reduction (Johnson *et al.*, 1997). Indeed, it has been suggested that for any plant–AMF association, there is a continuum of plant responses, varying from positive to negative (Gange and Ayres, 1999). If this is so, then the outcome of any AMF–plant–insect interaction may be positive for the insect in some situations, but negative in others. An increase in photosynthetic rate should translate into the fixation of more carbon, and the changes in plant chemistry as a result is another mechanism by which AMF can affect foliar-feeding insects. Indeed, Gange and West (1994) found that AMF colonization of *Plantago lanceolata* resulted in an increase in the carbon/nitrogen ratio in the leaves, which in turn led to increases in levels of two carbon-based iridoid glycosides, aucubin and catalpol. These increases were suggested to have resulted in a lowering of the growth of the generalist feeding lepidopteran, *Arctia caja,* observed on colonized plants. An increase in the carbon/nitrogen ratio in response to AMF colonization was also recorded in creeping thistle (*Cirsium arvense*) by Gange and Nice (1997). In this case, performance of the gall-inducing fly, *Urophora cardui,* was also reduced on the colonized plants. As AMF can result in a host of chemical changes in foliage (Smith and Read, 1997), it is likely that this is the main mechanism by which these fungi can alter the growth of foliar-feeding insects. However, exceptions may occur, as Rabin and Pacovsky (1985) could find no host plant chemical explanation for the reduction in performance of two chewing moth larvae on AMF colonized soybean (*Glycine max*) plants.

While the preceding examples all document reduced performance of foliage-chewing insects on mycorrhizal plants, all experiments with sap-feeding insects have demonstrated that these tend to perform better on colonized plants (Gange and West, 1994; Borowicz, 1997; Bower, 1997; Gange *et al.*, 1999). While the plant vigor hypothesis may explain these observations, it is possible that changes in host plant physiology may also play a role. For example, Krishna *et al.* (1981) found large increases in the size of vascular bundles in *Eleusine coracana* (finger millet) when plants were mycorrhizal. Therefore, the increase in aphid performance on mycorrhizal plants observed in the above studies may be due to the insects being more successful in locating the phloem elements when probing. Certainly, it is known that a major cause of poor aphid performance on "resistant" crop cultivars is their inability to locate the phloem most of the time (e.g., Gabrys and Pawluk, 1999).

In summary, the relatively scant literature to date suggests that AMF colonization of plants results in reduced performance of foliage-chewing insects, but increased performance of foliage-sucking species. This statement must be treated with some caution, as all experiments to date have used forbs as the host plant. Experiments involving grasses have yet to be published.

Of course, the interaction between these mycorrhizal fungi and foliar insects must be a two-way one, in that it is quite possible for the removal of foliage by grazers to have a negative effect on the fungus. Removal of photosynthetic tissue may mean the fixation of reduced amounts of carbon and hence a reduced carbon supply for the fungus in the roots (Daft and El-Ghiami, 1981). Evidence suggests that this is indeed the case; however, there are extremely few studies. Gehring and Whitham (1994) list a number of examples in which foliage removal resulted in reduced AMF colonization; however, these were virtually all instances in which the removal was artificial or by vertebrates. Bower (1997) and Gange and Bower (1997) have shown that insect herbivory results in lowered AMF colonization in roots and suggest that this may be one cause of the often patchy distribution of mycorrhizal colonization seen in populations of plants in natural communities. For example, Bower (1997) found that natural populations of ragwort (*Senecio jacobaea*) had highly variable levels of colonization, but that those plants that suffered chronic foliage removal by cinnabar moths (*Tyria jacobaeae*) were generally the least colonized. As plants in natural communities may be colonized by a variety of fungal species (Clapp *et al.*, 1995), it is important to understand if foliage removal results in a simple reduction of abundance of all mycorrhizal species or whether only some of the associates disappear. If the latter is true, then this could have very important consequences for plant community structure. For example, van der Heijden *et al.* (1998) have shown that increasing AMF diversity leads to an increase in plant diversity, in mesocosms. However, increased plant diversity may mean greater insect colonization (Knops *et al.*, 1999) that in turn would reduce the mycorrhizal effect, providing a feedback mechanism. As suggested by Gange *et al.* (1999), insect effects on mycorrhizas may be an important negative feedback mechanism, which may regulate the stability of the symbiosis (Allen, 1991).

Arbuscular mycorrhizal fungi and root-feeding insects

Only three papers have been published on the interactions between AMF and rhizophagous insects (Gange *et al.*, 1994; Gange, 1996; Gange, in press). In all cases, the presence of an AMF species colonizing the roots led to the reduction in growth of black vine weevil (*Otiorhynchus sulcatus*) larvae. However, there is evidence that the interaction is not straightforward, as Gange (1996, in press) found that while colonization by one species of fungus led to growth reductions, inoculating roots with two species had no effect at all. Multiple colonization of roots (as is presumed to be the case in the field, cf. Clapp *et al.*, 1995) may thus have little effect on root-chewing insects. However, again this statement must be treated with caution as it is based on one insect, attacking two different host plants.

We do not understand the mechanism by which mycorrhizal fungi affect root-feeding insects. It is possible that AMF alter the quantity of root available to the insects which could be important if these insects are more limited by the amount of root available than its nutritional quality (which is very low) (Brown and Gange, 1990). However, in the majority of studies, root biomass is higher in mycorrhizal plants (Smith and Read, 1997) as it was in the experiments of Gange *et al.* (1994) and Gange (1996). Thus, if rhizophagous insects are limited by the quantity of root, one might expect their growth to be enhanced on mycorrhizal plants.

However, this does not occur and so it is more likely that AMF alter the chemistry of roots (as they do foliage) and that this change has detrimental consequences for the root herbivore.

AMF can result in a wide variety of chemical changes in roots. Indeed, some of the chemicals that appear in response to colonization have potent anti-insect effects. Some examples are phenolics (Morandi, 1996), terpenoids (Peipp *et al.*, 1997), and isoflavonoids (Vierheilig *et al.*, 1998). All of these compounds have been shown to exhibit activity against phytophagous insects (Harborne, 1988). Furthermore, AMF are well known as bioprotectants of roots against soil-borne plant pathogens (Azcón-Aguilar and Barea, 1996), and defenses in roots elicited by mycorrhizas against pathogen invasion could also have activity against insects.

In summary, it is possible that AMF may protect plants against root-chewing insects. However, the consequences of this in field situations may be limited, if multiple colonization of roots by fungi is the norm. Of more interest perhaps is that the nature of the interaction may depend on the identity of the fungal species involved. In one of the few studies to examine temporal changes in AMF species colonization of a plant, Merryweather and Fitter (1998) found that the roots of bluebell (*Endymion non-scriptus*) are colonized by different species of fungi at different times during a season. It is therefore quite likely that the effects of AMF on rhizophagous insects may vary according to the time of year. Extending this idea to natural communities of plants, we can speculate that as AMF species are known to change during succession (Johnson *et al.*, 1991), the outcome of any interaction will depend on the successional age of the community.

Complex multitrophic interactions between mycorrhizal fungi, root- and foliar-feeding insects

Negative effects of AMF on insects are obviously important for the insect in question, but these interactions may have consequences for other insects, which feed on the same host plant, but are spatially separated from the observed species. As documented above, root- and foliar-feeding insects appear to interact in a +/− fashion, termed contramensalism. However, we do not understand what mechanism regulates such contramensalistic relations in natural situations and how foliar- and root-feeding insects coexist in natural communities. For example, if foliar-feeding insect populations were high, then this could lead to local

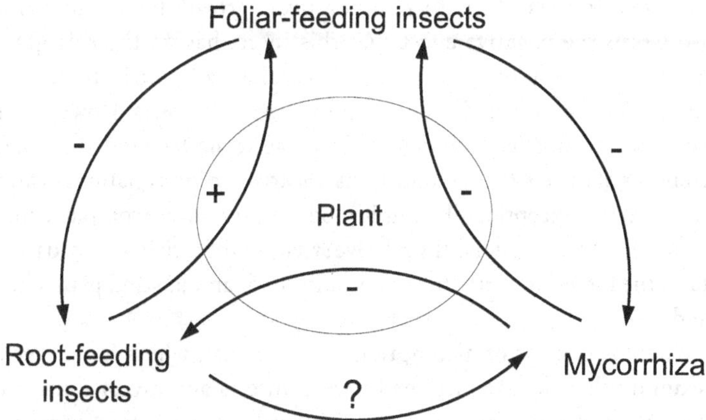

Fig. 9.2. A simplified model of how tritrophic interactions may occur between root- and foliar-feeding insects and arbuscular mycorrhizal fungi in natural plant communities.

extinction of root-feeding species. High populations of root feeders might be unlikely to occur, given their negative feedback mechanism with foliar feeders, yet clearly these situations do exist (Brown and Gange, 1990).

We suggest that the presence of a mycorrhiza, by having a negative effect on both root- and foliar-feeding insects, could reduce the strength of the insect–insect interaction and thereby assist in the maintenance of an equilibrium situation. Figure 9.2 elaborates on the model proposed by Masters *et al.* (1993) and illustrates this idea in a very simple model ecosystem. Here, there are two simple two-way interactions, that of the contramensalistic (+/−) insect–insect and the competitive (−/−) mycorrhiza–foliar insect. We assume that the competition in the latter case is for carbon, as this is directed to the fungus in the roots, but also lost to the insect when foliage is eaten. We do not know the effect of root herbivory on mycorrhizal colonization, hence the question mark, though we would expect this to be negative, thereby making the mycorrhiza–root insect another competitive interaction. Consumption of root may mean less available to be colonized and there may also be disruption of the mycorrhizal mycelium, as an occupational hazard of root feeding.

Of greater interest are the three-way interactions in this system. We predict that reduced growth of the rhizophage caused by AMF may reduce the strength of the positive effect that this insect has on a foliar-feeding species. In this way, the contramensalistic reaction may be stabilized. In turn, the reduction of folivore growth by the fungus will also lessen the negative effect that this insect has on the rhizophage. Therefore, the mycorrhiza acts as the regulatory factor in these multitrophic links between above- and below-ground insects. However, currently we do not have enough knowledge to understand the relative strengths of each of these interactions. Clearly, the interrelations of all the species in this exceptionally simple model ecosystem are complex, but we must begin to understand the relative strengths of each if we are to understand the forces that structure communities of animals and plants in the field.

There is one major assumption in our model, which is that we are assuming that the interactions between insects and mycorrhizal fungi seen in laboratory experiments do, indeed, occur in field situations. Circumstantial evidence suggests that this is a reasonable assumption (the experiment of Gange and West (1994) took place in a field situation and the work of Gange and Nice (1997) was in a semi-controlled "garden experiment"). However, long-term experiments, studying the roles of these interactions in plant community structure, have not previously been attempted. The remainder of this chapter is devoted to one such experiment, which we have recently begun, in which the roles of AMF and root- and foliar-feeding insects are being assessed in an early successional plant community (see Fig. 9.1).

A test of multitrophic interactions in an early successional community

An area of land measuring 450 m^2 was treated with weedkiller ("Roundup," containing glyphosate) in fall and plowed in January. The site was divided into plots, each 2.5×2.5m, and separated from each other by 1.5 m. Eight treatments were established, in which three different pesticides were applied in a fully factorial combination. The three products applied were: Dursban 5G (containing chlorpyrifos), to reduce levels of subterranean insect numbers; Dimethoate 40, to reduce foliar-feeding insects; and Rovral (containing iprodione), to limit mycorrhizal abundance. There were four replicates of each treatment, arranged in a randomized block design. Application of pesticides began in March and continued at six-weekly intervals for two growing seasons. Tests of phytotoxicity for the three compounds are described in Gange *et al.* (1992).

The developing vegetation was sampled at three-weekly intervals

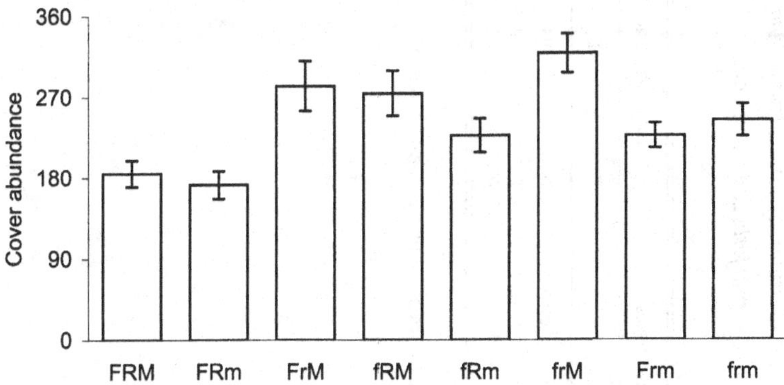

Fig. 9.3. Seasonal mean cover abundance of all vegetation in natural plant communities subjected to various exclusion treatments. (F) Foliar-feeding insects present, (f) foliar-feeding insects reduced, (R) root-feeding insects present, (r) root-feeding insects reduced, (M) arbuscular mycorrhizas present, (m) arbuscular mycorrhizas reduced.

during the first season and at monthly intervals during the second. On each occasion, a 38-cm linear steel grid, containing 10 3-mm point quadrat pins was placed randomly in each plot. A total of 50 pins per plot were sampled, and all touches of living vegetation recorded on each pin. Data were condensed to give a value for the total touches of each plant species and each plant life-history group in each plot. The data presented here are for the second growing season. Each data set was analyzed by a repeated-measures analysis of variance and seasonal means of the five monthly samples in year 2 are summarized here.

Field data

There was a significant effect of the three pesticide treatments on the total amount of vegetation in the communities (Table 9.1; Fig. 9.3). Reduction of root- and foliar-feeding insects increased the total vegetative cover, while reduction of mycorrhizal colonization reduced it. Although there were no significant statistical interactions between the treatments, interesting differences can be observed by visually inspecting pairs of means in Fig. 9.3. For example if we compare treatments FRM (control, with all insects and mycorrhiza present) and FRm (i.e., the application of fungicide when all insects were present) then there was little effect of mycorrhizal reduction. However, when all insects were reduced in number, application of fungicide resulted in a large reduction in cover abundance

Table 9.1. *Summary of ANOVA results, testing for the effects of excluding foliar- and root-feeding insects and arbuscular mycorrhizal fungi from an early successional plant community. Parameters analyzed were cover abundance of the total community, perennial forbs and grasses, and grass/forb ratio*

	Total		Perennial forbs		Perennial grasses		Grass/forb ratio	
	F	P	F	P	F	P	F	P
Foliar-feeding insects	6.68	<0.05	8.52	<0.01	6.79	<0.05	0.44	ns
Root-feeding insects	7.76	<0.05	9.75	<0.01	4.88	<0.05	4.61	<0.05
Arbuscular mycorrhizas	6.05	<0.05	4.96	<0.05	9.46	<0.01	0.79	ns

Note:
Repeated Measures analysis was performed, only main effects are given here. Interaction terms are omitted due to lack of significance. All degrees of freedom: 1,24.

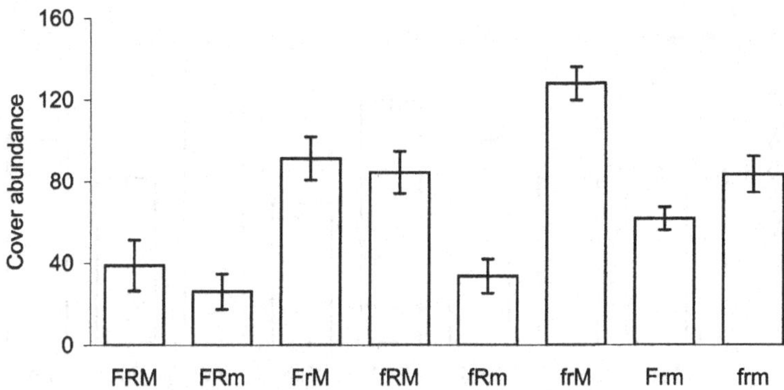

Fig. 9.4. Seasonal mean cover abundance of perennial forbs in natural plant communities subjected to various exclusion treatments. (F) Foliar-feeding insects present, (f) foliar-feeding insects reduced, (R) root-feeding insects present, (r) root-feeding insects reduced, (M) arbuscular mycorrhizas present, (m) arbuscular mycorrhizas reduced.

(treatments frM versus frm). Therefore, these data suggest that the effect of mycorrhizal colonization in increasing the total amount of vegetation was only seen when insect herbivory was also reduced. Gange and Brown (1992) suggested that insect attack on plants may be one reason as to why mycorrhizal effects on plant growth are rarely seen in field experiments, compared to laboratory situations, and these results would appear to support this suggestion.

In this second year of succession, the plant community was mainly composed of perennial forbs and perennial grasses (Fig. 9.1), the remainder being annual forbs and a few biennial forbs and sedges. Data for perennial forbs are presented in Fig. 9.4. Application of either insecticide had a highly significant effect in enhancing the abundance of this life-history group (Table 9.1). Subterranean insects are known to be major determinants of forb growth and hence species richness in these early successional communities (Brown and Gange, 1989b). Application of fungicide caused a significant reduction in the abundance of perennial forbs, and although no interaction terms were statistically significant, it again appeared that the effect of reducing mycorrhizal abundance was more clearly seen when insects were also reduced. This is seen by comparing pairs of means FRM versus FRm (little difference) and frM versus frm (large difference).

Data for perennial grasses are depicted in Fig. 9.5. Here, application of

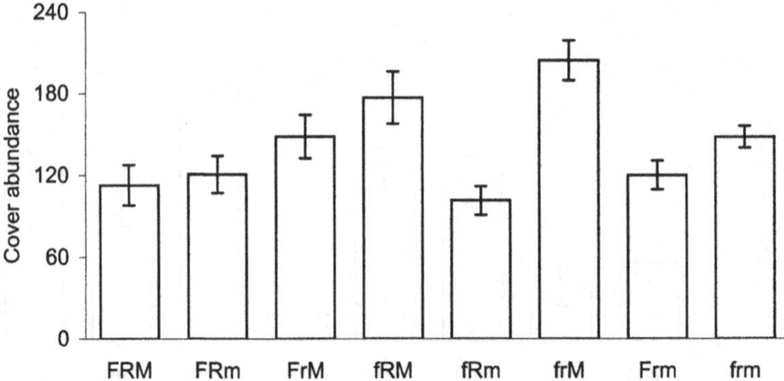

Fig. 9.5. Seasonal mean cover abundance of perennial grasses in natural plant communities subjected to various exclusion treatments. (F) Foliar-feeding insects present, (f) foliar-feeding insects reduced, (R) root-feeding insects present, (r) root-feeding insects reduced, (M) arbuscular mycorrhizas present, (m) arbuscular mycorrhizas reduced.

each of the three treatments was significant, with both insecticides increasing grass abundance and fungicide decreasing it. In this case, there was also a significant interaction between foliar insecticide and fungicide ($F_{1,24} = 5.1$, $P < 0.05$). This was because when foliar-feeding insects were present there was no effect of mycorrhizal reduction, but when these insects were excluded, mycorrhizal reduction was seen to have a large effect. This result is extremely interesting and suggests that at least one of the interactions seen in laboratory experiments may translate into significant community effects in the field. It is has been shown that the presence of mycorrhizal colonization tends to increase the growth and reproduction of sucking insects (Gange and West, 1994; Gange *et al.*, 1999). As stated above, it is also true that in these early successional communities, the bulk of the foliar-feeding insect community is composed of sucking insects, feeding on perennial grasses (Brown and Gange, 1999). Therefore, if mycorrhizal presence increases the growth of these insects, we should expect to see the greatest effect of insecticide application on plant production when the fungus is present, rather than when it is reduced. This can be seen by a comparison of the ratios of treatment means of foliar insecticide (treatment fRM) versus control (FRM), which is 1.58, with that of foliar insecticide + fungicide (fRm) versus fungicide (FRm), which is 0.84. The former is much greater, indicating that the effect of insecticide application on plant production was much greater when the mycorrhiza was present.

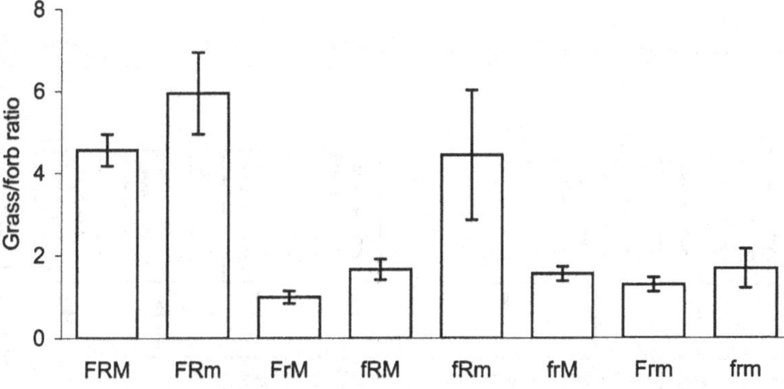

Fig. 9.6. Seasonal mean grass/forb ratio in natural plant communities subjected to various exclusion treatments. (F) Foliar-feeding insects present, (f) foliar-feeding insects reduced, (R) root-feeding insects present, (r) root-feeding insects reduced, (M) arbuscular mycorrhizas present, (m) arbuscular mycorrhizas reduced.

We believe that an important mechanism in the structuring of these early successional communities by insects and fungi is the grass/forb ratio, depicted in Fig. 9.6. Application of soil insecticide significantly decreased this ratio, i.e., communities in which root-feeding insects were excluded had much higher amounts of perennial forbs relative to perennial grasses. It is interesting that there was a clear difference between some of the treatments, pointing to the importance of plant competition as another structuring force. As an example, the grass/forb ratio was very high in plots only treated with fungicide (treatment FRm). This occurred because very few perennial forbs established in these plots, as seedlings were non-mycorrhizal and eaten by insects. However, the plots became dominated by perennial grasses, which were released from competition with the forbs. Although the overall total vegetative cover was relatively low in these plots, it was composed mainly of grasses. A second treatment which also had a high grass/forb ratio was that of the foliar insecticide + fungicide treatment (fRm). We explain this as follows: application of fungicide resulted in low numbers of forb and grass seedlings establishing. Root-feeding insects were present, thus reducing the forb abundance still further. However, foliar-feeding insects were absent, thus allowing enhanced grass growth. The result would be a situation in which grasses would be easily able to outcompete the forbs and the result would be a high grass/forb ratio.

The above statements imply that the three important agents of

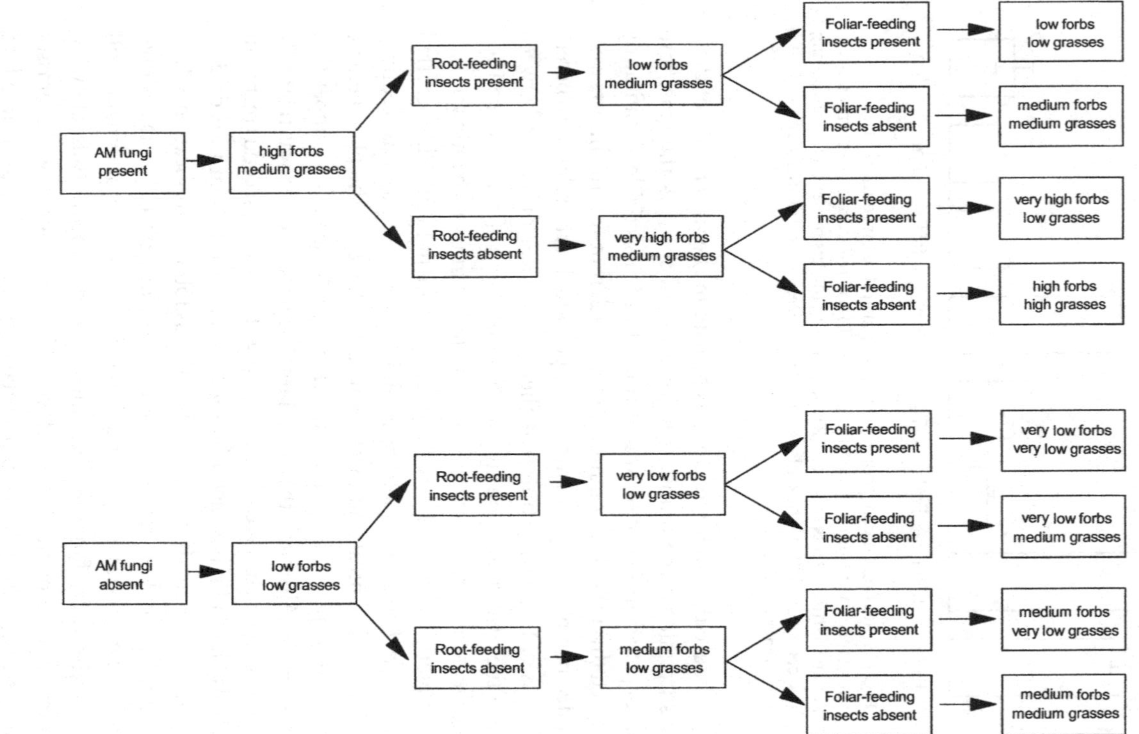

Fig. 9.7. A simple model of how arbuscular mycorrhizal (AM) fungi and phytophagous insects may act to structure early successional communities. Effects are assumed to occur in a linear temporal sequence. Although this is probably grossly oversimplified, the model does predict the abundance of vegetation observed remarkably well (see Fig. 9.3 and text).

seedling establishment act in a linear manner through time. We assume that the radicle of the seedling will first encounter a mycorrhizal hyphal network, and plugging into this network will allow enhanced establishment probability. Next in order of effect is likely to be the subterranean insect herbivore. These will attack the radicle once it has reached a sufficient size and before the seedling has appeared above ground. Indeed, it has been shown that the majority of seedling mortality, attributable to root-feeding insects, occurs below ground (Gange *et al.*, 1991). Once the seedling has appeared above ground, then the effect of excluding foliar-feeding insects is likely to be seen. We have assembled these ideas into a very simple model of community structuring, depicted in Fig. 9.7. For simplicity in this model we have used the terms low, medium, and high as indicators of the abundance of the two life-history groups. As an example, let us follow through one of the threads in this model, namely, that of the treatment of foliar insecticide + fungicide (fRm) outlined above. If AM fungi are absent, this will lead to low forb and grass establishment. If root-feeding insects are present, forb number will be reduced to a very low level, but grass abundance will be little affected and remain low. If foliar-feeding insects are then excluded, this will allow the emerging grass seedlings to increase in number to medium abundance. The result is a community with very low forb abundance and medium grass abundance and a high grass/forb ratio, as seen in Fig. 9.7. If we follow through the seedling establishment and subsequent competition idea, this model would predict a certain ascending order of abundance of our treatments as follows (in brackets after each one is the abundance of grasses/forbs): (1) Fungicide – FRm (very low/very low); (2) Control – FRM (low/low); (3) soil insecticide + fungicide – Frm (very low/medium); (4) foliar insecticide + fungicide – fRm (medium/very low); (5) all three pesticides – frm (medium/medium); (6) foliar insecticide – fRM (medium/medium); (7) soil insecticide – FrM(low/very high); (8) soil + foliar insecticide – frM (high/high). Although this is a very simple model, we are extremely encouraged by it, as this is precisely the order of cover abundance in each treatment seen in Fig. 9.3.

It therefore appears that multitrophic interactions recorded in laboratory experiments do translate into important community effects in the field. Our challenge for the future is to expand these interaction studies to include more species or combinations of species. Only then will we achieve a situation in which we can claim to understand the complex process of succession, and to manage it properly.

Conclusions and future directions

We have aimed to demonstrate the complexity of soil-mediated tritrophic interactions. The very complexity of the plant–mycorrhizal-fungal–insect–herbivore interactions is such that we have had to focus. We accept there are many gaps in our review, but we hope these gaps will serve as a stimulus for further research. Such research needs to be experimental, it needs to be manipulative and most importantly long term. However, we assert that AMF and root- and foliar-feeding insects are three important determinants of plant community structure. It is possible to demonstrate two- and three-way interactions between them in field studies, suggesting that an understanding of the effects of any one group is not possible without a consideration of the other. Our experiment has begun to unravel some of the intricacies of these interactions and we have been able to speculate on mechanisms of multitrophic interactions in relation to successional dynamics. Working within the context of a grassland successional sere we know (Fig 9.1), we are able to suggest the roles of the various organisms, their relative importance and that of the different trophic interactions. We have briefly touched on the influence of plant competition and consider this as an important mechanism by which the interactions are found to occur. We appreciate that other organisms, such as plant pathogens, nematodes, decomposers, and of course vertebrates all have important roles to play in the successional process. Further work on these is also needed (but see Mortimer *et al.*, 1999). We know the succession naturally proceeds to a sward dominated by perennial grasses. Therefore, those treatments in our field experiment and model (Fig. 9.7) which favour a high grass/forb ratio may be seen to accelerate the rate of succession. We suggest that restoration programs take into consideration results, such as those presented in this chapter, when aiming to reclaim land, either reduced in biodiversity value by agricultural practice or lost to a change in land use. The message is clear, in that future studies of plant community structure must be multidisciplinary and involve those interested or trained in soil microbiology, entomology, mycology, and mammalogy.

Acknowledgments

The authorship of this paper is merely alphabetical, with the chapter representing equal input from the authors. We wish to acknowledge the support of the UK Natural Environment Research Council that has funded much of the research over a number of years. The experiments would not have been possible without the assistance of a large number of field assistants, over the years, to whom we are extremely grateful.

REFERENCES

Allen, M. F. (1991) *The Ecology of Mycorrhizae*. New York: Cambridge University Press.

Azcón-Aguilar, C. and Barea, J. M. (1996) Arbuscular mycorrhizas and biological control of soil-borne plant pathogens: an overview of the mechanisms involved. *Mycorrhiza* **6**: 457–464.

Borowicz, V. A. (1997) A fungal root symbiont modifies plant resistance to an insect herbivore. *Oecologia* **112**: 534–542.

Bower, E. (1997) Interactions between arbuscular mycorrhizal fungi and foliar-feeding insects. PhD thesis, University of London.

Brown, V. K. (1986) Life cycle strategies and plant succession. In *The Evolution of Insect Life Cycles*, ed. F. Taylor and R. Karban, pp. 105–124. New York: Springer-Verlag.

Brown, V. K. (1990) Insect herbivores, herbivory and plant succession. In *Insect Life Cycles: Genetics, Evolution and Co-ordination*, ed. F. Gilbert, pp. 183–196. London: Springer-Verlag.

Brown, V. K. and Gange, A. C. (1989a) Differential effects of above- and below-ground insect herbivory during early plant succession. *Oikos* **54**: 67–76.

Brown, V. K. and Gange, A. C. (1989b) Herbivory by soil-dwelling insects depresses plant species richness. *Functional Ecology* **3**: 667–671.

Brown, V. K., and Gange, A. C. (1990) Insect herbivory below ground. *Advances in Ecological Research* **20**: 1–58.

Brown, V. K. and Gange, A. C. (1992) Secondary plant succession: how is it modified by insect herbivory? *Vegetatio* **101**: 3–13.

Brown, V. K., and Gange, A. C. (1999) Plant diversity in successional grasslands: how is it modified by foliar insect herbivory? In *Biodiversity in Ecosystems*, ed. A. Kratochwil, pp. 133–146. Amsterdam, Netherlands: Kluwer Academic Publishers.

Brown, V. K., and Southwood, T. R. E. (1987) Secondary succession: patterns and strategies. In *Colonization, Succession and Stability*, ed. A. J. Gray, M. J. Crawley and P. J. Edwards, pp. 315–337. Oxford: Blackwell Science.

Carson, W. P. and Root, R. B. (1999) Top-down effects of insect herbivores during early succession: influence on biomass and plant dominance. *Oecologia* **121**: 260–272.

Clapp, J. P., Young, J. P. W., Merryweather, J. and Fitter, A. H. (1995) Diversity of fungal symbionts in arbuscular mycorrhizas from a natural community. *New Phytologist* **130**: 259–265.

Clements, R. O, Bentley, B. R. and Nuttall, R. M. (1987) The invertebrate population and response to pesticide treatment of two permanent and temporary pastures. *Annals of Applied Biology* **105**: 129–145.

Daft, M. J. and El-Ghiami, A. A. (1981) Effect of arbuscular mycorrhiza on plant growth VII. Effect of defoliation and light on selected hosts. *New Phytologist* **80**: 365–372.

Davidson, D. W. (1993) The effects of herbivory and granivory on terrestrial plant succession. Oikos **68**: 23–35.

Edwards-Jones, G. and Brown, V. K. (1993) Successional trends in insect herbivore population densities: a field test of a hypothesis. *Oikos* **66**: 463–471.

Egler, F. E. (1954) Vegetation science concepts I. Initial floristic composition, a factor in old-field vegetation development. *Vegetatio* **4**: 412–417.

Gabrys, B. and Pawluk, M. (1999) Acceptability of different species of Brassicaceae as hosts for the cabbage aphid. *Entomologia Experimentalis et Applicata* **91**: 105–109.

Gange, A. C. (1996) Reduction in vine weevil larval growth by arbuscular mycorrhizal fungi. *Mitteilungen aus der Biologischen Bundesanstalt fur Land- und Forstwirtschaft* **316**: 56–60.

Gange, A. C. (2001) Species-specific responses of a root- and shoot-feeding insect to arbuscular colonization of its host plant. *New Phytologist* **150**: 615–618.

Gange, A. C. and Ayres, R. L. (1999) On the relation between mycorrhizal colonization and plant "benefit." *Oikos* **87**: 615–621.

Gange, A. C. and Bower, E. (1997) Interactions between insects and mycorrhizal fungi. In: *Multitrophic Interactions in Terrestrial Systems*, ed. A. C. Gange and V. K. Brown, pp. 115–132. Oxford: Blackwell Science.

Gange, A. C. and Brown, V. K. (1989) Effects of root herbivory by an insect on a foliar-feeding species, mediated through changes in the host plant. *Oecologia* **81**: 38–42.

Gange, A. C. and Brown, V. K. (1992) Interactions between soil-dwelling insects and mycorrhizas during early plant succession. In *Mycorrhizas in Ecosystems*, ed. I. J. Alexander, A. H. Fitter, D. H. Lewis and D. J. Read, pp.177–182. Wallingford: CAB International.

Gange, A. C., and Brown, V. K. (eds) (1997) *Multitrophic Interactions in Terrestrial Systems.* Oxford: Blackwell Science.

Gange, A. C. and Nice, H. E. (1997) Performance of the thistle gall fly, *Urophora cardui*, in relation to host plant nitrogen and mycorrhizal colonization. *New Phytologist* **137**: 335-343.

Gange, A. C. and West, H. M. (1994) Interactions between arbuscular mycorrhizal fungi and foliar-feeding insects in *Plantago lanceolata* L. *New Phytologist* **128**: 79–87.

Gange, A. C., Brown, V. K. and Farmer, L. M. (1990) A test of mycorrhizal benefit in an early successional plant community. *New Phytologist* **115**: 85–91.

Gange, A. C., Brown, V. K. and Farmer, L. M. (1991) Mechanisms of seedling mortality by subterranean insect herbivores. *Oecologia* **88**: 228–232.

Gange, A. C., Brown, V. K. and Farmer, L. M. (1992) Effects of pesticides on the germination of weed seeds: implications for manipulative experiments. *Journal of Applied Ecology* **29**: 303–310.

Gange, A. C., Brown, V. K. and Sinclair, G. S. (1993) VA mycorrhizal fungi: a determinant of plant community structure in early succession. *Functional Ecology* **7**: 616–622.

Gange, A. C., Brown, V. K. and Sinclair, G. S. (1994) Reduction of black vine weevil larval growth by vesicular–arbuscular mycorrhizal infection. *Entomologia Experimentalis et Applicata* **70**: 115–119.

Gange, A. C., Bower, E. and Brown, V. K. (1999) Positive effects of an arbuscular mycorrhizal fungus on aphid life history traits. *Oecologia* **120**: 123–131.

Gehring, C. A. and Whitham, T. G. (1994) Interactions between above-ground herbivores and the mycorrhizal mutualists of plants. *Trends in Ecology and Evolution* **9**: 251–255.

Gibson, C. W. D. and Brown, V. K. (1992) Grazing and vegetation change: deflected or modified succession? *Journal of Applied Ecology* **29**: 120–131.

Gibson, C. W. D., Brown, V. K. and Jepsen, M. (1987) Relationships between the effects of insect herbivory and sheep grazing on seasonal changes in an early successional plant community. *Oecologia* **71**: 245–253.

Grime, J. P., Hodgson, J. G. and Hunt, R. J. (1988) *Comparative Plant Ecology: A Functional Approach to Common British Species.* London: Unwin Hyman.

Harborne, J. (1988) *Introduction to Ecological Biochemistry.* London: Academic Press.

Haukioja, E. (1980) On the role of plant defences in the fluctuation of herbivore populations. *Oikos* **35**: 202–213.

Hunter, M. D. (1987) Opposing effects of spring defoliation on late season oak caterpillars. *Ecological Entomology* **12**: 373–382.

Ingham, R. E. and Detling, J. K. (1984) Plant–herbivore interactions in a North American mixed-grass prairie III. Soil nematode populations and root biomass on *Cynomys ludovicianus* colonies and adjacent uncolonized areas. *Oecologia* **63**: 307–313.

Johnson, N. C., Zak, D. R., Tilman, D. and Pfleger, F. L. (1991) Dynamics of vesicular–arbuscular mycorrhizae during old-field succession. *Oecologia* **86**: 349–358.

Johnson, N. C., Graham, J. H. and Smith, F. A. (1997) Functioning of mycorrhizal associations along the mutualism-parasitism continuum. *New Phytologist* **135**: 575–585.

Kidd, N. A. C., Lewis, G. B. and Howell, C. A. (1985) An association between two species of pine aphid, *Schizolachnus pineti* and *Eulachnus agilis. Ecological Entomology* **10**: 427–432.

Knops, J. M. H., Tilman, D., Haddad, N. M., Naeem, S., Mitchell, C. E., Haarstad, J., Ritchie, M. E., Howe, K. M., Reich, P. B., Siemann, E. and Groth, J. (1999) Effects of plant species richness on invasion dynamics, disease outbreaks, insect abundances and diversity. *Ecology Letters* **2**: 286–293.

Krishna, K. R., Suresh, H. M., Syamsunder, J. and Bagyaraj, D. J. (1981) Changes in the leaves of finger millet due to VA mycorrhizal infection. *New Phytologist* **87**: 717–722.

Masters, G. J. (1995) The impact of root herbivory on aphid performance: field and laboratory evidence. *Acta Oecologia* **16**: 135–142.

Masters, G. J. and Brown, V. K. (1992) Plant-mediated interactions between two spatially separated insects. *Functional Ecology* **6**: 175–179.

Masters, G. J.and Brown, V. K. (1997) Interactions between spatially separated herbivores. In *Multitrophic Interactions in Terrestrial Ecosystems*, ed. A. C. Gange and V. K. Brown, pp. 217–237. Oxford: Blackwell Science.

Masters, G. J., Brown, V. K. and Gange, A. C. (1993) Plant mediated interactions between above- and below-ground insect herbivores. *Oikos* **66**: 148–151.

Masters, G. J., Jones, T. H. and Rogers, M. (in press) Host-plant mediated effect of root herbivory on insect seed predators and their parasitoids. *Oecologia*.

Mawdsley, J. L. and Bardgett, R. D. (1997) Continuous defoliation of perennial ryegrass (*Lolium perenne*) and white clover (*Trifolium repens*) and associated changes in the composition and activity of the microbial population of an upland grassland soil. *Biology and Fertility of Soils* **24**: 52–58.

Merryweather, J. W.. and Fitter, A. H. (1998) Patterns of arbuscular mycorrhiza colonization of the roots of *Hyacinthoides non-scripta* after disruption of soil mycelium. *Mycorrhiza* **8**: 87–91.

Mikola, J., Yeates, G. W., Wardle, D. A., Barker, G. M. and Bonner, K. I. (2001) Response of soil food-web structure to defoliation of different plant species combinations in an experimental grassland community. *Soil Biology and Biochemistry* **33**: 205–214.

Moore, J. C. and De Ruiter, P. C. (1991) Temporal and spatial heterogeneity of trophic interactions within below-ground food webs. *Agriculture, Ecosystems, Environment* **34**: 371–397.

Moore, J. C. and De Ruiter, P. C. (1997) Compartmentalization of resource utilization within soil ecosystems. In *Multitrophic Interactions in Terrestrial Systems*, ed. A. C. Gange and V. K. Brown, pp. 375–393. Oxford: Blackwell Science.

Moran, N. A. and Whitham, T. G. (1990) Interspecific competition between root-feeding and leaf-galling aphids mediated by host-plant resistance. *Ecology* **71**: 1050–1058.

Morandi, D. (1996) Occurrence of phytoalexins and phenolic compounds in endomycorrhizal interactions, and their potential role in biological control. *Plant and Soil* **185**: 241–251.

Mortimer, S. R., Van der Putten, W. H. and Brown, V. K. (1999) Insect and nematode herbivory below-ground: interactions and role in vegetation succession. In *Herbivores: Between Plants and Predators*, ed. H. Olff, V. K. Brown and R. H. Drent, pp. 205–238. Oxford: Blackwell Science.

Noy-Meir, I., Gutman, M. and Kaplan, Y. (1989) Responses of Mediterranean grassland plants to grazing and protection. *Journal of Ecology* **77**: 290–310.

Peipp, H., Maier, W., Schmidt, J., Wray, V. and Strack, D. (1997) Arbuscular mycorrhizal fungus-induced changes in the accumulation of secondary compounds in barley roots. *Phytochemistry* **44**: 581–587.

Prach, K. and Pyšek, P. (1999) How do species dominating in succession differ from others? *Journal of Vegetation Science* **10**: 383–392.

Prach, K., Pyšek, P. and Šmilauer, P. (1993) On the rate of succession. *Oikos* **66**: 343–346.

Price, P.W. (1991) The plant vigor hypothesis and herbivore attack. *Oikos* **62**: 244–251.

Rabin, L. B. and Pacovsky, R. S. (1985) Reduced larva growth of two Lepidoptera (Noctuidae) on excised leaves of soybean infected with a mycorrhizal fungus. *Journal of Economic Entomology* **78**: 1358–1363.

Seastedt, T. R., Ramundo, R. A. and Hayes, D. A. (1988) Maximization of densities of soil animals by foliage herbivory: empirical evidence, graphical and conceptual models. *Oikos* **51**: 243–248.

Smith, S. E. and Read, D. J. (1997) *Mycorrhizal Symbiosis*. San Diego, CA: Academic Press.

Staddon, P. L., Fitter, A. H. and Robinson, D. (1999) Effects of mycorrhizal colonization and elevated atmospheric carbon dioxide on carbon fixation and below-ground carbon partitioning in *Plantago lanceolata*. *Journal of Experimental Botany* **50**: 853–860.

van der Heijden, M. G. A., Klironomos, J. N., Ursic, M., Moutoglis, P., Streitwolf-Engel, R., Boller, T., Wiemken, A. and Sanders, I. R. (1998) Mycorrhizal fungal diversity determines plant biodiversity, ecosystem variability and productivity. *Nature* **396**: 69–72.

Vierheilig, H., Bago, B., Albrecht, C., Poulin, M. J. and Piche, Y. (1998) Flavonoids and arbuscular–mycorrhizal fungi. *Advances in Experimental Medicine and Biology* **439**: 9–33.

Watt, T. A. (1981) A comparison of grazed and ungrazed grassland in East Anglian Breckland. *Journal of Ecology* **69**: 499–508.

White, T. C. R. (1984) The abundance of invertebrate herbivores in relation to the availability of nitrogen in stressed food plants. *Oecologia* **63**: 90–105.

Wilson, A. D. and Shure, D. J. (1993) Plant competition and nutrient limitation during early succession in the Southern Appalachian mountains. *American Midland Naturalist* **129**: 1–9.

STEFAN SCHEU AND HEIKKI SETÄLÄ

10

Multitrophic interactions in decomposer food-webs

Introduction

Trophic interactions in soil form the basis of virtually all terrestrial life. Without the recycling of plant materials produced above the ground, and the mineralization of the nutrients therein, plant life would cease quickly and with it the whole above-ground food-web. It is surprising therefore that, in comparison to aquatic food-webs and food-webs above the ground, the below-ground community has received little attention. One of the major intentions in writing this review was to emphasize this gap and to outline that the lack of knowledge on food-web interactions in below-ground systems is a major constraint in current ecological thinking.

We will stress in this review that below-ground systems are unique in several respects and that, due to this uniqueness, the understanding of interactions in below-ground systems may significantly enrich the way we perceive nature. Since from the energetic perspective the below-ground decomposer system is far more important than the herbivore system above the ground, it is certainly necessary to include the peculiarities of below-ground food-webs into ecological thought. Although this has been realized by various people (Beare *et al.*, 1995; Wardle and Giller, 1996; Bengtsson, 1998; Young and Ritz, 1998; Scheu *et al.*, 1999b; Villani *et al.*, 1999; Wall and Moore, 1999), the bias in terrestrial ecology towards above-ground systems has experienced little change. However, there are very promising signs (Wardle *et al.*, 1997; Bardgett *et al.*, 1998a; Huhta *et al.*, 1998; Laakso and Setälä, 1999a; Scheu *et al.*, 1999a; Ponsard and Arditi, 2000; Scheu and Falca, 2000) and we will highlight recent achievements below.

Price (1988) stressed that it was soil ecologists who deal with the most essential part of terrestrial life: "All started in soil: present food-webs reflect ancient associations (the earliest systems started from bacteria, through mutualistic interactions, to eukaryotes, including plants). The first food-webs were composed of bacteria, Protozoa and fungi, which even nowadays form the basis for all food-webs." More recently, May (1997) underlined this by stressing that "A full understanding of the causes and consequences of biological diversity, in all its richness, cannot be had until the contribution made by decomposers to the structure and functioning of ecosystems is fully understood."

We hope to convince ecologists working above the ground that their research essentially relies on below-ground food-web interactions and that the two systems are much more closely linked than commonly realized. The links are largely based on multitrophic interactions below ground, but also on those between components of both systems. The linkages often are complex, in most cases including indirect effects, and often involve abiotic resources, such as nutrients and root exudates.

To present the peculiarities of below-ground food-webs, we emphasize the complexity of trophic interactions among soil organisms and between these and primary production. Since the unique features of below-ground interactions follow from peculiarities of the soil habitat, and the distribution of the resources therein, we first deal with these before turning to the organisms themselves. The major part of this chapter then assesses the ways in which below-ground and above-ground systems are linked. Finally, we will stress that trophic interactions are only part of the story; in soil systems they may even be a minor part of the story.

Special attributes of the soil system in respect to trophic interactions

Trophic interactions involve consumers and resources. Confronted predominantly with plant–herbivore and herbivore–predator systems, animal ecologists are accustomed to think of both as being living organisms. However, in detritus systems, basal resources are the remains of living organisms and therefore, one of the components is dead organic matter. This is of prominent importance for trophic interactions and has shaped the peculiarities of soil organisms.

Habitat and resources

The soil habitat fundamentally differs in various ways from habitats above the ground, and this fact has important implications for trophic interactions. The basis of the decomposer food-web is dead organic matter, which has attributes that make it a unique resource.

Food-web implications of dead organic matter as basal resource

A unique feature of the resource–consumer relationship between dead organic matter and detritivores is that the interactors do not coevolve. This is in contrast to herbivores and plants and also to predator–prey systems. Major consequences of the lack of coevolution are that (1) detritivores tend to be less specialized than herbivores and (2) the gamma diversity of detritivores is comparatively low. A low degree of specialization implies that detritivores tend to be food generalists, i.e., there is no close association between plant species and species of detritivores, and the overlap in the resources used by detritivores is high. In combination with the well-accepted view of detritivores as being food-limited (Hairston *et al.*, 1960; Pimm, 1982; Hairston, 1989), one may expect strong competition between decomposer organisms. Surprisingly, however, there is little experimental evidence that competition really is a major structuring force in soil animal communities (Hairston, 1989; Scheu *et al.*, 1999b).

A marked feature of feeding on detritus is that the food substrate is packed with microorganisms. In fact, it has been commonly assumed that detritivorous animals primarily rely on microorganisms as food, with the ingestion of dead organic matter being unavoidable in order to exploit this more nutritious resource. This is reflected by the analogy of Cummins (1974) of detritivorous animals being feeders of peanut butter on undigestible biscuit. Translating this to trophic level interactions, three trophic levels (or even more, see below) appear to be involved in resource–detritivore interactions (Fig. 10.1). Using food-web terminology, this implies that most detritivores may better be viewed as trophic level omnivores. However, the extent to which detritivores digest microorganisms is still poorly understood. In fact, there is increasing evidence that most of the microorganisms in the food substrate of detritivores survive the gut passage and flourish in casts (Fischer *et al.*, 1995; Maraun and Scheu, 1996; Wolter and Scheu, 1999; Tiunov and Scheu, 2000).

Viewing detritivores as trophic level omnivores is even more evident when considering that they may also function as predators (Fig. 10.1). It has been shown that earthworms may digest nematodes, protozoans, and possibly even enchytraeids (Gorny, 1984; Roesner, 1986; Bonkowski and Schaefer, 1997). Earthworms search selectively for microsites rich in Protozoa

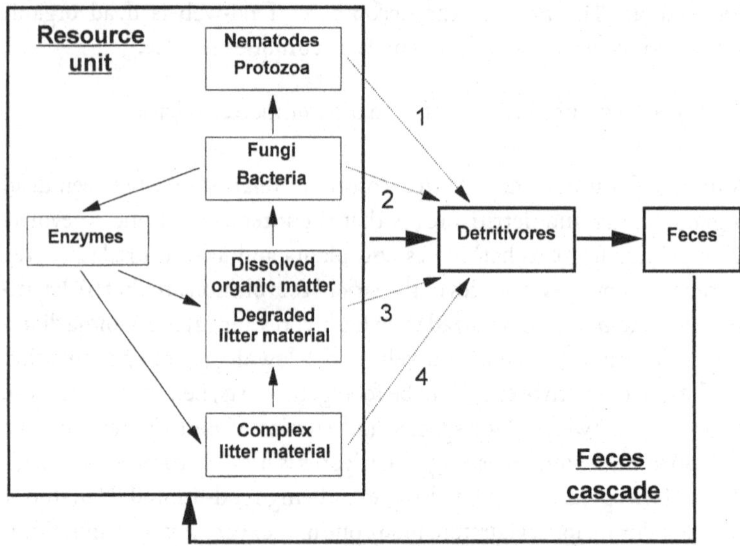

Fig. 10.1. Detritivore–resource relationships. Numbers indicate major potential food substrates exploited by detritivores from ingested material ("resource unit"). Depending on the resource exploited detritivores function as predators (1), microbivores (2), competitors for microbially degraded substrates (3) or as primary/secondary decomposers (4); when exploiting a mixture of resources they function as omnivores. Note that in the latter case detritivores may feed simultaneously on three trophic levels.

and benefit considerably when the food material ingested contains Protozoa (Miles, 1963; Flack and Hartenstein, 1984; Bonkowski and Schaefer, 1997).

The basis of the view outlined so far is that detritivorous animals ingest dead organic matter composed of a variety of resources (the "resource unit" in Fig. 10.1). Even more selectively feeding species, like Collembola and oribatid mites, which are commonly viewed as fungal feeders, ingest large quantities of dead organic matter (Anderson and Healey, 1972; Chen *et al.*, 1996). An important mechanism contributing to resource exploitation by these microbi-detritivores is the fragmentation of organic matter. In fact, microbi-detritivores uniformly appear to shred plant residues. This even applies to species which lack mandibles like earthworms and enchytraeids. In the former it has been shown that the combined ingestion of litter materials and sand grains significantly increases litter fragmentation (Schulmann and Tiunov, 1999).

Considering the ubiquity of resources composed of complex plant compounds, like cellulose and lignin, it is surprising that the majority of detritivorous animals appear to be enzymatic amateurs compared to soil microorganisms. This raises the question of how detritivorous animals manage to exploit these resources. Surprisingly, the solution to this problem, at least in part, is that detritivorous animals benefit from the superior enzymatic capabilities of litter and soil microorganisms. Microorganisms exploit resources by liberating enzymes which in the vicinity of the cell digest complex substrates thereby making resources more easily available. These resources presumably constitute a major portion of the diet of many detritivores (Fig. 10.2). This mechanism emphasizes that detritivore–resource interactions involve more than two trophic levels. If detritivorous animals rely on the enzymatic action of microorganisms, the acquisition of resouces by detritivorous animals may simply be a consequence of the limited ability of microorganisms to combine digestion and uptake of food resources.

This model includes the view that detritivorous animals live in an external rumen from which they get their resources (Swift *et al.*, 1979). In this case, microorganisms function as external mutualists of detritivores. A striking consequence of this model is that it explains why most detritivorous animals have not evolved the capability to digest complex plant litter compounds such as cellulose and lignin, but became and remain enzymatic amateurs.

Linkage between habitat and food resources and its implications

A further striking feature of detrital systems is that habitat and food are closely interconnected; in terms of the external rumen concept, habitat and food are two aspects of the same entity. Microbi-detritivores therefore may be viewed as organisms living in food resources – a situation reminiscent of paradise. However, the unique feature of food resources is that they contain large substrates predigested externally by microorganisms and internally by other animals, i.e., fecal material (Fig. 10.1). Indeed, re-ingestion of organic matter is one of the fundamental mechanisms responsible for habitat formation in soil. This is most obvious in soils with thick organic layers on top of the mineral soil (moder), where the lower layers constitute differentially fragmented plant residues. Ultimately, the lowermost organic horizon is formed of humus material which is mainly composed of fecal pellets of minute soil animals like enchytraeids, Collembola, and oribatid mites (Rusek, 1985). In soils where large detritivores are present, plant residues are decomposed and mixed

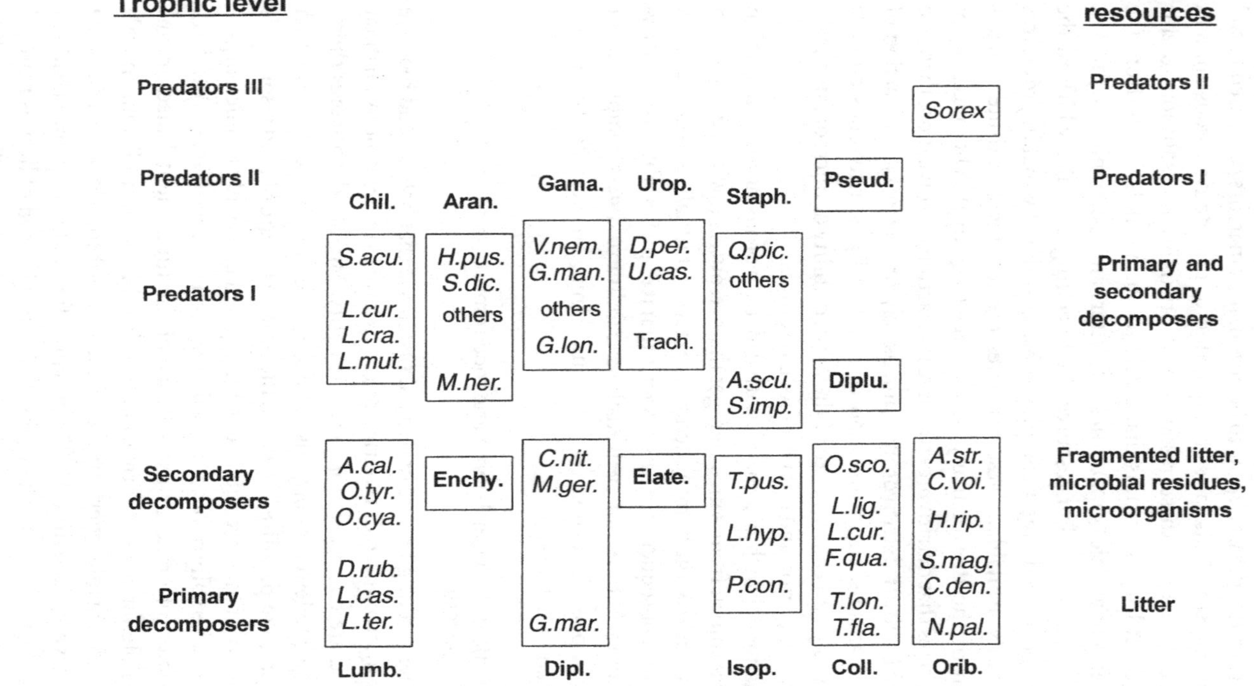

Fig. 10.2. Trophic structure of the decomposer community of a beech wood on limestone ("Göttinger Wald") as indicated by ^{15}N analysis (from Scheu & Falca, 2000). Aran: Araneida; Chil: Chilopoda; Coll: Collembola; Dipl: Diplopoda; Diplu: Diplura; Elate: elaterid larvae (Coleoptera); Enchy: Enchytraeidae; Gama: Gamasina; Isop: Isopoda; Lumb: Lumbricidae; Orib: Oribatida; Pseud: Pseudoscorpionida; Staph: Staphylinidae (Coleoptera); Urop: Uropodina; for full species names see Scheu & Falca (2000).

with mineral soil material (mull) by cascading through the guts of litter-feeding earthworms, millipedes, and isopods.

Re-ingestion of an organism's own fecal material has been shown to stimulate growth of diplopod and isopod species and this intraspecific coprophagy has received considerable attention (McBrayer, 1973; Hassall and Rushton, 1985; Gunnarsson and Tunlid, 1986). However, a large number of detritivorous species, the secondary decomposers, rely on fecal material of other detritivores. Surprisingly, this unique feature of decomposer food-webs has received little attention despite its paramount importance for soil formation and trophic interactions in soil (Szlavecz and Pobozsny, 1995). Affecting another species, via feces, is a kind of indirect interaction which may function as micro-scale facilitation. Particularly earthworms beneficially affect other soil invertebrates, and earthworms also benefit from the production of fecal pellets by other detritivores. For example, the litter-feeding earthworm *Lumbricus terrestris* has been shown to affect other soil animal species by providing fecal material enriched in organic residues (Szlavecz, 1985; Maraun *et al.*, 1999). Similarly, it has been shown in laboratory and field experiments that endogeic earthworms such as *Octolasion tyrtaeum* benefit from the production of fecal pellets by millipedes (Scheu and Sprengel, 1989; Bonkowski *et al.*, 1998).

Intra- and interspecific coprophagy add considerably to the complexity of decomposer food-webs. The common distinction of primary and secondary decomposers certainly is oversimplistic. Incorporating the cascading of detritus through microbi-detritivores in food-web models is difficult, since it is an indirect interaction. Therefore, interactions among detrivores add to the view that indirect interactions are of paramount importance for structuring soil communities (Bengtsson *et al.*, 1996; Scheu *et al.*, 1999b). The fact that detritivores, for various reasons, do not fit conventional food-web categories hampers the modeling of decomposer communities and approaches which attempt to combine population- and process-based views of decomposer communities (Bengtsson *et al.*, 1995; De Ruiter *et al.*, 1996; Persson *et al.*, 1996).

For experimental soil ecologists, the close interrelationship between habitat and resources in soil systems poses particular problems, since it is hard to manipulate one without changing the other. This is unfortunate because, to understand the dominant structuring forces in animal communities, it is essential to separate effects caused by habitat structure and by food resources. Another key question in understanding the structuring forces in animal communities is whether bottom-up or top-down forces predominate (Hairston *et al.*, 1960; Strong, 1992; Hairston and Hairston, 1993, 1997; Polis and Winemiller, 1996; Persson, 1999). Most experiments which addressed this issue for soil animal communities suffered from the difficulty that the manipulation of food resources also alters habitat structure (Judas, 1990; Hövemeyer, 1992; Ponge *et al.*, 1993). However, some experiments succeeded in manipulating food resources without concomitantly changing habitat structure. Chen and Wise (1997, 1999) manipulated food resources of the soil arthropod community of a deciduous forest by adding high-quality substrate (chopped mushrooms, potatoes, and instant fruit-fly medium) to the forest floor and demonstrated substantial bottom-up control. The increase in the resource base propagated through the food-web and resulted in a twofold density of strictly predaceous arthropods like centipedes, pseudoscorpions, and spiders. Thus, the resource manipulations cascaded up the food-web for at least two trophic levels. In contrast, in the experiment of Scheu and Schaefer (1998), which manipulated basal resource supply by repeatedly adding liquid carbon and nutrient resources to the floor of a deciduous forest, the density of detritivores increased but that of major predators, centipedes, declined. Presumably, this decline resulted from indirect effects due to habitat modification caused by an increased earthworm density. Since centipedes have been shown to feed on earthworms, the implication is that by habitat modification, prey species in soil such as earthworms may effectively control one of their own major predators (see section "Beyond trophic interactions: habitat modifications," below).

The organisms

Soil animals are extremely densely packed: underneath one footprint of forest soil there may be billions of Protozoa, hundreds of thousands of nematodes, thousands of Collembola and mites, and a large number of isopods, spiders, beetles, etc. Obviously, the majority of soil animals are small; most species are <1 mm in length. Small size enables the exploitation of resources accessible only via soil pores, while the porous nature of soils can be used as refuge from predators.

The porous nature of soils has major implications for predator–prey interactions. There must have been a strong selection for predators to reduce their body diameter to the size of their prey. For prey species the small size of predators may have increased their ability to defend against predator attacks. In fact, one of the most diverse microbi-detritivorous soil animal group, the oribatid mites, may have evolved such effective defense mechanisms that adults currently live in a predator-free space (Norton, 1994). However, this may not apply to juveniles (Laakso *et al.*, 1995; Laakso and Setälä, 1999a).

Both the ubiquity of refuges and the effective defense mechanisms of prey species suggest that the influence of predators on prey populations in soil is limited. This argument also applies to the interaction between microbial grazers and their bacterial or fungal prey. There is very limited evidence that microorganisms in soil are controlled by microbivorous species (Brussaard *et al.*, 1995; Mikola and Setälä, 1998a; Schlatte *et al.*, 1998; Kandeler *et al.*, 1999; Laakso and Setälä, 1999b; but see Scholle *et al.*, 1992; Scheu and Parkinson, 1994a; Alphei *et al.*, 1996; Hendrix *et al.*, 1998). Contrary to the normal functioning of predators in reducing the number of prey, in soil the presence of fungal and bacterial grazers often results in an increase in prey population density and biomass (Verhoef *et al.*, 1989; Bardgett *et al.*, 1998b). This is commonly ascribed to compensatory growth due to indirect effects of the grazers on the mobilization of nutrients (Visser, 1985; Wolters, 1991; Lussenhop, 1992). Also, there is very limited evidence that bacterivorous and fungivorous soil animals are controlled by predators, but the issue remains controversial and certainly needs further investigation (Schaefer, 1995; Hyvönen and Persson, 1996; Kajak, 1997). Generally, however, it is likely that in soil top-down trophic cascades are of limited importance for community organization (Mikola and Setälä, 1998a; Laakso and Setälä, 1999a, b). This implies that the dominant food-web models, i.e., trophic dynamics models (Oksanen *et al.*, 1981; Polis, 1999), are of limited use for understanding the major structuring forces in soil communities. Omnivory may exert strong regulation of other trophic levels in ways not predicted by cascading trophic interactions (Morin and Lawler, 1995; Persson *et al.*, 1996; Pace *et al.*, 1999).

It has been repeatedly argued that the prevalence of top-down trophic cascades declines from aquatic to terrestrial habitats (Strong, 1992; Persson, 1999), and we assume soil communities to be at the lowermost level of this gradient. Generally, this is consistent with the view that trophic cascades predominate in simple food-webs (Strong, 1992; Polis and Strong, 1996) since, due to the very dense species packing, the prevalence of generalist feeders and the ubiquity of trophic level omnivory (see above), soil communities are exceptionally complex. Stressing that trophic cascades presumably are of limited importance in soil communities does not mean that top-down control may not be important under certain conditions and for regulating certain soil animal species or groups. In fact, there is evidence that fungivores, nematodes and detritivores are controlled by predators (Kajak *et al.*, 1993; Schaefer, 1995; Hyvönen and Persson, 1996; Setälä *et al.*, 1996; Yeates and Wardle, 1996; Mikola and Setälä, 1998b), and predator control of microbi-detritivores has even been shown to slow down decomposition processes (Kajak, 1997; Lawrence and Wise, 2000).

Due to the complexity of the soil food-web, and the difficulties of ascribing decomposer organisms to specific trophic levels, information on the structure of soil food-webs is difficult to obtain. A methodology which may improve this situation is the analysis of variation in the abundance of stable isotopes in soil animals. Using $\delta^{15}N$ data, it has been shown recently that soil organisms form two clusters consisting of predators and microbi-detritivores which span a range equivalent to approximately four trophic levels (excluding detritus: Ponsard and Arditi, 2000; Scheu and Falca, 2000; Fig. 10.2). Both microbi-detritivores and predators spanned over two trophic levels, suggesting that microbi-detritivores do indeed comprise a gradient from primary to secondary decomposers. Interestingly, in microbi-detritivorous groups, such as Collembola, oribatid mites, diplopods, and earthworms, species spanned over a very similar range, indicating that species of very different groups exploit similar resources (Scheu and Falca, 2000). A major implication of this is that higher taxonomic units are inadequate for depicting trophic levels, only reflecting very general trophic groups, such as microbi-detritivores and predators. This questions the usefulness of soil food-web models in which higher taxonomic units are used to depict trophic relationships and energy flow in below-ground communities (Hunt *et al.*, 1987; Bengtsson *et al.*, 1996; De Ruiter *et al.*, 1996).

In a similar way to microbi-detritivores, predators in soil uniformly appear to be generalist feeders. At first sight, this is surprising considering that the size of predators in soil commonly only slightly exceeds that of their prey. Generally, small size differences between predators and prey favor the evolution of specialist predators (Peters, 1983; Futuyma and Moreno, 1988). Obviously, this is not the case in soil. The fact that predators in soil have remained generalist feeders presumably is related to the high density of potential prey and the difficulties of locating specific prey in the opaque and porous soil habitat. The latter is the major reason why the parasitic Hymenoptera, which is a dominant element of the predator community above the ground, is rare in soil; in some of the major soil animal taxa (earthworms, millipedes, isopods, Collembola, oribatid mites) hymenopteran parasitoids do not exist at all (Ulrich, 1988). A striking feature supporting the view that in soil generalist predators predominate is that one of the dominant groups are centipedes which are "sit-and-wait" predators (Poser, 1991; Schaefer, 1991; Ekschmitt *et al.*, 1997). Most important soil microarthropods are so ubiquitous and mobile that for many predators it does not pay to search for them. This situation is reminiscent of sit-and-wait predators or even suspension feeders on the floor of aquatic habitats foraging on zooplankton. The difference is that the prey in soil are actively mobile, whereas in aquatic habitats they are transported by the water current. Being confronted with a multitude of different predator species, it is likely that prey species in soil have been unable to evolve traits to avoid predation by specific predators. In above-ground systems, the implications of multiple predator effects on prey species are receiving closer scrutiny (Sih *et al.*, 1998); in soil, where the topic is of prime importance, it has hardly been touched.

The indiscriminate feeding on mobile prey by many soil predators implies that they almost uniformly function as trophic level omnivores, since the trophic range of their prey, microbi-detritivorous species, extends over at least two trophic levels (see above). In addition, predators are likely to feed within their own trophic level; intraguild predation (IGP) and cannibalism is certainly widespread in decomposer food-webs (Polis, 1991; Gunn and Cherrett, 1993). In their study on the trophic structure of the soil fauna community of two beech forests, Scheu and Falca (2000) reported that, based on differences in $\delta^{15}N$, predator species on average differ from prey species by more than one trophic level. They assumed that this resulted from a mixed diet composed of primary and secondary decomposers with a bias towards the latter. However, the larger difference in trophic levels between predators and microbi-detritivores may also be caused by IGP and, in fact, this is how Ponsard and Arditi (2000) explained the trophic level difference in excess of one between predators and microbi-detritivores in their study on the soil fauna community of three deciduous forests.

IGP and cannibalism have recently received considerable attention and, in contrast to previous assumptions (Pimm and Lawton, 1977; Pimm, 1982), are now thought to play a major role in structuring communities (Polis and Holt, 1992; Holt and Polis, 1997; Rosenheim, 1998). Certainly, IGP and cannibalism are very widespread in soil communities. This can be inferred as most predators in soil are generalists. In the soil, therefore, top-down effects of predators on prey, but also of "microbivores" on microorganisms, cannot be understood without considering IGP and predator switching. Unfortunately, except for epigeic predators such as lycosid spiders and carabid beetles (Wise and Chen, 1999; Snyder and Wise, 1999), the implications of IGP for soil community structure are poorly known. That IGP and cannibalism are widespread among soil predators and even in "fungivores" has been stressed by Walter (1987). The dominance of generalist predators and the high incidence of IGP and cannibalism in soil communities add to the intriguing complexity of below-ground foodwebs. It is likely that the exceptionally high local diversity of soil organims at least in part results from the fact that soil predators are so closely intermingled. The very large overlap in prey (among predators) may prevent competitively superior microbi-detritivores from outcompeting inferior species, despite close resource overlap among them. Predator switching and suppression of competitively superior prey species has been shown to be a key process in maintaining species diversity in aquatic and aboveground terrestrial systems (Paine, 1966; Menge and Sutherland, 1976), but has yet to be shown in soil communities.

The similarity of trophic positions of species in very different taxonomic groups within microbi-detritivores and predators supports the assumption that functional redundancy among soil animals is high (Wardle *et al.*, 1997, 1998; Setälä *et al.*, 1998). A detailed discussion of this issue is beyond the scope of this paper. However, we emphasize that to date only gross ecosystem properties like respiration and primary production have been investigated; other functional attributes of ecosystems, such as stability, resilience, constancy, and resistance, certainly need further attention. In contrast to the long-standing assumption that more complex systems are less stable (May, 1972; Pimm, 1980), recent experimental evidence suggests that the opposite might be true (McGrady-Steed *et al.*, 1997). We believe decomposer communities to be ideal systems to explore these issues, since very complex communities can be established and manipulated even in small experimental units. The few investigations undertaken have indicated that more complex soil communities recover more quickly from perturbations than simpler systems (Maraun *et al.*, 1998; B. S. Griffiths, unpublished data), although some counterevidence also exists (H. Setälä *et al.*, unpublished data).

Plants as players in multitrophic interactions in soil

The growing interest in exploring the interface of population-level and ecosystem-level approaches to ecology (Jones and Lawton, 1995) requires acknowledgement of the significance of both the above- and below-ground compartments of terrestrial ecosystems. This is because above-ground systems are responsible for most of the production (carbon input) in an ecosystem, whereas below-ground systems are responsible for most of the decomposition (carbon loss) in the system. The above-ground and below-ground compartments are dependent upon one another because of the role of plants as a carbon source for soil biota, and because soil biota in turn release nutrients bound up in relatively recalcitrant compounds into simpler forms that are more readily taken up by the plant (Wardle, 1999). Despite the obvious interdependence between soil food-webs and above-ground grazer pathways, the knowledge on the interplay between the two worlds, mediated through nutrient dynamics, is scanty.

The resources supplied by plants not only consist of above-ground residues, as the input via the root system may even exceed the litter input from above. Below-ground residues comprise a mixture of resources consisting of dead roots, cells lost by growing roots, and a multitude of liquid compounds exuded by roots (Curl and Truelove, 1986; Lynch, 1990). Due to the continuous and high resource input in the vicinity of roots, the rhizosphere is a focus of intense microbial and animal activity. The close spatial association between primary production (plant roots in soil) and decomposition is a marked difference between terrestrial and aquatic systems (Wagener *et al.*, 1998). The association between plants and decomposer organisms in the rhizosphere and the fact that the relationship is beneficial to both result in complex mutualistic interactions between plants and soil organisms, of which the plant–mycorrhiza mutualism is most prominent. However, this is only one aspect of the close relationship between the below-ground and the above-ground systems; there is a multitude of ways in which plants and decomposers interact and this is what we focus on in this chapter.

Certainly, it is the rhizosphere where decomposer organisms and plants interact most intensively. However, plants and decomposers also interact via plant material produced above ground, albeit with the interactors being spatially separated. By the decomposition of plant litter and the mineralization of the nutrients therein, decomposer organisms may affect their own resources via modification of the nutrient content of plants and therefore of the dead organic matter entering the detrital system. The tightly connected positive feedback system in which decomposers can improve the quality and quantity of their resource through recycling of nutrients has been proposed to be characteristic of terrestrial

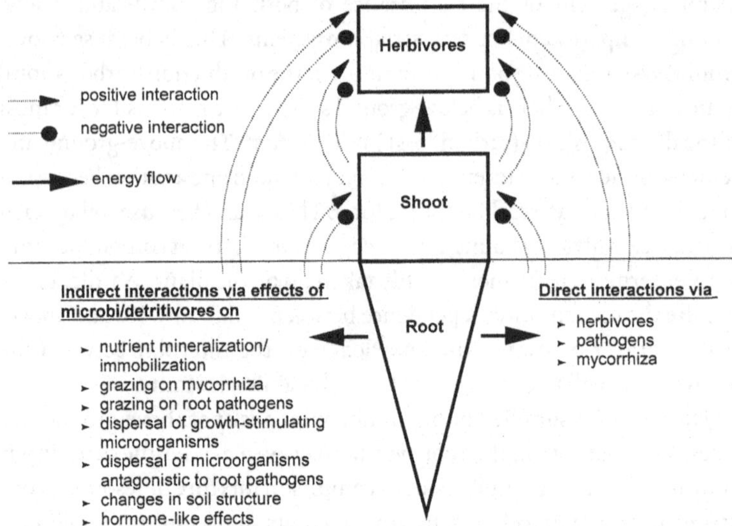

Fig. 10.3. Ways by which soil animals affect plant growth and thereby herbivore performance. Both direct and indirect interactions between roots and soil biota may stimulate or reduce plant growth and herbivore development. The activity of rhizosphere biota is driven by energy transfer from plant roots (root exudates).

systems (Bengtsson *et al.*, 1996). The intimate reciprocal interactions between the above-ground and below-ground systems warrant treating the producer–decomposer system as a single food-web with an array of interactive components (May, 1997; Wardle *et al.*, 1998).

Here, we specifically explore the implications of these interactions for the structure of the soil food-web and for plant growth. In doing this we focus on effects from below (soil biota community) to above (plants and herbivores: Fig. 10.3), others have stressed that the structure of the above-ground system, i.e., the composition of the plant community and plant–herbivore interactions, also strongly modifies the structure of the below-ground system and the interactions therein (Seastedt, 1985; Bardgett *et al.*, 1998a).

The multitude of plant–decomposer interactions

As for above-ground residues, the basis of plant–decomposer interactions in the rhizosphere involves non-living resources, root deposits for detritivores and mineral nutrients for plants. Since mineral nutrients are also essential resources for the majority of decomposers (bacteria and fungi), plants and decomposers may compete via consumption of abiotic resources. Therefore, interactions between soil biota and plants are often not direct feeding interactions but are mediated through abiotic resources (Fig. 10.3). In fact, the interaction between primary decomposers (soil bacteria and fungi) largely depends on the availability of nutrients (Tamm, 1991; Norton and Firestone, 1996). It is well documented, for example, that if carbon resources are in ample supply, microorganisms are superior competitors for nutrients, causing plants to suffer from nutrient deficiency (Zak *et al.*, 1990; Harte and Kinzig, 1993; Schmidt *et al.*, 1997; Lipson *et al.*, 1999). This raises the question of how plants manage to acquire resources when nutrients are in short supply. The fact that plants manage to do so is surprising considering that, in the vicinity of roots, microorganisms gain carbon resources from the plant as root exudates and other rhizosphere deposits. As is well known, plants rely on other organisms in the rhizosphere to improve their competitive ability against saprophytic miroorganisms. These organisms might be viewed in total as plant mutualists, of which mycorrhizal fungi are the most prominent. However, other components of the soil food-web also function as plant mutualists, although they are not directly connected to plant roots.

Competition for abiotic resources is not limited to plant–microbe interactions, but is also a common feature for other distantly related soil-inhabiting organisms and these interactions are likely to modify plant growth. There is increasing evidence that saprophytic soil bacteria and mycorrhizal fungi can inhibit each other (Garbaye, 1991; Fitter and Garbaye, 1994). Wilson *et al.* (1989) showed that spore germination of vesicular–arbuscular (VA) fungi was dramatically reduced by soil bacteria. On the other hand, the presence of ectomycorrhizal (EM) fungi can inhibit the growth of rhizosphere bacteria, which has been ascribed to reduced root exudation (Olsson *et al.*, 1996b; Bonkowski *et al.*, in press a) or to the production of fungal antibiotics (Grayston *et al.*, 1996; Nurmiaho-Lassila *et al.*, 1997). Despite the existence of various negative interactions in the rhizosphere, there appears to be consensus, however, that mutualistic relationships predominate which mitigate the competition for nutrients in the rhizosphere of plants (Ingham and Molina, 1991; Grayston *et al.*, 1996).

The function of most of the multitrophic interactions in soil is the facilitation of the acquisition of nutrients by plants via indirect effects (Clarholm, 1985; Kuikman and van Veen, 1989; Setälä, 1995; Bever *et al.*,

1997; Bonkowski *et al.*, in press a; Fig. 10.3). Excretion of nutrients often is considered as a mechanism by which microbi-detritivores directly alter plant growth. However, since both microbi-detritivores and plants rely on the same nutrient pools strong indirect interactions occur. More complex mechanisms which involve trophic interactions between microbi-detritivores and microorganisms include changes in soil nutrient pools by grazing on saprophytic miroorganisms (Setälä and Huhta, 1991; Haimi *et al.*, 1992; Alphei *et al.*, 1996), grazing on mycorrhizal fungi (Klironomos and Kendrick, 1995, Setälä, 1995), and grazing on plant pathogens (Curl *et al.*, 1988; Pussard *et al.*, 1994; Lartey *et al.*, 1994). Other indirect ways, which relate to trophic interactions, include the dispersal of plant growth stimulating microorganisms such as rhizobia, mycorrhiza, and mutualistic rhizosphere bacteria (Gange, 1993; Stephens *et al.*, 1993, 1995; Harinikumar and Bagyaraj, 1994; Lussenhop, 1996), and microorganisms antagonistic to root pathogens (Stephens and Davoren, 1997). Furthermore, microbi-detritivores indirectly modify plant growth by changes in soil structure (Hoogerkamp *et al.*, 1983; Boyle *et al.*, 1997) and by hormone-like effects (Jentschke *et al.*, 1995; Muscolo *et al.*, 1996, 1999). Other more direct effects, such as plant seed dispersal by soil animals (Thompson *et al.*, 1993; Willems and Huijsmans, 1994), and plant damage due to root herbivores and pathogens, mainly nematodes (Masters *et al.*, 1993; Sarathchandra *et al.*, 1996; Strong *et al.*, 1996; Zunke and Perry, 1997), add to the spectrum of links between soil animals and plants. As emphasized by Strong (1999), root herbivores are imbedded in a complex predator–parasite community including organisms as different as parasitic nematodes (as antagonists of insect root herbivores) and nematode-trapping fungi (as antagonists of root-feeding nematodes) which may exert top-down effects on the plant root systems and therefore on plant growth.

The prevelance of indirect interactions among biota also has important implications for the functioning of food-webs. In contrast to direct interactions via feeding on other organisms, effects of indirect interactions via consumption of a common resource, such as nutrients in soil, may propagate through the web without changing the size of other web components. This is a fundamentally different way in which food-web components are interconnected from the conventional view of trophic dynamics in which effects only propagate through the web by changing the size of adjacent trophic levels (Hairston *et al.*, 1960; Oksanen *et al.*, 1981). Since primary production is regulated by the availability of nutrients and the decomposer community relies on this availability, indirect interactions via abiotic resources, i.e., nutrients in soil, are essential for understanding the functioning of food-webs.

As already stated, saprophytic microorganisms compete for nutrients with mycorrhizal fungi (Wilson *et al.*, 1989; Brundrett, 1991) and plants (Jamieson and Killham, 1994; Cheng *et al.*, 1967). The implications of these competitive interactions, however, can only be understood by considering all three groups simultaneously. The growth of saprophytic microorganisms and the resulting immobilization of nutrients into microbial biomass facilitate mycorrhizal infection of plant roots, thereby improving the plant's ability to exploit heterogeneously distributed resources (nutrients and water: Smith and Read, 1996), but also strengthening the plant's resistence against root pathogens (Duchesne *et al.*, 1987, 1988). It is well established that low nutrient concentrations promote infection of plants by mycorrhizal fungi, which consistently results in enhanced growth and survival of the plants (Linderman, 1988; Allen, 1991).

The interrelationship between saprophytic microorganisms, mycorrhizal fungi, and plants is further complicated by another soil food-web component, the fungal-feeding fauna (Chakraborty *et al.*, 1985; Shaw, 1985; Ingham, 1988; Klironomos *et al.*, 1999; Fig. 10.3). Grazing on rhizosphere fungi (mycorrhizal and saprophytic) has been shown to control the growth of mycorrhizal fungi, thereby modifying the plant–fungus association (Warnock *et al.*, 1982; Finlay, 1985; Harris and Boerner, 1990; Ek *et al.*, 1994; Setälä, 1995; Setälä *et al.*, 1997). The interplay between these four food-web components is mediated by abiotic resources, i.e., soil nutrients and plant root exudates. The following scenario may be commonplace: plant nutrient acquisition and growth are stimulated by mycorrhizal fungi in soils where nutrients are in short supply, whereas the opposite may occur in fertile soils. In the latter case, plants may be disadvantaged by mycorrhizal infection due to the fact that when nutrients are freely available the investment of resources in root symbionts may not pay (Bowen, 1980; Allen, 1991) so turning symbiosis into parasitism (Smith and Smith, 1996). Similarly, albeit more locally, soil fauna, by increasing concentration of nitrogen and phosphorus in the rhizosphere (Ingham *et al.*, 1985; Setälä and Huhta, 1991; Alphei *et al.*, 1996), can indirectly affect the cost–benefit balance of the plant–fungus symbiosis (Harris and Boerner, 1990; Ek *et al.*, 1994; Setälä *et al.*, 1997). By feeding upon saprophytic microorganisms, soil fauna speed up the cycling of nutrients, an effect known to induce the plant to reduce allocation of carbon resources to its fungal partner (Gemma and Koske, 1988; Wallander, 1992). Thus, by feeding on soil fungi, the soil fauna indirectly affect both the performance of mycorrhizal fungi and the stability of the plant–fungus symbiosis via nutrient mobilization. This complex scenario illustrates interactions among distantly related organisms which have been assumed to be a characteristic element of plant–decomposer systems (Swift *et al.*, 1979; Coleman, 1996). However, an even more characteristic feature is the close interplay between the biotic and abiotic components. Understanding the indirect interactions mediated via abiotic resources is a prerequisite for fully capturing the feedbacks that control decomposition, nutrient cycling, and plant growth in terrestrial ecosystems (Bengtsson *et al.*, 1996).

Mutual benefits

The close spatial and temporal relationship between primary producers and decomposers in terrestrial environments (Wagener *et al.*, 1998) and the continuous shortage of nutrients in soils (Tamm, 1991) has resulted in many coevolutionary associations between plants and decomposer organisms (Perry *et al.*, 1992). This contrasts the lack of coevolution in the decomposer–dead organic matter system. As stressed above, many of the decomposer–primary producer associations are indirect and function under a wide range of conditions as mutualistic relationships (Lavelle *et al.*, 1995; Fig. 10.3). The direct forms of mutualistic associations are the best known, such as the symbiosis between mycorrhizal fungi and plants, between legumes and their nitrogen-fixing bacteria (*Rhizobium*) in the root nodules, and between actinomycetes (*Frankia*) and their host plants (such as alder). Obviously, mutualistic associations between plants and microorganisms evolved independently at different times and with different microbial partners (VA fungi, EM fungi, rhizobia, actinomycetes), each capable of providing their host with nutrients, particularly nitrogen and phosphorus. These independent evolutionary lines demonstrate that severe competition for nutrients in the rhizosphere was a major selection agent in the evolution of plants. Certainly, the main function of these mutualistic associations in the plant–decomposer system is to minimize competition for mineral nutrients between decomposers and plants and this presumably contributed to the world becoming so impressively green.

Each of the mutualistic associations presented above represents classical forms of direct mutualistic symbioses (*sensu* Boucher *et al.*, 1982), in which the two players evolved intimate physiological links with each other. In soil, there is a number of other mutualistic relationships in which the partners are not in direct physical contact with each other but interact indirectly via abiotic resources. Again, by mediating trophic interactions, the abiotic resources function as important components in the below-ground food-web. A prominent example is the link between microfaunal grazers and plant roots known as the microbial loop (Clarholm, 1985). In this tritrophic interaction, plant root exudates serve as the basal energy resource for saprophytic soil microorganisms (Grayston *et al.*, 1996). Rapid growth of microorganisms in the rhizosphere results in immobilization of mineral nutrients into microbial biomass. From the plant's point of view, the reduced availability of nutrients is problematic, unless a third party, soil fauna, remineralizes the microbially fixed nutrients. Soil Protozoa and nematodes are known to graze extensively on rhizosphere microorganisms thereby facilitating plant nutrient acquisition by liberating nutrients bound in microbial biomass for root uptake (Clarholm, 1989; Kuikman, 1990; Griffiths, 1994; Zwart *et al.*, 1994). Thus, Protozoa and nematodes function as bacteria-mediated mutualists (Bonkowski *et al.*, 2000a, in press a). In the mediated mutualism, plant resources are released as exudates, stimulating bacterial and subsequent protozoan growth; nitrogen is released due to grazing on bacteria by Protozoa. In the direct mutualism, plant carbon resources are directly transferred to mycorrhizal fungi in exchange for phosphorus. The mediated mutualism between plants and Protozoa is associated with a more complex root system, consisting of more branches, and longer and finer roots (Jentschke *et al.*, 1995; Alphei *et al.*, 1996; Bonkowski *et al.*, in press a). In contrast, the plant–mycorrhiza mutualism commonly is associated with a less complex root system, i.e., with fewer and bigger roots (Alexander, 1981; Jentschke *et al.*, 1995; Bonkowski *et al.*, 2000b, in press a). Since the plant–mycorrhiza mutualism mainly facilitates plant phosphorus uptake, whereas the mediated mutualism between plants and Protozoa predominantly functions by increasing plant nitrogen uptake, the two mutualistic systems have been hypothesized to complement each other (Bonkowski *et al.*, in press a).

Although this tritrophic interaction resembles a trophic cascade, in which carnivores indirectly affect the biomass of primary producers by consuming herbivores (Hairston *et al.*, 1960), it functions as a mutualistic interaction between plants and Protozoa which is mediated by bacteria and involves root exudates and nutrients. It is due to these non-living resources that the biomass of the microorganisms, the second trophic level, does not necessarily decline due to predation (Brussaard *et al.*, 1995;

Mikola and Setälä, 1998b); rather, it may even increase (Abrams and Mitchell, 1980; Griffiths, 1986; Hedlund *et al.*, 1991; Mikola and Setälä, 1998a; see also section "Limitations of plant–decomposer interactions," below). The ability of prey populations to respond "positively" to grazing is thought to result from a grazer-mediated increase in the availability of nutrients (Coleman *et al.*, 1983; Ingham *et al.*, 1985; Mikola and Setälä, 1998a).

Abiotic components have also been found to be of crucial importance in a very different trophic interaction pathway in soil. It has been demonstrated that above the ground plants may "call for" predators or parasites of herbivores by producing volatile substances, thereby strengthening top-down trophic cascades (Turlings *et al.*, 1991, chapter 7, this volume; Dicke and Vet, 1999). Recently, Hall and Hedlund (1999) found a similar interaction to exist below ground, with soil fungi, collembolan grazers, and predators (mesostigmatid mites) as the players. Soil fungi have long been known to produce substances that inhibit the growth of other microorganisms (Wicklow, 1992). Hall and Hedlund (1999) observed that fungal-derived volatiles also function as a cue which attracts predatory mites and guides them to the place where Collembola, their major prey, graze on the fungus. As soils are opaque, three-dimensional habitats where visual cues are of limited value, tritrophic indirect interactions of this kind may well be a common feature throughout the plant–decomposer food-web, in comparison to above ground where volatiles may be used for long-distance communication. Such communication in soil via volatile chemical cues will clearly be restricted to short distances.

The tight connection between abiotic resources and biotic interactions in soils is likely to be the key factor responsible for the evolution of the great variety of positive feedbacks in soil food-webs (Perry *et al.*, 1992; Bengtsson *et al.*, 1996; Bonkowski *et al.*, 2000b, in press a).

Limitations of plant–decomposer interactions

Soil food-webs are regarded as classic textbook examples of donor-controlled food-webs in which the density of the resource (detritus) controls the density of the consumers, but not the reverse (Pimm, 1982). However, there is evidence that under certain circumstances, predators (including grazers) may control the number and growth rates of soil microbivores (Hyvönen and Persson, 1996; Yeates and Wardle, 1996; Mikola and Setälä, 1998a, b; Laakso and Setälä, 1999a) and microbi-detritivores (Kajak *et al.*, 1993; Schaefer, 1995; Setälä *et al.*, 1996; Laakso and Setälä, 1999a). Further, as described in context of the microbial loop, the growth and activity of soil microorganisms can be effectively controlled by soil micro- and mesofauna (cf. Bengtsson *et al.*, 1996). It therefore seems that, despite predominant bottom-up regulation, top-down forces can be strong enough to control lower trophic levels even at the very base of the detrital food-web. Indeed, there is evidence showing that exclusion of top predators may affect the activity of the soil microflora with concomitant changes in decomposition rate of organic matter (Santos *et al.*, 1981; Kajak *et al.*, 1993; Lawrence and Wise, 2000). However, very little is known about the relationship between top predators and plant growth, i.e., whether the effects due to removal of top soil carnivores would propagate down the food-web so as to be reflected in a change in primary production.

The microcosm experiment of Laakso and Setälä (1999a) explored the impact of the structure of the decomposer food-web on the growth of birch (*Betula pendula*) seedlings by manipulating both the diversity of trophic groups and the within-trophic-group complexity. The results of this study corroborated earlier findings that, through consumption of soil microflora, microbi-detritivores (Collembola, oribatid mites, enchytraeids, dipteran larvae) stimulated plant growth. This stimulation of primary production was due to an increase in nutrient mineralization, mainly nitrogen, in the rhizosphere. Presence of top predators (predatory mites) did not modify the microbi-detritivore induced stimulation of plant growth, despite significant reduction in prey populations.

A similar observation was made by Setälä *et al.* (1996). Although large predators (Coleoptera and centipedes) substantially reduced the biomass of their prey (small earthworms, millipedes, and others), neither the growth nor concentration of nitrogen in the leaves of poplar seedlings was affected. However, excluding fungal and bacterial feeding soil mesofauna from the systems resulted in reduced uptake of nitrogen by the seedlings. Therefore, it appears that top-down control over plant productivity is restricted to the base of the detrital food-web, for example to the control of fungi by fungivores, and is dampened at higher trophic levels, i.e., trophic cascades appear to be of limited importance (see also Mikola and Setälä, 1998a, b).

The reason for the apparent functional insignificance of top-down trophic cascades in soil food-webs is poorly known. Pace *et al.* (1999) compiled a list of conditions that promote or inhibit the transmission of predatory effects in food-webs: the relative productivity of ecosystems, presence of refuges, omnivory, and the potential for compensation. As stressed in the section "Special attributes of the soil system in respect to trophic interactions" (above), the ubiquity of omnivory and refuges diminishes the cascading of top-down effects in decomposer systems, with productivity and compensation likely to reduce further the incidence of top-down effects in soil. This particularly holds for the control of primary decomposers, i.e., bacteria and fungi, which can colonize minute soil pores (refuges) hardly accessible to their major predators, protozoans and nematodes (Elliott *et al.*, 1979; Rutherford and Juma, 1992, England *et al.*, 1993; Mikola and Setälä, 1998b). Moreover, soil bacteria and fungi are able to compensate for consumed biomass (Abrams and Mitchell, 1980; Ingham *et al.*, 1985; Brussaard *et al.*, 1995; Griffiths, 1986; Mikola and Setälä, 1998a) by accelerating their tunover rate which is known to be facilitated by enhanced consumer-driven nutrient mobilization (Mikola and Setälä, 1998a; Wardle, 1999).

Another feature typical of soil food-webs is the heterogeneity within trophic levels (Moore and Hunt, 1988). In soils, energy and nutrient pathways are compartmentalized into distinct channels based on bacteria, fungi, and plant roots. As these channels have been shown to differ in their response to for example microbial grazing (De Ruiter *et al.* 1993; Wardle and Yeates, 1993; Mikola and Setälä, 1998b), it is not surprising that propagation of predatory effects is likely to fade away under heterogeneity in structure and function.

Interestingly, the fungal-based energy channel may be much more prone to trophic cascades than the bacterial-based channel. The reason for this is the dissimilarity of the intermediate predators, the microbivores, between the two channels. In the bacteria-based energy–nutrient pathway, the major predators are Protozoa and nematodes, which represent not only the intermediate consumers but also seem to represent the top predators in the web. In the fungal channel, however, the diverse assemblage of fungal-feeding microarthropods forms the basis for an abundant and species-rich community of carnivores (Persson *et al.*, 1980; Schaefer, 1991; Berg, 1997). Consequently, the functional importance of top predators is likely to be more important in systems where energy and nutrients are transferred along the fungal-based channel. Soils of boreal forest evidently fulfill this criterion, whereas improved grasslands and agricultural systems with predominantly bacterial-based energy–nutrient channels do so to a lesser extent.

Forests and grass-dominated ecosystems differ not only in the type and structure of primary producers, but also in the plant–fungal interaction, the mycorrhiza. Arbuscular mycorrhizae (AM) are typical of grasslands and agroecosystems, whereas ectomycorrhizae predominate in forest soils (Allen, 1991). Importantly, the two types of mycorrhiza differ in respect to the way the fungi affect their host plant and the rhizosphere (Brundrett, 1991). AM plants typically are grasses and herbs with a low shoot-to-root ratio (1/1). Due to the substantial amount of photosynthates allocated below ground, the production of root exudates (commonly sugars) is extensive and can be as high as 40% of the net primary production (Holland *et al.*, 1996; Bardgett *et al.*, 1998a). In grasslands and agroecosystems, with the constant input of easily utilizable energy resource, bacteria may dominate over fungi (Alexander, 1977; Swift *et al.*, 1979). In such systems, bacterial feeding microfauna, Protozoa in particular, appear to be the only trophic group to control efficiently the growth and biomass of primary decomposers (Clarholm, 1985; Alphei *et al.*, 1996; Coleman, 1996; Bonkowski *et al.*, in press a). Such systems with a strong bacterial component are best described as "fast cycle systems" (*sensu* Coleman *et al.*, 1983) in which the rapid rate of energy and nutrient transfer through microorganisms, via microfaunal consumers, to plants reflects the rapid turnover rate of the organisms of the soil food-web (Coleman, 1996). Consequently, as there is virtually no evidence for effective control of bacterial grazers, the effects due to removal of top carnivores are likely to be of limited value in grass-dominated systems.

In forest ecosystems, on the other hand, predators of microbial feeding fauna are abundant (Persson *et al.*, 1980; Huhta *et al.*, 1986; Schaefer, 1991; Berg, 1997) and likely to have a strong influence on prey density (Usher, 1985). Moreover, in boreal forests ectomycorrhizal plants predominate and, as compared with grasses, these plants allocate a substantially smaller proportion of energy below ground (Jackson *et al.*, 1996). In addition, EM fungi form a thick fungal mantle on the short roots of their host plant, which can prevent root exudates from reaching the soil (Marx, 1972; see also Setälä *et al.*, 2000). Consequently, the biomass of rhizosphere bacteria has been reported to be less in the presence than in the absence of EM fungi (Olsson *et al.*, 1996b; Nurmiaho-Lassila *et al.*, 1997; Bonkowski *et al.*, in press a), whereas in AM systems no such decline has been observed (Olsson *et al.*, 1996a). Not surprisingly, saprophytic fungi typically dominate over bacteria in forest soils, particularly under more acidic conditions (Anderson and Domsch, 1975; Scheu and Parkinson, 1994b; Berg *et al.*, 1998), resulting in the dominance of the fungal-based energy channel in these systems (Moore and Hunt, 1988; Ingham *et al.*,

1989). As saprophytic fungi, in contrast to bacteria, have relatively long generation times ("slow-cycle" decomposition systems *sensu* Coleman *et al.* (1983)) and react slowly to consumption and other disturbances (Moore and De Ruiter, 1997), the fungal-based pathway is likely to be a persistent and predictable avenue providing nutrition to carnivores at the top of the food-web. Therefore, we expect trophic cascades to manifest themselves in forest soils and other fungal-dominated systems where fungivores as intermediate "predators" predominate.

As briefly described in the section "The multitude of plant–decomposer interactions" (above), mycorrhizal fungi not only benefit plants by enhancing water and nutrient uptake but can also serve various types of soil fauna as a food resource (Shaw, 1985; Thimm and Larink, 1995; Klironomos *et al.*, 1999). This direct trophic interaction between fungi and fungal-feeding soil fauna can induce fundamental changes in the performance of the plant–fungus association, and, importantly, limits the propagation of feeding effects across the plant–soil food-web. Depending on the severity of grazing, fungal-feeding fauna have been shown to retard plant growth (Warnock *et al.*, 1982; Finlay, 1985), have a negligible effect (Ek *et al.*, 1994), or increase primary production (Harris and Boerner, 1990; Setälä, 1995).

The outcome of interactions between fungivorous fauna, mycorrhizal fungi, and plants are more complex and less straightforward than one might assume. The importance of abiotic resources, nutrients in particular, in mediating the plant–decomposer food-web is mainly responsible for this complexity. Fertility of the soil, together with the intensity of grazing, can, in unexpected ways, determine whether grazing by fauna on mycorrhizal fungi is beneficial or detrimental for the plant. The relationship between plants and EM fungi can vary from parasitism via neutralism to mutualism depending on the availability of nutrients in the soil, i.e., the outcome of the interaction is conditional on abiotic conditions (see Bronstein, 1994; Smith and Smith, 1996). When nutrients are scarce fungal-feeding soil fauna can overconsume the beneficial mycorrhizal fungi and thus render the symbiosis ineffective (Setälä *et al.*, 1997). On the other hand, when nutrients are abundant and mycorrhizal fungi become parasitic, fungal grazing may be beneficial to the plant due to harming its parasite (Setälä *et al.*, 1997).

In conclusion, plant–decomposer interactions appear to be notoriously variable which is primarily due to the fact that abiotic components are involved. However, the pattern that trophic cascades are of limited importance in regulating primary productivity certainly holds for most detrital systems. Trophic cascades are much less important in soil than in aquatic habitats. Since nutrients are recycled almost exclusively by microorganisms which are able to compensate for losses due to grazing and other disturbances, the consumers are unlikely to affect indirectly the renewal rate of primary decomposers.

Links between decomposer and herbivore system

We stressed that plants are significantly affected by the activity and composition of the decomposer community. This certainly has implications for the whole above-ground community, i.e., plants link the decomposer and the herbivore system. By modifying the resource base of the above-ground community, this pathway is under bottom-up control. However, there is another pathway linking the below-ground and above-ground community which functions in a top-down way. Animals of the decomposer community are an important component of the diet of generalist predators above the ground and this may strengthen herbivore control by these predators (Settle *et al.*, 1996; Oksanen *et al.*, 1997; Snyder and Wise, 1999, 2001). We cover this aspect only briefly here; for a more detailed discussion see Scheu (2001).

So far, direct interactions between soil organisms and plants have dominated in studies on links between soil organisms and plant herbivores, i.e., root feeding, root pathogens, and mycorrhizal infection (Rabin and Pacovsky, 1985; Gange and Brown, 1989; Moran and Whitham, 1990; Masters *et al.*, 1993; Gange and West, 1994; Gange *et al.*, 1994, 1999; Borowicz, 1997; Gange and Nice, 1997; see also chapter 9, this volume). As we have outlined above, there is also a multitude of indirect interactions by which soil organisms modify plant growth thereby altering the food resource of herbivores (Fig. 10.3). Surprisingly, there are only two studies that have focused on how microbi-detritivore animals affect herbivore performance (Scheu *et al.*, 1999a; Bonkowski *et al.*, in press b). Certainly, these interactions need much more attention, as future work in this direction will highlight the close interrelationship between the below-ground and above-ground communities.

It is well documented that generalist predators, such as spiders and carabid and staphylinid beetles, essentially rely on prey species originating from the decomposer community (Kajak and Jakubczyk, 1976; Thiele, 1977; Wise, 1993; Lövei and Sunderland, 1996; Ekschmitt *et al.*, 1997). Since these taxa are among the most important predators of herbivores in many habitats they may be viewed as predators of the above-ground system which are subsidized by resources from the decomposer system (Polis and Strong, 1996; Oksanen *et al.*, 1997). Very few studies have explicitly addressed this topic (but see Settle *et al.*, 1996; Snyder and Wise, 1999, 2001). More research will no doubt demonstrate the close interconnection between the below-ground and above-ground communities.

Beyond trophic interactions: habitat modifications

Since trophic interactions are a fundamental entity of any community, they tend to be at the forefront of attempts to understand structuring forces in animal, plant, and microbial communities. However, trophic interaction is only one of the ways by which organisms interact with each other; another is by modifying environmental conditions, thereby facilitating or inhibiting the living conditions of other organisms. In this, the physical modifications of the environment are of paramount importance ("ecosystem engineering": Jones *et al.*, 1994, 1997; Waid, 1999). This is certainly the case in below-ground communities where organisms live in a habitat formed by the action of others (see section "Introduction"). Since the action of organisms is vital for soil formation, ecosystem engineering is a fundamental force structuring virtually all soil communities. Here, we only emphasize that the composition of the soil animal community and ecosystem engineering are closely interdependent.

While trophic interactions necessarily are associated with energy flow, ecosystem engineering *per se* is not, but via modifications in the resource supply to other organisms engineering effects modify trophic interactions and thereby energy flow. In contrast to the processing of material in soil, which is mainly due to the action of minute organisms such as bacteria, fungi, and Protozoa, modifications of the physical habitat are mainly due to the action of large soil animals. The most important are shredders and organisms mixing organic and inorganic soil components. The focus on indirect soil faunal–microbial interactions (Anderson, 1988; Wolters, 1991; Lussenhop, 1992) may therefore simply be inferred from the complementary action of small organisms as energy transfer agents and large organisms as agents forming the habitat in which the energy is transferred (Fig. 10.4).

Despite the fundamental importance of large soil invertebrates as ecosystem engineers, their effects on other soil organisms via habitat

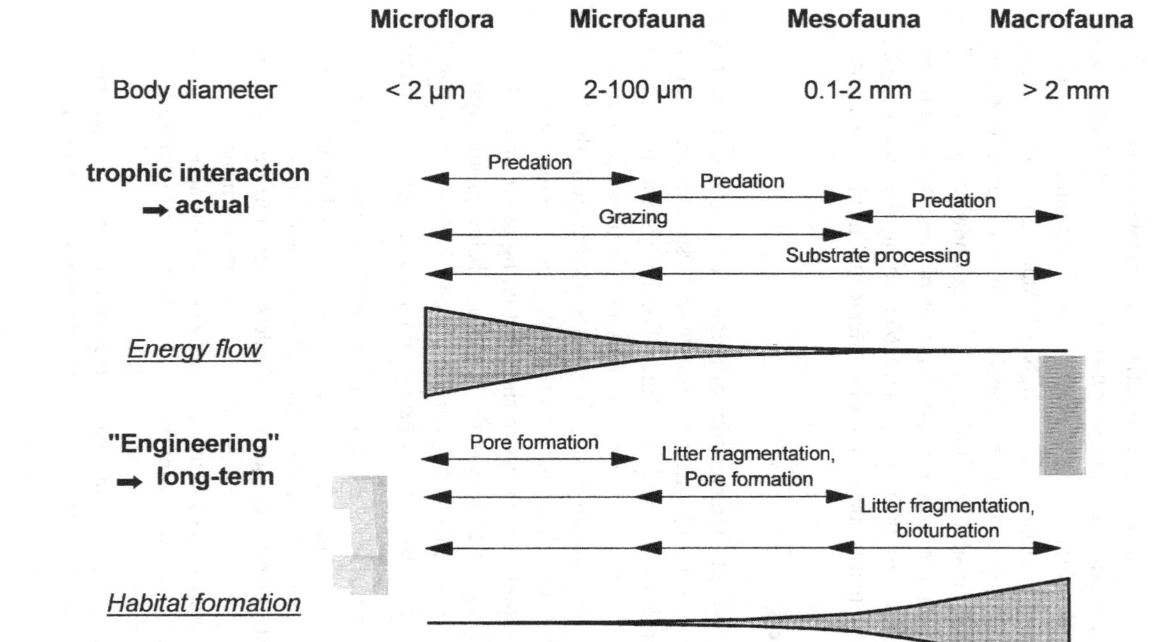

Fig. 10.4. Size dependent interactions among soil organisms. Trophic interactions and interactions caused by "engineering" are separated; both are indicated by arrows. Note that trophic interactions and interactions caused by engineering are strongly size dependent but complement each other (tapering and widening triangles). Both function at different scales: trophic interactions drive the current energy flow, engineering sets the conditions for the existence of the soil biota community in the long-term.

modifications is still poorly studied (Lavelle *et al.*, 1997; Parkinson and McLean, 1998). Earthworms have been the focus of attention. The mixing of plant residues with mineral soil components by earthworms (bioturbation) transforms forest soils with thick organic layers (ectorganic matter; moder soils) to soils where organic materials are intimately mixed into the mineral soil (entorganic matter; mull soils). Mull and moder soils dramatically differ in the composition of the soil animal community (Schaefer, 1991; Ponge, 1999). Currently, forests in North America which are free of earthworms are being invaded by European earthworm species, which situation offers the possibility of following in detail how ecosystem engineers in soil alter the structure and function of ecosystems and even landscapes (Scheu and Parkinson, 1994a, c; McLean and Parkinson, 1998a, b). There is much to learn beyond trophic interactions from soil systems.

Prospects

We have highlighted characteristics of below-ground systems and the prevalence of multitrophic interactions therein. Despite its great importance for decomposition and recycling of nutrients the interactions among soil biota and between soil animals and plants are little understood. This is unfortunate since, due to its peculiarities, the below-ground system has much to offer to widen ecological thinking and theory. For a more comprehensive view on the functioning of life on earth the efforts in studying the ecology of the above- and below-ground systems have to become more balanced. Particularly promising lines of ecological research are the links between the below- and the above-ground systems:

1. Interactions in the rhizosphere of plants and their implications for plant growth; study of the integration of Protozoa and nematodes as antagonists of rhizosphere bacteria and of Collembola as antagonists of mycorrhizal fungi is essential for understanding these interactions.
2. Links between the decomposer and the herbivore system via modifications of plant performance, a theme with an enormous scope; which soil organisms are involved, which mechanisms are at work, which above-ground processes are affected, etc.
3. Links between the decomposer and the herbivore system via generalist predators and the strengthening of top-down trophic cascades above the ground.

Topics (2) and (3) are particularly exciting since they may contribute to pest control in agroecosystems. Certainly, the links between the below-

and above-ground systems need considerable further attention and to become a pivotal field in ecological research. We hope that this review will help in achieving this goal.

Acknowledgments

Comments by two anonymous reviewers are gratefully acknowledged. We give thanks to Teja Tscharntke and Bradford A. Hawkins for comments and encouragement.

REFERENCES

Abrams, B. I. and Mitchell, M. J. (1980) Role of nematode–bacterial interactions in heterotrophic systems with emphasis on sewage sludge decomposition. *Oikos* **35**: 404–410.

Alexander, M. (1977) *Introduction to Soil Microbiology*, 2nd edn. New York: John Wiley.

Alexander, M. (1981) *Picea sitchensis* and *Lactarius rufus* mycorrhizal association and its effects on seedling growth and development. *Transactions of the British Mycological Society* **76**: 417–423.

Allen, M. F. (1991) *The Ecology of Mycorrhizae*. Cambridge: Cambridge University Press.

Alphei, J., Bonkowski, M. and Scheu, S. (1996) Protozoa, Nematoda and Lumbricidae in the rhizosphere of *Hordelymus europaeus* (Poaceae): faunal interactions, response of microorganisms and effects on plant growth. *Oecologia* **106**: 111–126.

Anderson, J. M. (1988) Spatiotemporal effects of invertebrates on soil processes. *Biology and Fertility of Soils* **6**: 216–227.

Anderson, J. M. and Healey, J. N. (1972) Seasonal and interspecific variation in major components of the gut contents of some woodland Collembola. *Journal of Animal Ecology* **41**: 359–368.

Anderson, J. P. E. and Domsch, K. H. (1975) Measurement of bacterial and fungal contributions to respiration of selected agricultural and forest soils. *Canadian Journal of Microbiology* **21**: 314–322.

Anderson, O. R. and Bohlen, P. J. (1998) Abundances and diversity of Gymnamoebae associated with earthworm (*Lumbricus terrestris*) middens in a northeastern US forest. *Soil Biology and Biochemistry* **30**: 1213–1216.

Bardgett, R. D., Wardle, D. A. and Yeates, G. W. (1998a) Linking above-ground and below-ground interactions: how plant responses to foliar herbivory influence soil organism. *Soil Biology and Biochemistry* **30**: 1867–1878.

Bardgett, R. D., Keiller, S., Cook, R. and Gilburn, A. S. (1998b) Dynamic interactions between soil animals and microorganisms in upland grassland soils amended with sheep dung: a microcosm experiment. *Soil Biology and Biochemistry* **30**: 531–539.

Beare, M. H., Coleman, D. C., Crossley, D. A., Jr., Hendrix, P. F. and Odum, E. P. (1995) A hierarchical approach to evaluating the significance of soil biodiversity to biogeochemical cycling. *Plant and Soil* **170**: 5–22.

Bengtsson, J. (1998) Which species? What kind of diversity? Which ecosystem function? Some problems in studies of relations between biodiversity and ecosystem function. *Applied Soil Ecology* **10**: 191–199.

Bengtsson, J., Setälä, H. and Zheng, D. W. (1996) Food-webs and nutrient cycling in soils: interactions and positive feedbacks. In *Food-Webs: Integration of Patterns and Dynamics*, ed. G. A. Polis and K. O. Winemiller, pp. 30–38. New York: Chapman and Hall.

Bengtsson, J., Zheng, D. W., Agren, G. I. and Persson, T. (1995) Food-webs in soil: an interface between population and ecosystem ecology. In *Linking Species and Ecosystems*, ed. C. G. Jones and J. H. Lawton, pp. 159–165. New York: Chapman and Hall.

Berg, M. P. (1997) *Decomposition, Nutrient Flow and Food-Web Dynamics in a Stratified Pine Forest Soil*. Amsterdam, Netherlands: Vrije Universitet.

Berg, M. P., Kniese, J. P. and Verhoef, H. A. (1998) Dynamics and stratification of bacteria and fungi in the organic layers of a Scots pine forest soil. *Biology and Fertility of Soils* **26**: 313–322.

Bever, J. D., Westover, K. M. and Antonovics, J. (1997) Incorporating the soil community into plant population dynamics: the utility of the feedback approach. *Journal of Ecology* **85**: 561–573.

Bonkowski, M. and Schaefer, M. (1997) Interactions between earthworms and soil Protozoa: a trophic component in the soil food-web. *Soil Biology and Biochemistry* **29**: 499–502.

Bonkowski, M., Scheu, S. and Schaefer, M. (1998) Interactions of earthworms (*Octolasion lacteum*), millipedes (*Glomeris marginata*) and plants (*Hordelymus europaeus*) in a beechwood on a basalt hill: implications for litter decomposition and soil formation. *Applied Soil Ecology* **9**: 161–166.

Bonkowski, M., Griffiths, B. S. and Scrimgeour, C. (2000a) Substrate heterogeneity and microfauna in soil organic 'hotspots' as determinants of nitrogen capture and growth of rye-grass. *Applied Soil Ecology* **14**: 37–53.

Bonkowski, M., Cheng, W., Griffiths, B., Alphei, J. and Scheu, S. (2000b) Microbial faunal interactions in the rhizosphere and effects on plant growth. *European Journal of Soil Biology* **36**: 135–147.

Bonkowski, M., Jentschke, G. and Scheu, S. (in press a) Contrasting interests in the rhizosphere: interactions between Norway spruce seedlings (*Picea abies* Karst.), mycorrhiza (*Paxillus involutus* (Bartsch) Fr.) and naked amoebae (Protozoa). *Applied Soil Ecology*.

Bonkowski, M., Geoghegan, I. E., Birch, A. N. E. and Griffiths, B. S. (in press b) Effects of microbi-detritivores (earthworms and Protozoa) on an above-ground phytophagous insect (cereal aphid) mediated through changes in the host plant. *Oikos*.

Borowicz, V. A. (1997) A fungal root symbiont modifies plant resistance to an insect herbivore. *Oecologia* **112**: 534–542.

Boucher, D. H., James, S. and Keeler, K. H. (1982) The ecology of mutualism. *Annual Review of Ecology and Systematics* **13**: 315–347.

Bowen, G. D. (1980) Misconceptions, concepts and approaches in rhizosphere biology. In *Contemporary Microbial Ecology*, ed. D. C. Ellwood, J. N. Hedger, M. J. Latham, J. M. Lynch and J. M. Slater, pp. 283–304. London: Academic Press.

Boyle, K. E., Curry, J. P. and Farrell, E. P. (1997) Influence of earthworms on soil properties and grass production in reclaimed cutover peat. *Biology and Fertility of Soils* **25**: 20–26.

Bronstein, J. L. (1994) Our current understanding of mutualism. *Quarterly Review of Biology* **69**: 31–51.

Brundrett, M. (1991) Mycorrhizas in natural ecosystems. *Advances in Ecological Research* **21**: 171–313.

Brussaard, L., Noordhuis, R., Geurs, M. and Bouwman, L. A. (1995) Nitrogen mineralization in soil in microcosms with or without bacterivorous nematodes and nematophagous mites. *Acta Zoologica Fennica* **196**: 15–21.

Chakraborty, S., Theodorou, C. and Bowen, G. D. (1985) The reduction of root colonization by mycorrhizal fungi by mycophagous amoebae. *Canadian Journal of Microbiology* **31**: 295–297.

Chen, B., Snider, R. J. and Snider, R. M. (1996) Food consumption by Collembola from northern Michigan deciduous forest. *Pedobiologia* **40**: 149–161.

Chen, B. R. and Wise, D. H. (1997) Responses of forest-floor fungivores to experimental food enhancement. *Pedobiologia* **41**: 316–326.

Chen, B. R. and Wise, D. H. (1999) Bottom-up limitation of predaceous arthropods in a detritus-based terrestrial food-web. *Ecology* **80**: 761–772.

Cheng, W. X., Zhang, Q. L., Coleman, D. C. Carroll, C. R. and Hoffmann, C. A. (1996) Is available carbon limiting microbial respiration in the rhizosphere? *Soil Biology and Biochemistry* **28**: 1283–1288.

Clarholm, M. (1985) Interactions of bacteria, Protozoa and plants leading mineralization of soil nitrogen. *Soil Biology and Biochemistry* **17**: 181–187.

Clarholm, M. (1989) Effects of plant–bacterial–amoebal interactions on plant uptake of nitrogen under field conditions. *Biology and Fertility of Soils* **8**: 373–378.

Coleman, D. C. (1996) Energetics of detritivory and microbiovory in soil in theory and practice. In *Food-Webs: Integration of Patterns and Dynamics*, ed. G. A. Polis and K. O. Winemiller, pp. 39–50. New York: Chapman and Hall.

Coleman, D. C., Reid, C. P. P. and Cole, C. V. (1983) Biological strategies of nutrient cycling in soil systems. In *Advances in Ecological Research*, ed. A. Macfadyen and E. D. Ford, pp. 1–55. New York: Academic Press.

Cummins, K. W. (1974) Structure and function of stream ecosystems. *BioScience* **24**: 631–641.

Curl, E. A. and Truelove B. (1986) *The Rhizosphere*. New York: Springer-Verlag.

Curl, E. A., Lartey, R. and Peterson, C. M. (1988) Interactions between root pathogens and soil microarthropods. *Agriculture, Ecosystems, Environment* **24**: 249–261.

De Ruiter, P. C., van Veen, J. A., Moore, J. C., Brussaard, L. and Hunt, H. W. (1993) Calculation of nitrogen mineralization in soil food-webs. *Plant and Soil* **157**: 263–273.

De Ruiter, P. C., Neutel, A. M. and Moore, J. C. (1996) Energetics and stability in below-ground food-webs. In *Food-Webs: Integration of Patterns and Dynamics*, ed. G. A. Polis and K. O. Winemiller, pp. 201–210. New York: Chapman and Hall.

Dicke, M. and Vet, L. E. M. (1999) Plant–carnivore interactions: evolutionary and ecological consequences for plant, herbivore and carnivore. In *Herbivores: Between Plants and Predators*, ed. H. Olff, V. K. Brown and R. H. Drent, pp. 483–520. Oxford: Blackwell Science.

Duchesne, L. C., Peterson, R. L. and Ellis, B. E. (1987) The accumulation of plant-produced antimicrobial compounds in response to ectomycorrhizal fungi: a review. *Phytoprotection* **68**: 17–27.

Duchesne, L. C., Peterson, R. L. and Ellis, B. E. (1988) Pine root exudates stimulates antibiotic synthesis by the ectomycorrhizal fungus *Paxillus involutus*. *New Phytologist* **108**: 471–476.

Ek, H., Sjögren, M., Arnebrant, K. and Söderström, B. (1994) Extramatrical mycelial growth, biomass allocation and nitrogen uptake in ectomycorrhizal systems in response to collembolan grazing. *Applied Soil Ecology* **1**: 155–169.

Ekschmitt, K., Wolters, V. and Weber, M. (1997) Spiders, carabids, and staphylinids: the ecological potential of predatory macroarthropods. In *Fauna in Soil Ecosystems: Recycling Processes, Nutrient Fluxes, and Agricultural Production*, ed. G. Benckiser, pp. 307–362. New York: M. Dekker.

Elliott, E. T., Coleman, D. C. and Cole, C. V. (1979) The influence of amoeba on the uptake of nitrogen by plants in gnotobiotic soil. In *The Soil–Root Interface*, ed. J. L. Harley and R. S. Russel, pp. 221–229. London: Academic Press.

England, L. S., Lee, H. and Trevors, J. T. (1993) Bacterial survival in soil: effect of clays and Protozoa. *Soil Biology and Biochemistry* **25**: 525–531.

Finlay, R. D. (1985) Interaction between soil micro-arthropods and endomycorrhizal associations of higher plants. In *Ecological Interactions in Soil*, ed. A. H. Fitter, pp. 319–331. Oxford: Blackwell Science.

Fischer, K., Hahn, D., Amann, R. I., Daniel, O. and Zeyer, J. (1995) *In situ* analysis of the bacterial community in the gut of the earthworm *Lumbricus terrestris* by whole-cell hybridization. *Canadian Journal of Microbiology* **41**: 666–673.

Fitter, A. H. and Garbaye, J. (1994) Interactions between mycorrhizal fungi and other soil organisms. *Plant and Soil* **159**: 123–132.

Flack, F. M. and Hartenstein, R. (1984) Growth of the earthworm *Eisenia foetida* on microorganisms and cellulose. *Soil Biology and Biochemistry* **16**: 491–495.

Futuyma, D. J. and Moreno, G. (1988) The evolution of ecological specialization. *Annual Review of Ecology and Systematics* **19**: 207–233.

Gange, A. C. (1993) Translocation of mycorrhizal fungi by earthworms during early succession. *Soil Biology and Biochemistry* **25**: 1021–1026.

Gange, A. C. and Brown, V. K. (1989) Effects of root herbivory by an insect on a foliar feeding species, mediated through changes in the host plant. *Oecologia* **81**: 38–42.

Gange, A. C. and Nice, H. E. (1997) Performance of the thistle gall fly, *Urophora cardui*, in relation to host plant nitrogen and mycorrhizal colonization. *New Phytologist* **137**: 335–343.

Gange, A. C. and West, H. M. (1994) Interactions between arbuscular mycorrhizal fungi and foliar feeding insects in *Plantago lanceolata* L. *New Phytologist* **128**: 79–87.

Gange, A. C., Brown, V. K. and Sinclair, G. S. (1994) Reduction of black vine weevil larval growth by vesicular–arbuscular mycorrhizal infection. *Entomologia Experimentalis et Applicata* **70**: 115–119.

Gange, A. C., Bower, E. and Brown, V. K. (1999) Positive effects of an arbuscular mycorrhizal fungus on aphid life history traits. *Oecologia* **120**: 123–131.

Garbaye, J. (1991) Biological interactions in the mycorrhizosphere. *Experientia* **47**: 370–375.

Gemma, J. N. and Koske, R. E. (1988) Pre-infection between roots and the mycorrhizal fungus *Gigaspora gigantea*: chemotropism of germ-tubes and root growth response. *Transactions of the British Mycological Society* **91**: 123–132.

Gorny, M. (1984) Studies on the relationship between enchytraeids and earthworms. In *Soil Biology and Conservation of the Biosphere*, ed. J. Szegi, pp. 769–776. Budapest, Hungary: Academiai Kiado.

Görres, J. H., Savin, M. and Amador, J. A. (1997) Dynamics of carbon and nirogen mineralization, microbial biomass, and nematode abundance within and outside the burrow walls of anecic earthworms (*Lumbricus terrestris*). *Soil Science* **162**: 666–671.

Grayston, S. J., Vaughan, D. and Jones, D. (1996) Rhizosphere carbon flow in trees, in comparison with annual plants: the importance of root exudation and its impact on microbial activity and nutrient availability. *Applied Soil Ecology* **5**: 29–55.

Griffiths, B. S. (1986) Mineralization of nitrogen and phosphorus by mixed cultures of ciliate protozoan *Colpoda steinii*, the nematode *Rhabditis* sp. and the bacterium *Pseudomonas fluorescens*. *Soil Biology and Biochemistry* **18**: 637–641.

Griffiths, B. S. (1994) Microbial-feeding nematodes and Protozoa in soil: their effects on microbial activity and nitrogen mineralization in decomposing hotspots in the rhizosphere. *Plant and Soil* **164**: 25–33.

Gunn, A. and Cherrett, J. M. (1993) The exploitation of food resources by soil meso- and macro-invertebrates. *Pedobiologia* **37**: 303–320.

Gunnarsson, T. and Tunlid, A. (1986) Recycling of fecal pellets in isopodes: microorganisms and nitrogen comounds as potential food for *Onicus asellus* L. *Soil Biology and Biochemistry* **18**: 595–600.

Haimi, J., Huhta, V. and Boucelham, M. (1992) Growth increase of birch seedlings under the influence of earthworms: a laboratory study. *Soil Biology and Biochemistry* **24**: 1525–1528.

Hairston, N. G., Jr. (1989) *Ecological Experiments: Purpose, Design and Execution*. Cambridge: Cambridge University Press.

Hairson, N. G., Jr. and Hairston, N. G., Sr. (1993) Cause–effect relationships in energy flow, trophic structure, and interspecific interactions. *American Naturalist* **142**: 379–411.

Hairston, N. G. and Hairston, N. G. (1997) Does food-web complexity eliminate trophic-level dynamics? *American Naturalist* **149**: 1001–1007.

Hairston, N. G., Smith, F. E. and Slobodkin, L. B. (1960) Community structure, population control, and competition. *American Naturalist* **94**: 421–425.

Hall, M. and Hedlund, K. (1999) The predatory mite *Hypoaspis aculeifer* is attracted to food of its fungivorous prey. *Pedobiologia* **43**: 11–17.

Harinikumar, K. M. and Bagyaraj, D. J. (1994) Potential of earthworms, ants, millipedes, and termites for dissemination of vesicular–arbuscular mycorrhizal fungi in soil. *Biology and Fertility of Soils* **18**: 115–118.

Harris, K. K. and Boerner, R. E. J. (1990) Effects of belowground grazing by Collembola on growth, mycorrhizal infection, and P uptake by *Geranium robertianum*. *Plant and Soil* **129**: 203–210.

Harte, J. and Kinzig, A. P. (1993) Mutualism and competition between plants and decomposers: implications for nutrient allocation in ecosystems. *American Naturalist* **141**: 829–846.

Hassall, M. and Rushton, S. P. (1985) The adaptive significance of coprophagous behaviour in the terrestrial isopod *Porcellio scaber*. *Pedobiologia* **28**: 169–175.

Hedlund, K., Boddy, L. and Preston, C. M. (1991) Mycelial responses of the soil fungus, *Mortierella isabellina*, to grazing by *Onychiurus armatus* (Collembola). *Soil Biology and Biochemistry* **23**: 361–366.

Hendrix, P. F., Peterson, A. C., Beare, M. H. and Coleman, D. C. (1998) Long-term effects of earthworms on microbial biomass nitrogen in coarse and fine textured soils. *Applied Soil Ecology* **9**: 375–380.

Holland, J. N., Cheng, W. and Crossley, D. A., Jr. (1996) Herbivore-induced changes in plant carbon allocation: assessment of below-ground carbon fluxes using carbon-14. *Oecologia* **107**: 87–94.

Holt, R. D. and Polis, G. A. (1997) A theoretical framework for intraguild predation. *American Naturalist* **149**: 745–764.

Hoogerkamp, M., Rogaar, H. and Eijsackers, H. J. P. (1983) Effect of earthworms on grassland on recently reclaimed polder soils in the Netherlands. In *Earthworm Ecology from Darwin to Vermiculture*, ed. J. E. Satchell, pp. 85–105. London: Chapman and Hall.

Hövemeyer, K. (1992) Response of Diptera populations to experimentally modified leaf litter input in a beech forest on limestone. *Pedobiologia* **36**: 35–49.

Huhta, V., Hyvönen, R., Koskenniemi, A., Vilkamaa, P. K. P. and Sulander, M. (1986) Response of soil fauna to fertilization and manipulation of pH in coniferous forests. *Acta Forestalia Fennica* **195**: 1–30.

Huhta, V., Persson, T. and Setälä, H. (1998) Functional implications of soil fauna diversity in boreal forests. *Applied Soil Ecology* **10**: 277–288.

Hunt, H. W., Coleman, D. C., Ingham, E. R., Ingham, R. E., Elliott, E. T., Moore, J. C., Rose, S. L., Reid, C. P. P. and Morley, C. R. (1987) The detrital food-web in a shortgrass prairie. *Biology and Fertility of Soils* **3**: 57–68.

Hyvönen, R. and Persson, T. (1996) Effect of fungivorous and predatory arthropods on nematodes and tardigrades in microcosms with coniferous forest soils. *Biology and Fertility of Soils* **21**: 121–127.

Ingham, E. R. (1988) Interaction between nematodes and vesicular-arbuscular mycorrhizae. *Agriculture Ecosystems, Environment* **24**: 169–182.

Ingham, E. R. and Molina, R. (1991) Interactions among mycorrhizal fungi, rhizosphere organisms, and plants. In *Microbial Mediation of Plant–Herbivore Interactions*, ed. P. Barbosa, V. A. Krischik and C. G. Jones, pp. 169–198. New York: John Wiley.

Ingham, E. R., Coleman, D. C. and Moore, J. C. (1989) An analysis of food-web structure and function in a shortgrass prairie, a mountain meadow, and lodgepole pine forest. *Biology and Fertility of Soils* **8**: 29–37.

Ingham, R. E., Trofymow, J. A., Ingham, E. R. and Coleman, D. C. (1985) Interactions of bacteria, fungi and their nematode grazers: effects of nutrient cycling and plant growth. *Ecological Monographs* **55**: 119–140.

Jackson, R. B., Canadell, J., Ehleringer, J. R., Mooney, H. A., Sala, O. E. and Schulze, E. D. (1996) A global analysis of root distribution for terrestrial biomes. *Oecologia* **108**: 389–411.

Jamieson, N. and Killham, K. (1994) Biocide manipulation of nitrogen flow to investigate root microbe competition in forest soil. *Plant and Soil* **159**: 283–290.

Jentschke, G., Bonkowski, M., Godbold, D. L. and Scheu, S. (1995) Soil Protozoa and forest tree growth: non-nutritional effects and interaction with mycorrhizae. *Biology and Fertility of Soils* **19**: 263–269.

Jones, C. G. and Lawton, J. H. (eds.) (1995) *Linking Species and Ecosystems*. New York: Chapman and Hall.

Jones, C. G., Lawton, J. H. and Shachak, M. (1994) Organisms as ecosystem engineers. *Oikos* **69**: 373–386.

Jones, C. G., Lawton, J. H. and Shachak, M. (1997) Positive and negative effects of organisms as physical ecosystem engineers. *Ecology* **78**: 1946–1957.

Judas, M. (1990) The development of earthworm populations following manipulation of the canopy leaf litter in a beechwood on limestone. *Pedobiologia* **34**: 247–255.

Kajak, A. (1997) Effects of epigeic macroarthropods on grass litter decomposition in mown meadow. *Agriculture, Ecosystems, Environment* **64**: 53–63.

Kajak, A. and Jakubczyk, H. (1976) The effect of intensive fertilization on the structure and productivity of meadow ecosystems XVIII. Trophic relationships of epigeic predators. *Polish Ecological Studies* **2**: 219–229.

Kajak, A., Chmielewski, K., Kaczmarek, M. and Rembialkowska, E. (1993) Experimental studies on the effect of epigeic predators on matter decomposition processes in managed peat grasslands. *Polish Ecological Studies* **17**: 289–310.

Kandeler, E., Kampichler, C., Joergensen, R. G. and Molter, K. (1999) Effects of mesofauna in a spruce forest on soil microbial communities and nitrogen cycling in field mesocosms. *Soil Biology and Biochemistry* **31**: 1783–1792.

Klironomos, J. N. and Kendrick, W. B. (1995) Stimulative effects of arthropods on endomycorrhizas of sugar maple in the presence of decaying litter. *Functional Ecology* **9**: 528–536.

Klironomos, J. N., Bednarczuk, E. M. and Neville, J. (1999) Reproductive significance of feeding on saprobic and arbuscular mycorrhizal fungi by the collembolan, *Folsomia candida*. *Functional Ecology* **13**: 756–761.

Kuikman, P. (1990) Mineralization of nitrogen by protozoan activity in soil. PhD thesis, University of Wageningen, Netherlands.

Kuikman, P. J. and van Veen, J. A. (1989) The impact of Protozoa on the availability of bacterial nitrogen to plants. *Biology and Fertility of Soils* **8**: 13–18.

Laakso, J. and Setälä, H. (1999a) Sensitivity of primary production to changes in the architecture of below-ground food-webs. *Oikos* **87**: 57–64.

Laakso, J. and Setälä, H. (1999b) Population- and ecosystem-level effects of predation on microbial-feeding nematodes. *Oecologia* **120**: 279–286.

Laakso, J., Salminen, J. and Setälä, H. (1995) Effects of abiotic conditions and microarthropod predation on the structure and function of soil animal communities. *Acta Zoologica Fennica* **196**: 162–167.

Lartey, R. T., Curl, E. A. and Peterson, C. M. (1994) Interactions of mycophagous collembola and biological control fungi in the suppression of *Rhizoctonia solani*. *Soil Biology and Biochemistry* **26**: 81–88.

Lavelle, P., Lattaud, C., Trigo, D. and Barois, I. (1995) Mutualism and biodiverity in soils. *Plant and Soil* **170**: 23–33.

Lavelle, P., Bignell, D., Lepage, M., Wolters, V., Roger, P., Ineson, P., Heal, O. W. and Dhillion, S. (1997) Soil function in a changing world: the role of invertebrate ecosystem engineers. *European Journal of Soil Biology* **33**: 159–193.

Lawrence, K. L. and Wise, D. H. (2000) Spider predation on forest-floor Collembola and evidence for indirect effects on decomposition. *Pedobiologia* **44**: 33–39.

Linderman, R. G. (1988) Mycorrhizal interactions with the rhizosphere microflora: the mycorrhizosphere effect. *Phytopathology* **78**: 366–371.

Lipson, D. A., Schmidt, S. K. and Monson, R. K. (1999) Links between microbial population dynamics and nitrogen availability in an alpine ecosystem. *Ecology* **80**: 1623–1631.

Lövei, G. L. and Sunderland, K. D. (1996) Ecology and behavior of ground beetles (Coleoptera: Carabidae). *Annual Review of Entomology* **41**: 231–256.

Lussenhop, L. (1992) Mechanisms of microarthropod–microbial interactions in soil. *Advances in Ecological Research* **23**: 1–33.

Lussenhop, J. (1996) Collembola as mediators of microbial symbiont effects upon soybean. *Soil Biology and Biochemistry* **28**: 363–369.

Lynch, J. M. (1990) *The Rhizosphere*. Chichester: John Wiley.

Maraun, M. and Scheu, S. (1996) Changes in microbial biomass, respiration and nutrient status of beech (*Fagus sylvatica*) leaf litter processed by millipedes (*Glomeris marginata*). *Oecologia* **107**: 131–140.

Maraun, M., Visser, S. and Scheu, S. (1998) Oribatid mites enhance the recovery of the microbial community after a strong disturbance. *Applied Soil Ecology* **9**: 175–181.

Maraun, M., Alphei, J., Bonkowski, M., Buryn, R., Migge, S., Peter, M., Schaefer, M. and Scheu, S. (1999) Middens of the earthworm *Lumbricus terrestris* (Lumbricidae): microhabitats for micro- and mesofauna in forest soil. *Pedobiologia* **43**: 276–287.

Marx, D. H. (1972) Ectomycorrhizae as biological deterrents to pathogenic root infection. *Phytopathology* **10**: 429–454.

Masters, G. J., Brown, V. K. and Gange, A. C. (1993) Plant-mediated interactions between above- and below-ground insect herbivores. *Oikos* **66**: 148–151.

May, R. M. (1972) Will a large complex system be stable? *Nature* **238**: 413–414.

May, R. M. (1997) Complex animal interactions: introductory remarks. In *Multitrophic Interactions in Terrestrial Systems*, ed. A. C. Gange and V. K. Brown, pp. 305–306. Oxford: Blackwell Science.

McBrayer, J. F. (1973) Exploitation of deciduous leaf litter by *Apheloria montana* (Diplopoda: Eurydesmidae). *Pedobiologia* **13**: 90–98.

McGrady-Steed, J., Harris, P. M. and Morin, P. J. (1997) Biodiversity regulates ecosystem predictability. *Nature* **390**: 162–165.

McLean, M. A. and Parkinson, D. (1998a) Impacts of epigeic earthworm *Dendrobaena octaedra* on oribatid mite community diversity and microarthropod abundances in pine forest floor: a mesocosm study. *Applied Soil Ecology* **7**: 125–136.

McLean, M. A. and Parkinson, D. (1998b) Impacts of the epigeic earthworm *Dendrobaena octaedra* on microfungal community structure in pine forest floor: a mesocosm study. *Applied Soil Ecology* **8**: 61–75.

Menge, B. A. and Sutherland, J. P. (1976) Species diversity gradients: synthesis of the roles of predation, competition, and temporal heterogeneity. *American Naturalist* **110**: 351–369.

Mikola, J. and Setälä, H. (1998a) No evidence of trophic cascades in an experimental microbial-based soil food-web. *Ecology* **79**: 153–164.

Mikola, J. and Setälä, H. (1998b) Productivity and trophic-level biomasses in a microbial-based soil food-web. *Oikos* **82**: 158–168.

Miles, H. B. (1963) Soil Protozoa and earthworm nutrition . *Soil Science* **95**: 407–409.

Moore, J. C. and De Ruiter, P. C. (1997) Compartmentalization of resource utilization within soil ecosystems. In *Multitrophic Interactions in Terrestrial Systems*, ed. A. C. Gange and V. K. Brown, pp. 375–93. Oxford: Blackwell Science.

Moore, J. C. and Hunt, H. W. (1988) Resource compartmentation and the stability of real ecosystems. *Nature* **333**: 261–263.

Moran, N. A. and Whitham, T. G. (1990) Interspecific competition between root-feeding and leaf-galling aphids mediated by host-plant resistance. *Ecology* **71**: 1050–1058.

Morin, P. J. and Lawler, S. P. (1995) Food-web architecture and population dynamics: theory and empirical evidence. *Annual Review of Ecology and Systematics* **26**: 505–529.

Muscolo, A., Panuccio, M. R., Abenavoli, M. R., Concheri, G. and Nardi, S. (1996) Effect of molecular complexity and acidity of earthworm faeces humic fractions on glutamate dehydrogenase, glutamine synthetase, and phosphoenolpyruvate carboxylase in *Daucus carota* alpha II cells. *Biology and Fertility of Soils* **22**: 83–88.

Muscolo, A., Bovalo, F., Gionfriddo, F. and Nardi, S. (1999) Earthworm humic matter produces auxin-like effects on *Daucus carota* cell growth and nitrate metabolism. *Soil Biology and Biochemistry* **31**: 1303–1311.

Norton, J. M. and Firestone, M. K. (1996) Nitrogen dynamics in the rhizosphere of *Pinus ponderosa* seedlings. *Soil Biology and Biochemistry* **28**: 351–362.

Norton, R. A. (1994) Evolutionary aspects of oribatid mite life histories and consequences for the origin of the astigmata. In *Mites: Ecological and Evolutionary Analyses of Life History Patterns*, ed. M. A. Hank, pp. 99–135. New York: Chapman and Hall.

Nurmiaho-Lassila, E. L., Timonen, S., Haahtela, K. and Sen, R. (1997) Bacterial colonization patterns of intact *Pinus sylvestris* mycorrhizospheres in dry pine forest soil: an electron microscopy study. *Candian Journal of Microbiology* **43**: 1017–1035.

Oksanen, L., Fretwell, S. D., Arruda, J. and Niemelä, P. (1981) Exploitation ecosystems in gradients of primary productivity. *American Naturalist* **118**: 240–261.

Oksanen, L., Aunapuu, M., Oksanen, T., Schneider, M., Ekerholm, P., Lundberg, P. A., Armulik, T., Aruoja, V. and Bondestad, L. (1997) Outlines of food-webs in a low arctic tundra landscape in relation to three theories on trophic dynamics. In *Multitrophic interactions in terrestrial Systems*, ed. A. C. Gange and V. K. Brown, pp. 351–373. Oxford: Blackwell Science.

Olsson, P. A., Baath, E., Jakobsen, I. and Söderström, B. (1996a) Soil bacteria respond to presence of roots but not to mycelium of arbuscular mycorrhizal fungi. *Soil Biology and Biochemistry* **28**: 463–470.

Olsson, P. A., Chalet, M., Baath, E., Finlay, R. D. and Söderström, B. (1996b) Ectomycorrhizal mycelia reduce bacterial activity in a sandy soil. *FEMS Microbiology Ecology* **21**: 77–86.

Pace, M. J., Cole, J. J., Carpenter, S. R. and Kitchell, J. F. (1999) Trophic cascades revealed in diverse ecosystems. *Trends in Ecology and Evolution* **24**: 483–488.

Paine, R. T. (1966) Food-web complexity and species diversity. *American Naturalist* **100**: 65–75.

Parkinson, D. and McLean, M. A. (1998) Impacts of earthworms on the community structure of other biota in forest soils. In *Earthworm Ecology*, ed. C. A. Edwards, pp. 213–226. Boca Raton, FL: CRC Press.

Perry, D. A., Bell, T. and Amaranthus, M. P. (1992) Mycorrhizal fungi in mixed-species forests and other tales of positive feedback, redundancy and stability. *Special Publication of the British Ecological Society* **11**: 151–179.

Persson, L. (1999) Trophic cascades: abiding heterogeneity and the trophic level concept at the end of the road. *Oikos* **85**: 385–397.

Persson, L., Bengtsson, J., Menge, B. A. and Power, M. E. (1996) Productivity and consumer regulation: concepts, patterns and mechanisms. In *Food-Webs: Integration of Patterns and Dynamics*, ed. G. A. Polis and K. O. Winemiller, pp. 396–434. New York: Chapman and Hall.

Persson, T., Clarholm, M., Lundkvist, H., Söderström, B. and Sohlenius, B. (1980) Trophic structure, biomass dynamics and carbon metabolism in a Scots pine forest. In *Structure and Function of Northern Coniferous Forest: An Ecosystem Study*, ed. Persson, T. *Ecological Bulletins* **23**: 419–459.

Peters, R. H. (1983) *The Ecological Implications of Body Size*. Cambridge: Cambridge University Press.

Pimm, S. L. (1980) The properties of food-webs. *Ecology* **61:** 219–225.

Pimm, S. L. (1982) *Food-Webs*. New York: Chapman and Hall.

Pimm, S. L. and Lawton, J. H. (1977) Number of trophic levels in ecological communities. *Nature* **268:** 329–331.

Polis, G. A. (1991) Complex trophic interactions in deserts: an empirical critique of food-web theory. *American Naturalist* **138:** 123–155.

Polis, G. A. (1999) Why are parts of the world green? Multiple factors control productivity and the distribution of biomass. *Oikos* **86:** 3–15.

Polis, G. A. and Holt, R. D. (1992) Intraguild predation: the dynamics of complex trophic interactions. *Trends in Ecology and Evolution* **7:** 151–155.

Polis, G. A. and Strong, D. R. (1996) Food-web complexity and community dynamics. *American Naturalist* **147:** 813–846.

Polis, G. A. and Winemiller, K. O. (eds.) (1996) *Food-Webs: Integration Patterns and Dynamics*. New York: Chapman and Hall.

Ponge, J.-F. (1999) Interaction between soil fauna and their environment. In *Going Underground: Ecological Studies in Forest Soils*, ed. N. Rastin and J. Bauhus, pp. 45–76. Trivandrum, India: Research Signpost.

Ponge, J. F., Arpin, P. and Vannier, G. (1993) Collembolan response to experimental perturbations of litter supply in a temperate forest ecosystem. *European Journal of Soil Biology* **29:** 141–153.

Ponsard, S. and Arditi, R. (2000) What can stable isotopes (delta^{15}N and delta^{13}C) tell about the food-web of soil macroinvertebrates? *Ecology* **81:** 852–864.

Poser, G. (1991) Die Hundertfüßer (Myriapoda, Chilopoda) eines Kalkbuchenwaldes: Populationsökologie, Nahrungsbiologie und Gemeinschaftsstruktur. *Berichte des Forschungszentrums Waldökosysteme Göttingen A* **71:** 1–211.

Price, P.(1988) An overview of organismal interactions in ecosystems in evolutionary and ecological time. *Agriculture, Ecosystems, Environment* **24:** 369–377.

Pussard, M., Alabouvette, C. and Levrat, P. (1994) Protozoa interactions with soil microflora and possibilities for biocontrol of plant pathogens. In *Soil Protozoa*, ed. J. F. Darbyshire, pp. 123–146. Wallingford: CAB International.

Rabin, L. B. and Pacovsky, R. S. (1985) Reduced larva growth of two lepidoptera (Noctuidae) on excised leaves of soybean infected with a mycorrhizal fungus. *Journal of Economic Entomology* **78:** 1358–1363.

Roesner, J. (1986) Untersuchungen zur Reproduktion von Nematoden in Boden durch Regenwürmer. *Mededelingen van den Faculteit Landbouwwetenschappen Rijkuniversiteit Gent* **51:** 1311–1318.

Rosenheim, J. A. (1998) Higher-order predators and the regulation of insect herbivore populations. *Annual Review of Entomology* **43:** 421–447.

Rusek, J. (1985) Soil microstructures: contributions on specific soil organisms. *Questiones Entomologicae* **21:** 497–514.

Rutherford, P. M. and Juma, N. G. (1992) Influence of texture on habitable pore space and bacterial–protozoan populations in soil. *Biology and Fertility of Soils* **12:** 221–227.

Santos, P. F., Phillips, J. and Whitford, W. G. (1981) The role of mites and nematodes in early stages of buried litter decomposition in a desert. *Ecology* **62:** 664–669.

Sarathchandra, S. U., Watson, R. N., Cox, N. R., Dimenna, M. E., Brown, J. A., Burch, G. and Neville, F. J. (1996) Effects of chitin amendment of soil on microorganisms, nematodes, and growth of white clover (*Trifolium repens* L) and perennial ryegrass (*Lolium perenne* L). *Biology and Fertility of Soils* **22**: 221–226.

Schaefer, M. (1991) The animal community: diversity and resources. In *Ecosystems of the World: Temperate Deciduous Forests*, ed. E. Röhrig and B. Ulrich, pp. 51–120. Amsterdam, Netherlands: Elsevier.

Schaefer, M. (1995) Interspecific interactions in the soil community. *Acta Zoologica Fennica* **196**: 101–106.

Scheu, S. (2001) Plants and generalist predators as mediators between the decomposer and the herbivore system. *Basic and Applied Ecology* **2**: 3–13.

Scheu, S. and Falca, M. (2000) The soil food-web of two beech forests (*Fagus sylvatica*) of contrasting humus form: stable isotope analysis of a macro- and mesofauna-dominated community. *Oecologia* **123**: 285–296.

Scheu, S. and Parkinson, D. (1994a) Effects of earthworms on nutrient dynamics, carbon turnover, and microorganisms in soil from cool temperate forests of the Canadian Rocky Mountains: laboratory studies. *Applied Soil Ecology* **1**: 113–125.

Scheu, S. and Parkinson, D. (1994b) Changes in bacterial and fungal biomass carbon, bacterial and fungal biovolume and ergosterol content after drying, remoistening and incubation of different layers of cool temperate forest soils. *Soil Biology and Biochemistry* **26**: 1515–1525.

Scheu, S. and Parkinson, D. (1994c) Effects of invasion of an aspen forest (Alberta, Canada) by *Dendrobaena octaedra* (Lumbricidae) on plant growth. *Ecology* **75**: 2348–2361.

Scheu, S. and Schaefer, M. (1998) Bottom-up control of the soil macrofauna community in a beechwood on limestone: manipulation of food resources. *Ecology* **79**: 1573–1585.

Scheu, S. and Sprengel, T. (1989) Die Rolle der endogäischen Regenwürmer im Ökosystem Kalkbuchenwald und ihre Wechselwirkung mit saprophagen Makroarthropoden. *Verhandlungen der Gesellschaft für Ökologie* **17**: 237–243.

Scheu, S., Theenhaus, A. and Jones, T. H. (1999a) Links between the detritivore and the herbivore system: effects of earthworms and Collembola on plant growth and aphid development. *Oecologia* **119**: 541–551.

Scheu, S., Alphei, J., Bonkowski, M. and Jentschke, C. (1999b) Soil food-web interactions and ecosystem functioning: experimental approaches with systems of increasing complexity. In *Going Underground: Ecological Studies in Forest Soils*, ed. N. Rastin and J. Bauhus, pp. 1–32. Trivandrum, India: Research Signpost.

Schlatte, G., Kampichler, C. and Kandeler, E. (1998) Do soil microarthropods influence microbial biomass and activity in spruce forest litter? *Pedobiologia* **42**: 205–214.

Schmidt, I. K., Michelsen, A. and Jonasson, S. (1997) Effects of labile soil carbon on nutrient partitioning between an arctic graminoid and microbes. *Oecologia* **112**: 557–565.

Scholle, G., Wolters, V. and Jörgensen, R. G. (1992) Effects of mesofauna exclusion on the microbial biomass in two moder profiles. *Biology and Fertility of Soils* **12**: 253–260.

Schulmann, O. P. and Tiunov, A. V. (1999) Leaf litter fragmentation by the earthworm *Lumbricus terrestris* L. *Pedobiologia* **43**: 453–458.

Seastedt, T. R. (1985) Maximization of primary and secondary productivity by grazers. *American Naturalist* **126**: 559–564.

Setälä, H. (1995) Growth of birch and pine seedlings in relation to grazing by soil fauna on ectomycorrhizal fungi. *Ecology* **76**: 1844–1851.

Setälä, H. and Huhta, V. (1991) Soil fauna increase *Betula pendula* growth: laboratory experiments with coniferous forest floor. *Ecology* **72**: 665–671.

Setälä, H., Marshall, V. G. and Trofymow, J. A. (1996) Influence of body size of soil fauna on litter decomposition and ^{15}N uptake by poplar in a pot trial. *Soil Biology and Biochemistry* **28**: 1661–1675.

Setälä, H., Rissanen, J. and Markkola, A. M. (1997) Conditional outcomes in the relationship between pine and ectomycorrhizal fungi in relation to biotic and abiotic environment. *Oikos* **80**: 112–122.

Setälä, H., Laakso, J., Mikola, J. and Huhta, V. (1998) Functional diversity of decomposer organisms in relation to primary production. *Applied Soil Ecology* **9**: 25–31.

Setälä, H., Kulmala, P., Mikola, J. and Markkola, A. M. (2000) Influence of ectomycorrhiza on the structure of detrital food-webs in pine rhizosphere. *Oikos* **87**: 113–122.

Settle, W. H., Ariawan, H., Tri Asuti, E., Cahyana, W., Hakim, A. L., Hindayana, D., Sri Lestari, A. and Sartano, P. (1996) Managing tropical rice pests through conservation of generalist natural enemies and alternative prey. *Ecology* **77**: 1975–1988.

Shaw, P. J. A. (1985) Grazing preferences of *Onychiurus armatus* (Insecta: Collembola) for mycorrhizal and saprophytic fungi in pine plantations. In *Microbial Interactions in Soil*, ed. A. H. Fitter, pp. 333–337. Oxford: Blackwell Science.

Shultz, P. A. (1991) Grazing preferences of two collembolan species, *Folsomia candida* and *Proisotoma minuta*, for ectomycorrhizal fungi. *Pedobiologia* **35**: 313–325.

Sih, A., Englund, G. and Wooster, D. (1998) Emergent impacts of multiple predators on prey. *Trends in Ecology and Evolution* **13**: 350–355.

Smith, F. A. and Smith, S. E. (1996) Mutualism and parasitism: biodiversity in function and structure in the 'arbuscular' (VA) mycorrhizal symbiosis. *Advances in Botanical Research* **22**: 1–43.

Smith, S. E. and Read, D. J. (1996) *Mycorrhizal Symbiosis*, 2nd edn. London: Academic Press.

Snyder, W. E. and Wise, D. H. (1999) Predator interference and the establishment of generalist predator populations for biocontrol. *Biological Control* **15**: 283–292.

Snyder, W. E. and Wise, D. H. (2001) Temporal shift in the impact of generalist predators on fruit production: from parallel to counteractive effects. *Ecology* **82**: 1571–1583.

Stephens, P. M. and Davoren, C. W. (1997) Influence of the earthworms *Aporrectodea trapezoides* and *A. rosea* on the disease severity of *Rhizoctonia solani* on subterranean clover and ryegrass. *Soil Biology and Biochemistry* **29**: 511–516.

Stephens, P. M., Davoren, C. W. and Hawke, B. G. (1995) Influence of barley straw and the lumbricid earthworm *Aporrectodea trapezoides* on *Rhizobium meliloti* L5–30R, *Pseudomonas corrugata* 2140R, microbial biomass and microbial activity in a red-brown earth soil. *Soil Biology and Biochemistry* **27**: 1489–1497.

Stephens, P. M., Davoren, C. W., Ryder, M. H. and Doube, B. M. (1993) Influence of the lumbricid earthworm *Aporrectodea trapezoides* on the colonization of wheat roots by *Pseudomonas corrugata* strain 2140R in soil. *Soil Biology and Biochemistry* **25**: 1719–1724.

Strong, D. L., Kaya, H. K., Whipple, A. V., Child, A. L., Kraig, S., Bondonno, M., Dyer, K. and Maron, J. L. (1996) Entomopathogenic nematodes: natural enemies of root-feeding caterpillars on bush lupine. *Oecologia* **108**: 167–173.

Strong, D. R. (1992) Are trophic cascades all wet? Differentiation and donor-control in speciose ecosystems. *Ecology* **73**: 747–754.

Strong, D. R. (1999) Predator control in terrestrial ecosystems: the underground food chain of bush lupine. In *Herbivores: Between Plants and Predator*, ed. H. Olff, V. K. Brown and R. H. Drent, pp. 577–602. Oxford: Blackwell Science.

Swift, M. J., Heal, O. W. and Anderson, J. M. (1979) *Decomposition in Terrestrial Ecosystems*. Oxford: Blackwell Science.

Szlavecz, K. (1985) The effect of microhabitats on the leaf litter decomposition and on the distribution of soil animals. *Holarctic Ecology* **8**: 33–38.

Szlavecz, K. and Pobozsny, M. (1995) Coprophagy in isopods and diplopods: a case for indirect interaction. *Acta Zoologica Fennica* **196**: 124–128.

Tamm, C. O. (1991) Nitrogen in terrestrial ecosystems: questions of productivity, vegetational changes and ecosystem stability. *Ecological Studies* **81**: 1–116.

Thiele, H. U. (1977) *Carabid Beetles in their Environment*. Berlin, Germany: Springer-Verlag.

Thimm, T. and Larink, O. (1995) Grazing preferences of some collembola for ectomycorrhizal fungi. *Biology and Fertility of Soils* **19**: 266–268.

Thompson, L., Thomas, C. D., Radley, J. M. A., Williamson, S. and Lawton, J. H. (1993) The effect of earthworms and snails in a simple plant community. *Oecologia* **95**: 171–178.

Tiunov, A. and Scheu, S. (2000) Microbial biomass, biovolume and respiration in *Lumbricus terrestris* L. cast material of different age. *Soil Biology and Biochemistry* **32**: 265–276.

Turlings, T. C. J., Tumlinson, J. H., Eller, F. J. and Lewis, W. J. (1991) Larval damaged plants: source of volatile synomones that guide the parasitoid *Cotesia marginiventris* to the micro-habitat of its hosts. *Entomologia Experimentalis et Applicata* **58**: 75–82.

Ulrich, W. (1988) Welche Faktoren beeinflussen die Populationen und die Strukturen der Gemeinschaften von bodenlebenden Hymenopteren in einem Buchenwald. PhD thesis, University of Göttingen, Germany.

Usher, M. B. (1985) Population and community dynamics in the soil system. In *Ecological Interactions in Soil*, ed. A. H. Fitter, pp. 243–265. Oxford: Blackwell Science.

Verhoef, H. A., Dorel, F. G. and Zoomer, H. R. (1989) Effects of nitrogen deposition on animal-mediated nitrogen mobilization in coniferous litter. *Biology and Fertility of Soils* **8**: 255–259.

Villani, M. G., Allee, L. L., Diaz, A. and Robbins, P. S. (1999) Adaptive strategies of edaphic arthropods. *Annual Review of Entomology* **44**: 233–256.

Visser, S. (1985) Role of soil invertebrates in the determining the composition of soil microbial communities. In *Ecological Interactions in Soil*, ed. A. H. Fitter, pp. 297–317. Oxford: Blackwell Science.

Wagener, S. M., Oswood, M. W. and Schimel, J. P. (1998) Rivers and soils: parallels in carbon and nutrient processing. *BioScience* **48**: 104–108.

Waid, J. S. (1999) Does soil biodiversity depend upon metabiotic activity and influences? *Applied Soil Ecology* **13**: 151–158.

Wall, D. H. and Moore, J. C. (1999) Interactions underground: soil biodiversity, mutualism, and ecosystem processes. *BioScience* **49**: 109–117.

Wallander, H. (1992) *Regulation of Ectomycorrhizal Symbiosis in* Pinus sylvestris *L. Seedlings: Influence of Mineral Nutrition*. Uppsala, Sweden: Swedish University of Agricultural Sciences.

Walter, D. E. (1987) Trophic behavior of "mycophagous" microarthropods. *Ecology* **18**: 226–229.

Wardle, C. A. (1999) How soil food-webs make plants grow. *Trends in Ecology and Evolution* **14**: 418–420.

Wardle, D. A. and Giller, K. E. (1996) The quest for a contemporary ecological dimension to soil biology. *Soil Biology and Biochemistry* **28**: 1549–1554.

Wardle, D. A. and Yeates, G. W. (1993) The dual importance of competition and predation as regulatory forces in terrestrial ecosystem: evidence from decomposer food-webs. *Oecologia* **93**: 303–306.

Wardle, D. A., Zackrisson, O., Hornberg, G. and Gallet, C. (1997) The influence of island area on ecosystem properties. *Science* **277**: 1296–1299.

Wardle, D. A., Verhoef, H. A. and Clarholm, M. (1998) Trophic relationships in the soil microfood-web: predicting the responses to a changing global environment. *Global Change Biology* **4**: 713–727.

Warnock, A. J., Fitter, A. H. and Usher, M. B. (1982) The influence of a springtail *Folsomia candida* (Insecta, Collembola) on the mycorrhizal association of leek *Allium porrum* and the vesicular–arbuscular mycorrhizal endophyte *Glomus fasciculatus*. *New Phytologist* **90**: 285–292.

Wicklow, D. T. (1992) Interference competition. In *The Fungal Community: Its Organization and Role in the Ecosystem*, 2nd edn, ed. G. C. Carroll and D. T. Wicklow, pp. 265–274. New York: M. Dekker.

Willems, J. H. and Huijsmans, K. G. A. (1994) Vertical seed dispersal by earthworms: a quantitative approach. *Ecography* **17**: 124–130.

Wilson, G. W. T., Hetric, B. A. D. and Kitt, D. G. (1989) Suppression of vesicular–arbuscular mycorrhizal fungus spore germination by nonsterile soil. *Canadian Journal of Botany* **67**: 18–23.

Wise, D. H. (1993) *Spiders in Ecological Webs*. Cambridge: Cambridge University Press.

Wise, D. H. and Chen, B. R. (1999) Impact of intraguild predators on survival of a forest-floor wolf spider. *Oecologia* **121**: 129–137.

Wolter, C. and Scheu, S. (1999) Changes in bacterial numbers and hyphal lengths during the gut passage through *Lumbricus terrestris* (Lumbricidae, Oligochaeta). *Pedobiologia* **43**: 891–900.

Wolters, V. (1991) Soil invertebrates: Effects on nutrient turnover and soil structure: a review. *Zeitschrift für Pflanzenernaehrung und Bodenkunde* **154**: 389–402.

Yeates, G. W. and Wardle, D. A. (1996) Nematodes as predators and prey: relationships to biological control and soil processes. *Pedobiologia* **40**: 43–50.

Young, I. M. and Ritz, K. (1998) Can there be a contemporary ecological dimension to soil biology without a habitat? *Soil Biology and Biochemistry* **30**: 1229–1232.

Zak, D. K., Groffman, P. M., Pregitzer, K. S., Christensen, S. and Tiedje, J. M. (1990) The vernal dam: plant–microbe competition for nitrogen in northern hardwood forests. *Ecology* **71**: 651–656.

Zunke, U. and Perry, R. N. (1997) Nematodes: harmful and beneficial organisms. In *Fauna in Soil Ecosystems: Recycling Processes, Nutrient Fluxes, and Agricultural Production*, ed. G. Benckiser, pp. 85–133. New York: M. Dekker.

Zwart, K. B., Kuikman, P. J. and van Veen, J. A. (1994) Rhizosphere Protozoa: their significance in nutrient dynamics. In *Soil Protozoa*, ed. J. F. Darbyshire, pp. 93–121. Wallingford: CAB International.

Index